-8 NOV 1985
ASTON U.L.
ASTON UNIVERSITY
LIS
WITHDRAWN
FROM STOCK

ON-LINE MONITORING OF CONTINUOUS PROCESS PLANTS

ON-LINE MONITORING OF CONTINUOUS PROCESS PLANTS

Edited by

D. W. BUTCHER

School of Chemistry and Systems Engineering
N.E. London Polytechnic
Dagenham, Essex

Published for the
SOCIETY OF CHEMICAL INDUSTRY
London

by
ELLIS HORWOOD LIMITED
Publishers · Chichester

First published in 1983 by

ELLIS HORWOOD LIMITED
Market Cross House, Cooper Street, Chichester, West Sussex, PO19 1EB, England

The publisher's colophon is reproduced from James Gillison's drawing of the ancient Market Cross, Chichester.

Distributors:

Australia, New Zealand, South-east Asia:
Jacaranda-Wiley Ltd., Jacaranda Press,
JOHN WILEY & SONS INC.,
G.P.O. Box 859, Brisbane, Queensland 40001, Australia

Canada:
JOHN WILEY & SONS CANADA LIMITED
22 Worcester Road, Rexdale, Ontario, Canada.

Europe, Africa:
JOHN WILEY & SONS LIMITED
Baffins Lane, Chichester, West Sussex, England.

North and South America and the rest of the world:
Halsted Press: a division of
JOHN WILEY & SONS
605 Third Avenue, New York, N.Y. 10016, U.S.A.

British Library Cataloguing in Publication Data
On-line monitoring of continuous process plants.
1. Production management – Data processing
I. Butcher, D. W.
658.5'0028'54044 TS155

Library of Congress Card No. 83-18333

ISBN 0-85312-683-6 (Ellis Horwood Ltd.)
ISBN 0-470-27504-9 (Halsted Press)

Typeset in Press Roman by Ellis Horwood Ltd.
Printed in Great Britain by R. J. Acford, Chichester.

Table of Contents

Foreword

D. W. Butcher

The increasing size and complexity of manufacturing and processing plant necessary for competitive operation require high levels of component reliability and system integrity. With the application of new materials that are subjected to extreme operating conditions, traditional monitoring and preventative maintenance methods are often unacceptable. Fortunately the development of chemical and physical NDT methods and sophisticated measurement techniques has facilitated the continuous measurement of plant performance. When linked with microcomputer based control equipment, performance monitoring and predictive control of complete systems are made possible. Developments are taking place at such a rate that a normal text would rapidly become dated. The conference medium, however, provides the means by which the latest ideas over a wide range of topics can be readily disseminated.

Individual chapters of this volume are based upon papers presented at the International Conference on On-line Surveillance and Monitoring – ICOSM – organised by the Society of Chemical Industry in London during 1983, following a similar event held during 1980. The contributing authors represent a wide range of industrial, research and academic organisations throughout the UK and Europe. Both theory and practice are treated in sufficient detail and at such a level that the book will be found invaluable to engineers concerned with the monitoring and control of plant performance and to students at senior undergraduate and postgraduate levels.

The subject matter has been systematically developed from physical and chemical condition monitoring to components, process and system monitoring. Many of the authors describe research and development work that is continuing in a range of industries, so the reader is provided with a state-of-the-art review of current practice. Additionally, the details contained in many chapters should be directly applicable to other industrial situations in which on-line monitoring is required.

Neither the Conference nor this volume would have been possible without the unstinting effort of the staff of the Society of Chemical Industry and of the

Conference Organising Committee. It is perhaps fitting to pay a special tribute to Ken Urwin, the original Organising Committee Chairman, who died in tragic circumstances during 1982. This volume may therefore be considered as a memorial to his work and to his initial efforts in putting the organisation of the conference on a firm footing.

Part 1
PHYSICAL CONDITION MONITORING

CHAPTER 1

Detection and monitoring of propagating cracks in steel structures by acoustic emission analysis

L. M. Rogers

The chapter reviews progress made in the application of acoustic emission analysis to in-service monitoring of the integrity of offshore steel structures. The detection and location of propagating fatigue cracks in the node joints and crane pedestals will be considered. On-line monitoring of natural propagating fatigue cracks by acoustic emission analysis has provided new insight into the mechanism of crack initiation and extension in structural steel weldments. It appears that subcritical extension of the principal crack front is by a rapid succession of discrete steps, each step corresponding to the fracture of up to a few grains of material in a time interval of the order of 0.1 to 1 μsec. The principal factor determining the overall rate of crack extension is the relatively long 'incubation time' between successive stages of localised extension of the crack front, the mean value of which decreases with increasing stress intensity.

1. INTRODUCTION

The detection, location and monitoring of propagating subcritical cracks in the node joints, risers, bulkheads and cranes of offshore production platforms by the method of acoustic emission analysis, could have a major influence on the inspection, maintenance and repair of such structures. For example, the method

(i) Allows automatic, continuous monitoring of crack growth rate as a function of environmental, structural and process parameters, from the safety and comfort of the platform control room.
(ii) Gives global coverage of all welds in complicated joints, including normally inaccessible areas.

(iii) Measures rate of crack extension directly for a known energy input (sea state) and hence defect significance with respect to joint integrity without need for accurate sizing of the defect, measurement of the strain field and reverting to the mathematics of finite analysis and fracture mechanics.

(iv) Is considerably more sensitive than conventional forms of non-destructive testing for the detection, location and measurement of the small changes in defect state occurring during subcritical fatigue crack growth in structural steel weldments.

(v) Gives assurance of the quality of repairs by locating defects during their initial 'shake-down' and shows that shake-down has occurred.

(vi) In the case of node joints simply involves attachment to the chord (during the summer weather window) of between 4 and 8 transducers depending on node complexity, and the running of a single cable to the surface.

(vii) Does not obstruct the welds or interfere with weld inspection.

The UK [1–5], France [6], Italy [7] and Norway [8–10] report good progress in the application of AE technology offshore. Recent results from CISE [7] support the authors earlier work on similar specimens [3]. The measurements show more clearly than earlier reported, the predominance of crack growth emission during the initial stages of crack propagation, progressing to both crack growth emission (under tension) and friction noise (under compression) during the latter stages of testing. This is a result of increased compliance in the joint.

It should be noted that measurements of corrosion fatigue cracking [11] in compact tension specimens have shown a predominance of crack closure emission from the first sign of cracking. We would, however, adivse caution when extrapolating results for small scale uniaxial stressing to the behaviour of full-scale tubular welded joints. The unique nature of crack growth emission compared with emission due to frictional rubbing of the crack faces is considered further in the following sections.

2. THE CASE FOR USING ACOUSTIC EMISSION ANALYSIS OFFSHORE

The maintenance and repair of offshore structures is a sensitive subject which few operators will openly discuss. The standard practice is to repair immediately all crack indications, once confirmed, by either grinding out or if necessary rewelding. However, present methods of underwater inspection, principally visual and fluorescent magnetic particle [12] are of questionable adequacy for detecting all fatigue cracks at all node joints. Fatigue damage is sometimes so extensive that in time these joints require stiffening using massive purpose-built

30-tonne-plus steel clamps. Maintenance and repairs on this scale present considerable problems and accompanying high cost [13]. We would pose the following questions:

(1) Why, with our present codes and guide lines [14, 15] have such problems arisen so early in the design lives of these platforms?
(2) Why wasn't the fatigue cracking detected much earlier, e.g. when the cracks were a few millimetres deep and could simply be ground out without need for repairs?
(3) How certain are we that the significant cracks in the structurally important joints have been detected and that personnel safety is not at risk?

Even in cases where the present inspection procedures have proved adequate for detecting all potential defects and cracks, there still remains the problem of deciding what to do about them once detected. The question is posed: to what extent is the fatigue life and hence overall structural integrity threatened by these defects? In other words do we repair, e.g. grind out the crack indication, or do we leave well alone? Is the crack growing or dormant? Can we catagorise the defect indications in order of increasing significance? Acoustic emission monitoring is capable of answering these questions.

A portable and completely recoverable single array, comprising four sensors and one pulser, would suffice for simple cylindrical geometries, nozzles, stub joints and four-branch nodes. The transducers would be readily installed on a node or region of hot spot stress without having to worry too much about their precise location relative to the welds. Conduit would be clamped to the appropriate jacket leg to protect the main cable as it passes through the splash zone. Use of, for example, a J-tube or obsolute riser or conduit passing through redundant slots in the conductor guide frames are reasonable alternatives to attaching conduit to the jacket leg.

It is paramount that the underwater components of the installation are of high reliability. It would be counter-productive to cut corners in this area. We believe the effort involved would be acceptable to operators once the value of the results is realised.

3. THE MECHANISM OF ACOUSTIC EMISSION DURING SUBCRITICAL CRACK GROWTH IN STRUCTURAL STEEL WELDMENTS

Acoustic emission or stress wave emission is the transient vibrational energy released during crack growth. This is a very fast discrete process producing pulses of ultra-sound which spread through the structure and can be detected and located unambiguously using four or more transducers.

The discrete pulses of acoustic emission correspond to advancement of the principal crack front in discrete steps with relatively long periods of apparent inactivity between stages of active extension. From experience gained monitoring cracks in full-scale structures, it has been observed that each period of active crack extension through newly created plastic zone comprises numerous shorter steps (between 50 and 1000) occurring in rapid succession. In the case of fatigue crack growth in the HAZ of major structural steel weldments, each of these discrete steps has been found to be typically 0.01 to 0.1 mm^2 [5] producing a relatively large amplitude acoustic emission event (70 to 80† re. 1 μV at a point 20 cm from source). There is growing evidence that with increasing crack depth the mean time between successive periods of crack extension (for creation of new plastic zone) decreases steadily, but the major steps (distance between 'beach marks') remain similar in size. However, as the crack approaches through thickness and criticality, the distance between beach marks, i.e. successive increments of crack extension, is expected and observed to increase dramatically [16].

4. RESULTS OF AE MONITORING OF AN OFFSHORE PEDESTAL CRANE

The base of an offshore pedestal crane, see Fig. 1, with known weld defects, was monitored continuously for five days during normal slewing operations. Subsequently, the crane was monitored under controlled loading conditions. Strain and vibration displacement were also measured. A schematic of the test instrumentation is shown in Fig. 2.

The objective of acoustic emission monitoring was to determine whether the known weld defects were growing and whether they were the only structurally significant defects present.

4.1 Set-up, calibration and test procedure

This was carried out in accordance with the EWGAE recommended 'Code for Acoustic Emission Examination – Location of Source of Discrete Events' [18] and the procedure described in reference [17].

Two quad arrays of transducers were used to obtain the required coverage of diametrically opposite sectors of the cylindrical pedestal base, see development drawing of one sector, Fig. 3. Also shown here are the defects detected by magnetic particle and ultrasonic inspection prior to acoustic emission analysis and the regions identified subsequently as 'acoustically active'.

4.2 Results during normal operation of crane

The most important routine day-time operations were the unloading of support vessels. The crane would lift from the west side of the platform and slew north to unload in the north-east quadrant. During night-time there was no such unloading, the crane being used principally in support of construction work in the north-east quadrant.

---

†The peak electronic noise of the detection equipment, referred back to the output of the crystal detector, is typically 20 dB or 10 μV on this scale.

(a)

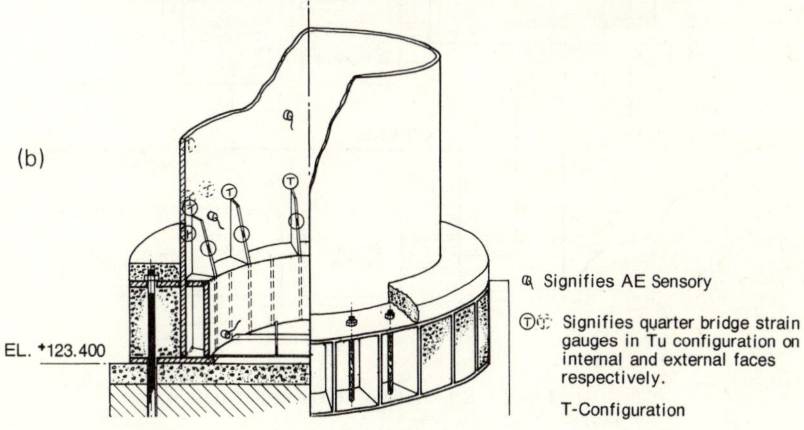

Fig. 1 – (a) Photograph of pedestal base showing external strain gauges. (b) Cut-away drawing of base showing position of the AE sensors and strain gauges attached internally.

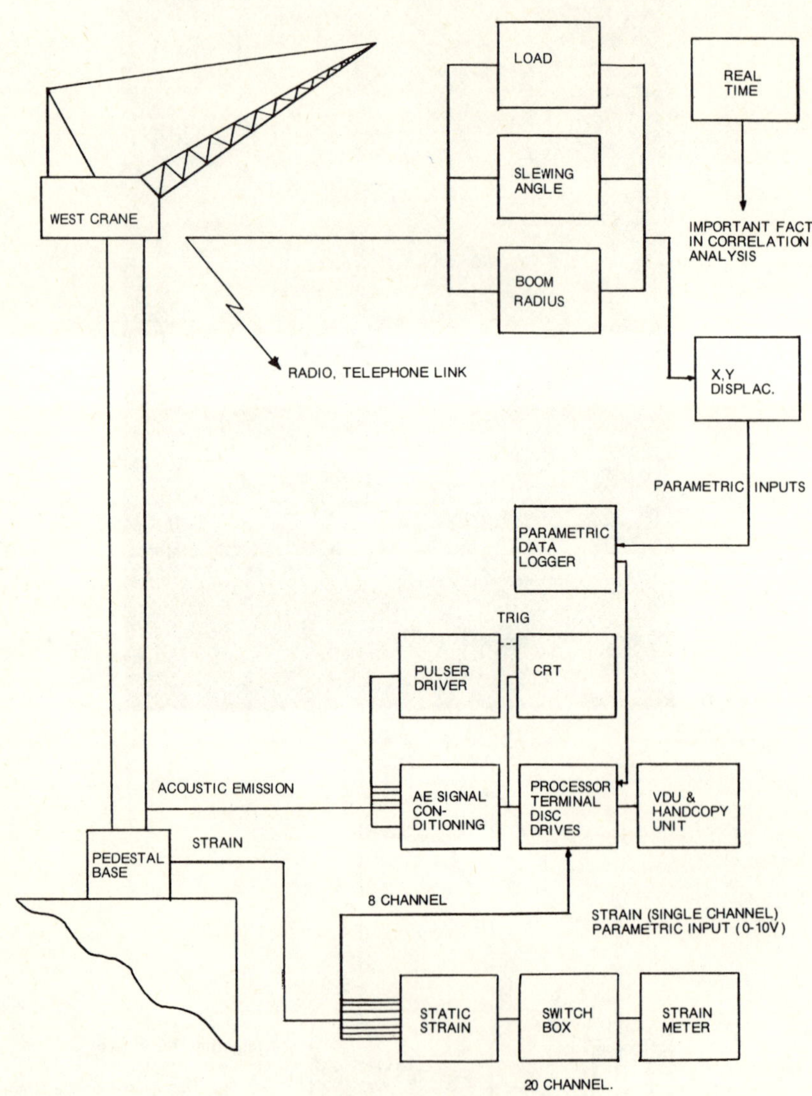

Fig. 2 – Schematic of test instrumentation.

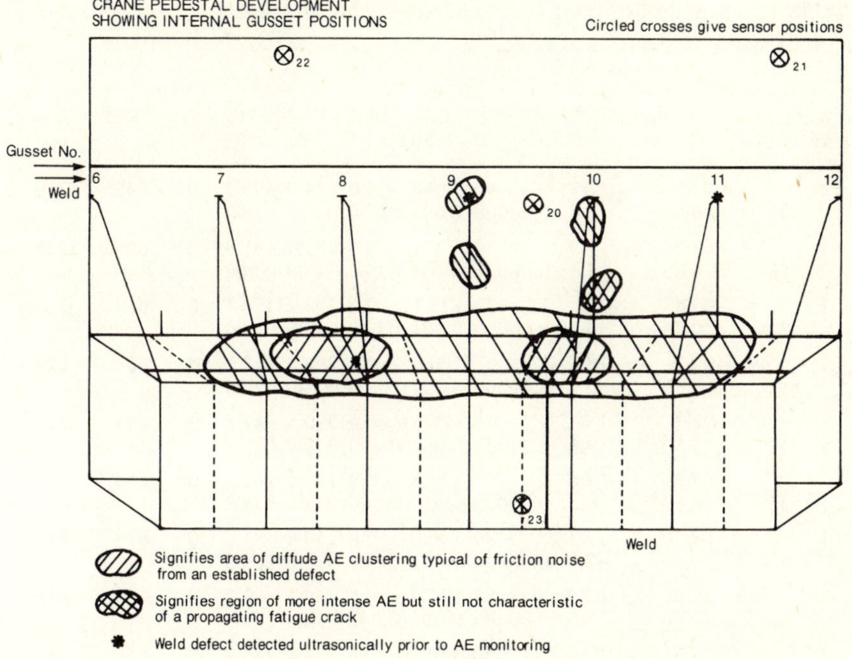

Fig. 3 – Summary of AE results for array 2 (before repairs).

Two types of acoustic emission were detected during these operations:

Type 1
Periodic intense 'beacon type' emission with highly reproducible source co-ordinates as a result of consistent signal amplitudes and arrival times, see Figs. 4 (a), (b) and (c), and listing, Table 1. This emission was particularly strong during the unloading of workboats.

Type 2
Emission with much greater variance in signal amplitude and arrival times producing diffuse clusters which developed principally when the crane was operating over the north-east quadrant, i.e. during night-time and when the principal source was not emitting. This type of emission was most prolific on array 2, in the region of the pedestal almost directly opposite to the sharply defined AE sources, see Fig. 3. It was also detected above the principal cluster, see Fig. 4(b).

Table 1 – Sample listing of event characteristics in primary cluster. For definition of symbols, see foot of table.

Ar	A	Location (X, Y)		N	DT0	DT1	DT2	DT3	C	Min.	Event no. into test
		D	R	U	P0	P1	P2	P3			
1		−0.66	−1.12	90	139.8	0.0	294.9	495.2	0	649	220
	53	0.65	0.2	0.00000	0.001	0.000	0.000	0.000			
1		−0.62	−1.09	74	126.4	0.0	281.4	481.8	0	649	221
	53	0.60	4.1	0.00000	0.001	0.000	0.000	0.000			
1		−0.64	−1.05	70	126.5	0.0	290.4	478.3	0	654	226
	52	0.60	0.2	0.00000	0.001	0.000	0.000	0.000			
1		−0.63	−1.05	67	126.6	0.0	290.5	482.1	0	654	227
	52	0.60	50.3	0.00000	0.001	0.000	0.000	0.000			
1		−0.63	−1.05	66	126.6	0.0	290.5	482.2	0	654	228
	52	0.60	49.8	0.00000	0.001	0.000	0.000	0.000			
1		−0.63	−1.05	67	126.6	0.0	290.5	482.2	0	654	229
	51	0.60	6.5	0.00000	0.001	0.000	0.000	0.000			
1		−0.63	−1.43	80	140.0	0.0	238.4	491.6	0	654	231
	53	0.60	62.5	0.00000	0.001	0.000	0.000	0.000			
1		−0.60	−1.39	86	131.2	0.0	229.3	482.9	0	654	232
	53	0.60	17.0	0.00000	0.001	0.000	0.000	0.000			
1		−0.66	−1.34	90	140.1	0.0	253.3	486.0	0	656	233
	53	0.14	4.1	0.00000	0.001	0.000	0.000	0.000			
1		−0.65	−1.43	95	140.1	0.0	238.5	486.3	0	658	235
	54	0.60	62.1	0.00000	0.001	0.000	0.000	0.000			
1		−0.64	−1.23	79	139.6	0.0	271.8	495.1	0	658	236
	53	0.60	4.0	0.00000	0.001	0.000	0.000	0.000			
1		−0.65	−1.43	86	140.0	0.0	238.3	486.2	0	659	239
	54	0.60	0.1	0.00000	0.001	0.000	0.000	0.000			
1		−0.64	−1.43	92	140.0	0.0	238.3	491.1	0	659	240
	54	0.60	16.7	0.00000	0.001	0.000	0.000	0.000			
1		−0.66	−1.12	85	139.9	0.0	295.0	495.4	0	659	241
	53	0.60	50.2	0.00000	0.001	0.000	0.000	0.000			
1		−0.66	−1.12	77	139.5	0.0	294.7	495.0	0	659	242
	53	0.16	4.2	0.00000	0.001	0.000	0.000	0.000			
1		−0.64	−1.23	81	139.6	0.0	272.1	495.0	0	659	243
	53	0.60	16.9	0.00000	0.001	0.000	0.000	0.000			
1		−0.66	−1.12	82	139.6	0.0	294.8	495.1	0	660	244
	53	0.16	3.9	0.00000	0.001	0.000	0.000	0.000			

Ar = Array number.
X, Y = Computed coordinates of AE source in m.
DT = Measured Delta T's at each sensor in μs re first hit.
A = Signal amplitude in dB re. 1 μV.
D = Duration of event in ms.
R = Rise time of event in μs.
U = Event energy (not measured), C = stress cycles (not printed).
P = Instantaneous values of parametrics (not printed).

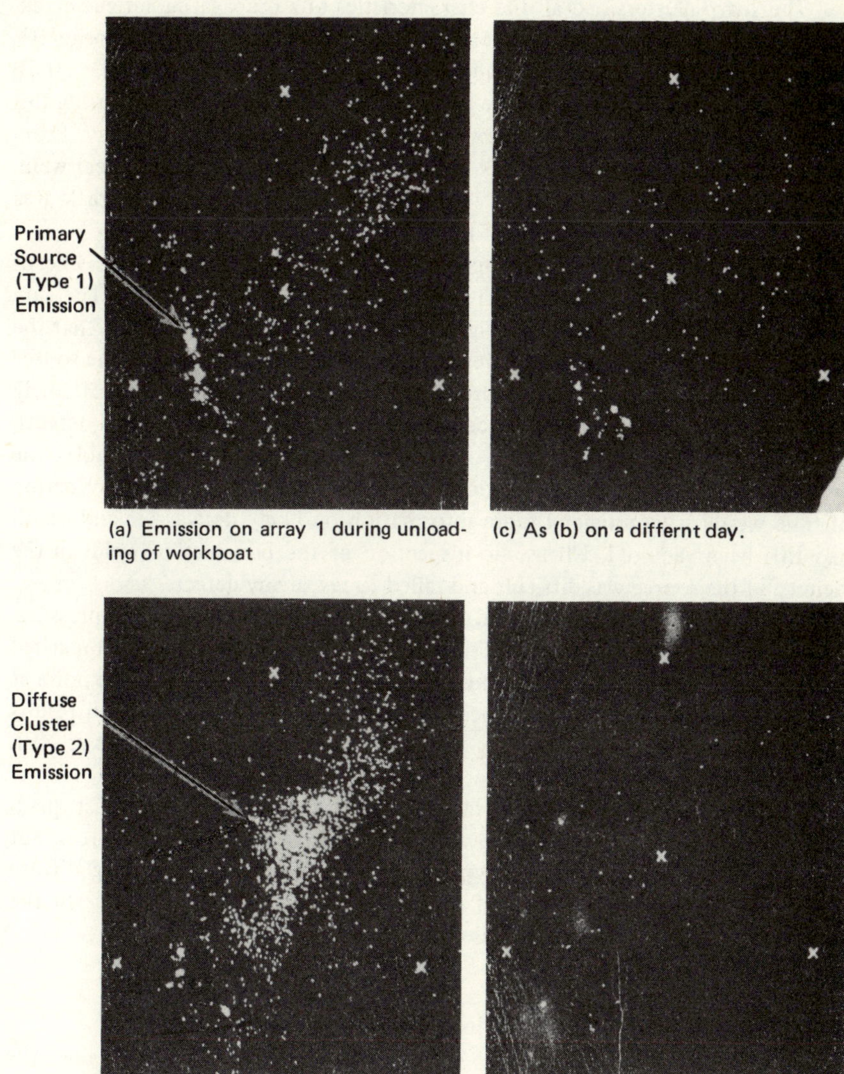

(a) Emission on array 1 during unloading of workboat

(c) As (b) on a differnt day.

(b) Emission on array 1 during general day-time slewing operation, principally over NE quadrant of platform.

(d) Emission on array 1 during static load tests.

Fig. 4 – AE event location plots (the crosses signify sensor positions).

The sharp clusters had all the characteristics of a propagating fatigue crack. During the unloading of one workboard (23 containers, total tonnage 93, maximum container weight 10 tonne) 316 large amplitude AE events ($>$ 60 dB at a point 20 cm from source) were recorded in the principal cluster. During this period of intense AE activity, the friction noise in comparison was low. From results for subcritical fatigue crack growth in large-scale structural steel weldments [3] we concluded that between 3 mm^2 and 30 mm^2 of new crack area was created during the unloading of this workboat.

The principal AE source comprised at least 10 discrete sources, some of which emitted more regularly than others, but rarely simultaneously. The sum total of this emission (combining Figs. 4 (a), (b) and (c)) would indicate that the defect was fairly large in plan. However, because of the uncertainty in the source computations, see §4.4, this is probably a significant exaggeration of likely defect size. Alternatively, the source could comprise a number of small defects, since a brief period of extension of an established crack front resembles the behaviour of several small cracks along the front. Monitoring for a longer period of time would have provided more information on likely defect size and depth had this been required. Ultrasonic inspection of the box section welds in the vicinity of the source was difficult and failed to reveal any defect.

The principal source produced no friction noise when under compression, from which we concluded that cracking was at an early stage and had not reached the point where there was sufficient compliance in the joint to produce noise as a result of rubbing of the crack faces. This behaviour is consistent with results obtained during the fatigue testing of large-scale tubular welded joints at NEL [3] and CISE [7].

During night-time when the crane was operating over the north-east quadrant, there was negligible crack growth emission from the principal sources, but the steady build-up of 'friction-type noise' elsewhere into the strong diffuse clusters observed on arrays 1 and 2 suggested appreciable compliance in the defective welds coincident with these sources.

4.3 Results obtained during static load tests

These tests were carried out using a 10-tonne water bag on the whip line (the maximum allowed load). The purpose of the tests was to measure the strain field around the base of the pedestal for a known load and different slewing and boom angles (radii).

During this period (one $+270°$ and one $-270°$ rotation and two boom movements between radius 50 ft and radius 110 ft at the due west and due south positions, a total of just 75 AE events were recorded, distributed approximately equally between arrays 1 and 2. The primary cluster, outlined by just 15 emissions, was discernible only with previous knowledge of its location, see Fig. 4(d). The small number of emissions was consistent with the fatigue expected from a single ±270° rotation of the crane under normal loading, for example the unloading of a 5–10-tonne container from a work boat in a reasonably calm sea.

We concluded that monitoring during controlled load testing does not produce sufficient cyclic stressing to allow unambiguous detection of all the structurally significant defects present. The latter requires continuous monitoring for a minimum of two to three days during normal operation of the crane and this must include monitoring during unloading of one or more work boats to achieve conditions representative of high dynamic stress.

4.4 Source location accuracy

The pedestal base was of complicated three-dimensional geometry resulting in numerous possible routes for sound propagation to each sensor. This effect was studied using a pulser transducer.

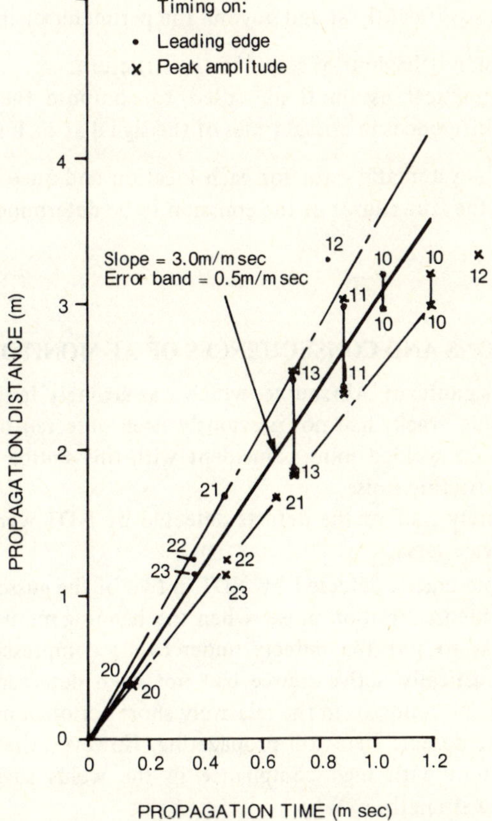

Fig. 5 – Time of flight of pulser signal measured from (i) leading edge and (ii) peak amplitude versus shortest propagation distance for all sensors.

Interference from background mechanical noise originating from the crane slewing machinery was low as a result of the remoteness of the pedestal base. A detection threshold of 6 dB above the peak electronic noise level of each channel was therefore used.

Timing of the event was on the leading edge (first threshold crossing). The wave velocity used to compute the source coordinated was 3.0 m/msec. This was a good representative value falling midway between the leading edge and peak amplitude values taking the shortest propagation distance between source and sensor, see Fig. 5.

The variation in wave velocity with propagation distance is due to dispersion and mode conversion at welds. The location accuracy of the arrays was measured using Hsu–Nielson standard pencil lead source [18]. Pencil leads were broken at the intersection of the horizontal lines (A, B, C, D) and the vertical lines through the gussets, see Fig. 6. The inaccuracies in location are shown as discontinuous lines joining actual and computed locations. The location error, which increases significantly at and beyond the perimeter of the array is due to:

(1) The dispersion behaviour of sound in the structure.
(2) The mathematical assumptions† used to compute the source from the measured differences in arrival times of the signal at each sensor.

It is, however, a systematic error for each location and once quantified for that location allows the true source of the emission to be determined.

5. CONCLUSIONS AND CONSEQUENCES OF AE MONITORING

(1) The most significant AE source, which was strongly indicative of a propagating fatigue crack, had not previously been detected by NDT. The compliance of the welded joint coincident with this source was not sufficient to produce friction noise.
(2) Approximately half of the defects detected by NDT were not acoustically active and vice versa.
(3) Large fatigue cracks detected by NDT in two of the gusset horizontal welds produced intense 'friction noise' when the bending moment of the pedestal was such as to put the defects under cyclic compressive stress. A third similar acoustically active source had not been detected by NDT. It was difficult to determine from the relatively short period of monitoring whether or not these defects were still propagating. However, the observed emission was consistent with high compliance in the welds suggesting significant reduction in strength.

---

†FLAIR software for DE 1032 system.

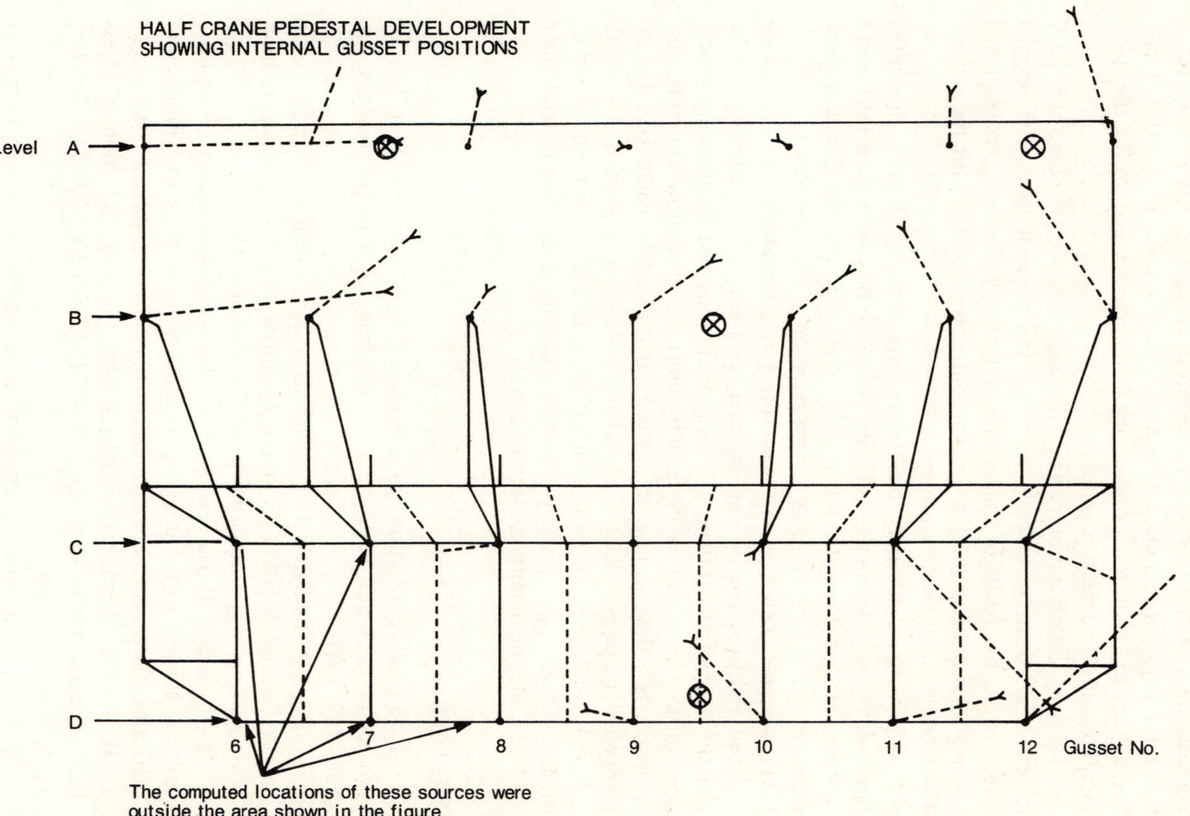

Fig. 6 – Results of pencil lead break tests showing errors in computed location of articial source.

(4) During static load testing for different boom and slewing angles relatively little acoustic emission was produced compared with that observed during the unloading of work boats. We therefore concluded that a minimum of two/three days continuous monitoring, while the crane is operating under normal service conditions, is necessary to detect and characterise the structurally significant defects present.

(5) As a result of the AE monitoring and the uncertain effectiveness of the NDT it was decided to completely refurbish the pedestal base. The original effects were the result of inadequate and poor weldments. An improved welding procedure with tight quality control was specified for the repairs. This work was completed recently.

It is intended to carry out further AE monitoring after the initial 'shake-down' of any minor weld defects which may have been introduced and not detected during the repair work. It is hoped that the results of these measurements will be available for presentation at the Conference.

(6) Using portable 8-channel AE monitoring instrumentation which can be installed and operated by one engineer, assurance of the integrity of the pedestal base of two offshore cranes, or the base and slewing ring attachment of one crane, can be obtained by short period continuous monitoring. A further two days should be allowed for installation, calibration and recovery of equipment. Larger 32-channel equipment is available for simultaneously monitoring of more cranes if required. Such monitoring would be carried out at approximately 6-month intervals, depending on the crane workload and the extent of damage found.

REFERENCES

[1] L. M. Rogers, Applications of acoustic emission source location, Inst. of Physics Conf. Series No. 44, Cr. 4., *Machine-Aided Image Analysis,* pp. 156–167, 1978.

[2] L. M. Rogers, The application of vibration signature analysis and acoustic emission source location to on-line condition monitoring of anti-friction bearings, *Tribology International,* pp. 51–59, April 1979.

[3] L. M. Rogers, J. P. Hansen, T. J. C. Webborn, Application of acoustic emission analysis to the integrity monitoring of offshore steel production platforms, *Materials Evaluation,* **38** No. 8, pp. 39–49, August 1980.

[4] L. M. Rogers, Acoustic emission system for early detection of stress corrosion cracking in steel, *Proc. ASNT 1980 FALL CONF. 5–9 October, 1980, Houston, Texas.*

[5] L. M. Rogers, Monitoring fatigue and stress corrosion cracking in process installations by acoustic emission, *Proc. Eurotest Int. Conf. on New Trends in NDT,* Brussels, 24–26 March 1982. Secretariat: Rue due Commerce 20-22-Bre 7, B-1040 Brussels.

[6] P. F. Dumousseau (CETIM), P. Laffone SNEA (P) and J. M. Thebault, CNEXO-COB, Experimental study of acoustic emission monitoring of crack propagation in offshore steel tubular joints, 11th Annual OCT, Houston, 1979.

[7] D. Bozzetti, E. Fontana, C. Panzoni, F. Tonolini, G. Villa and P. Bandi. Acoustic emission in fatigue of nodes for offshore platforms, CISE 1812, Paper Presented at the 6th International AE Symposium, SUSONO City, Tokyo, November 1982.

[8] C. Thaulow, NTH-SINTEF, Trondheim University, and O. Forli, Det. Norske Veritas, Oslo, Norway, private communication.

[9] Norwegian Petroluem Directorate, *Condition Monitoring of Process Plant on Offshore Installation,* Oljedirektoratet, ISBN 82-7257-043-2, April 1980.

[10] Norwegian Petroleum Directorate, *A State of the Art Review of Integrity Monitoring Systems Being Offered or Under Development for Use on Offshore Structures.* Oljedirekloratet, ISBN-82-7275-064-5, April 1981.

[11] P. Jax and B. Richter, Detection of corrosion fatigue by acoustic emission, 10th EWGAE Conference, Winterhur, Switzerland, October 1981.

[12] M. B. Moncaster, Underwater inspection of welds – an assessment of techniques and reliability, *Journal of the Society for Underwater Technology,* **8,** No. 3, Autumn 1982.

[13] *Proceedings of Short Course on Condition Monitoring of Offshore Installations,* Cranfield Institute of Technology, December, 1981.

[14] *Rules for the Design, Construction and Inspection of Offshore Steel Structures,* 1977, Det. Norske Veritas.

[15] BS6235: 1982. *Code of Practice for Fixed Offshore Structures,* Department of Energy Guidance Notes; (i) Design and Building of Platforms. (ii) Inspection, Maintenance and Repair of Offshore Structures. (iii) Fracture.

[16] Norwegian Public Reports NOU 1981: 11. The 'Alexander L. Kielland' Accident.

[17] L. M. Rogers and R. G. Monk, Monitoring of structural integrity by acoustic emission analysis, 2nd Nat. Conf. on Condition Monitoring in the Process Industries, 10–11 May, Cafe Royal, London.

[18] Code Sub-Group of the European Working Group on Acoustic Emission: *The EWGAE Code for Acoustic Emission Examination Location of Sources of Discrete Acoustic Events,* NDT International, August 1981, pp. 181–184.

CHAPTER 2

On-line monitoring using an acoustic emission testing service – what's really involved?

P. T. Cole

1. INTRODUCTION

Acoustic emission (AE) monitoring of petrochemical plant is a relatively new technique, increasingly practised world-wide by a number of companies, amongst them major insurance and inspection groups as well as the specialist AE companies such as Dunegan/Endevco.

AE is a technique for 'listening to' deterioration in a structure by detecting high-frequency acoustic signals which travel out from the site of defect growth and which can be detected by sensors often placed metres away. One of the largest, continuous monitoring, on-line acoustic emission systems is installed monitoring the blast furnace stove domes of the UK Redcar steelworks for stress corrosion cracking. This uses 128 detection sensors and a computer to process the data. It is a permanent installation and is reported in detail elsewhere [1, 2]. This chapter deals with the use of 'periodic' inspection by a testing service to find deterioration occurring in-service.

2. MATERIAL CHARACTERISTICS AND COMPARISON WITH CONVENTIONAL TECHNIQUES

It is fundamental to the success of using AE that the problems to be found result in detectable stress waves under the test conditions. Fibre reinforced plastic (FRP) materials emit so much when being damaged that it requires little skill to identify a problem and its location. The routine use of AE for checking the integrity of FRP structures is now common and a sub-group (called CARP) of the Corrosion Resistant Structures Committee of the Society of the Plastics Industry has issued a 'Recommended practice for acoustic emission testing of fibreglass tanks/vessels' based on experience from more than 1400 vessels. The increasing use of AE on composite structures is evident in that this year the first (four-day) international conference dedicated to AE from composites is to be held. On-line inspection of FRP tanks and pipework is straightforward and will be described in detail later.

The acoustic emission which comes from metallic structures when being damaged covers a very wide range of detectability depending upon the source, material, microstructure and effect of environment [3] (hydrogen, ammonia, temperature changes, etc.) and the main challenge when applying AE here is knowing over what distances the likely signals will be detected from any particular structure.

A large number of academic papers have been published showing that there are few emissions from undefective, low-strength metals when uniaxially loaded to failure. The point which is often missed is that problems in service (or during pre-service hydrotest) would not occur if the material were perfect. Problems are usually the result of faulty welding, heat treatment, or environmental factors which lead to material changes. Researchers at the Lawrence Livermore Laboratory [4] have also demonstrated that the stress ratio can play an important part in emissivity; since most laboratory work is under uniaxial conditions and pressure vessel shells see a biaxial stress, this goes some way to explaining why 'quiet' materials emit when made into pressure vessels (and particularly if in an embrittled condition). Anyone who has conducted AE tests on real structures knows that the problem is how to analyse the thousands of emissions in an economic timescale rather than not having any.

The other point which leads to so much confusion is that acoustic emission is a DYNAMIC technique which tells about *microstructural* changes occurring under the test conditions. It DOES NOT find STATIC defects (except by 'accident' if the crack faces rub together, a technique which may be used offshore in future to measure the length of cracks). Direct comparisons of AE results with conventional techniques such as ultrasonics, X-ray, etc., is therefore pointless since the techniques are complementary to each other.

AE should be used to monitor defects found by conventional methods to assess whether they are growing (and under what conditions; possibly the process can be controlled to prevent it).

Conventional techniques should be used to investigate areas of high acoustic activity. Sometimes there will be 'nothing there' if the AE was due to yielding or cracking which is too small at present to find conventionally (brittle intergranular crack growth in the micrometre size range can be heard using AE [5]).

Another major difference between AE and conventional methods is that AE can give global, volumetric monitoring rather than localised or surface inspection. This makes it an ideal technique for 'steering' conventional methods to critical areas (the alternative may be a limited percentage inspection). Often the fracture of a pencil lead against the surface of a thick-walled vessel can be detected on the vessel shell 30 metres away, giving some idea of the detection range (this does depend upon the structure and contents, so it is usually measured for each situation).

Acoustic emission can often be used 'on-line', resulting in obvious cost savings. Fig. 1 sums up the comparison between the techniques.

ACOUSTIC EMISSION IS COMPLEMENTARY TO CONVENTIONAL TESTING METHODS

Acoustic emission	*Conventional methods*
Sensitive to dynamic growth	Detects static presence
Volumetric coverage	Local coverage
Can be used in-service	Not usually possible to use in-service
Can be correlated to process parameters	Not usually possible to correlate with process parameters

Fig. 1

3. IN-SERVICE INSPECTION OF FIBRE REINFORCED PLASTIC TANKS USING ACOUSTIC EMISSION

The test starts before any instrumentation or sensors are installed, since the tank needs to be conditioned for a time leading up to the test at a level lower than maximum. The choice ranges from 12 hours at 10% up to seven days at 60% or less. If the process cannot accommodate these changes, then the test itself has to become 'semi-continuous', a more costly alternative which involves monitoring continuously for a number of days (including maximum level periods) to listen for creep damage and strain corrosion.

Provision has to be made for controlling the level in the vessel during the test by allowing process fluid in and out, using a submerged nozzle to avoid fill noise. The flow rate should be sufficient to allow the tank to be filled in a realistic time scale (2–4 hours is normal).

High frequency (150 kHz) sensors are attached to the tank using adhesive clips and ultrasonic couplant at all critical areas, particularly manways, nozzles and the bottom knuckle region. Usually 14 are used. Each sensor monitors an area of 0.5–1 metre radius (the signal attentuation in composite structures is high at these frequencies). Two low-frequency sensors are mounted to the main tank wall to provide coverage over a large area. The time taken for sensor installation and cabling depends upon access, but is usually 2–4 hours for a four-metre tank.

The instrumentation used is portable and may be used outside from the back of an estate car, for example, but mains power needs to be supplied. Once the tank is instrumented, a background noise check is started, lasting 30 minutes, in order to identify and quantify (or eliminate) any signals not the result of damage propagation. Occasionally, if the tank is already under load (it may be up to 60% full at the start of the test), active strain corrosion is detected as regular emissions from particular locations. After the noise check, the tank is filled in steps to 100% of maximum operating level and this level is held for 30 minutes to listen for creep damage. The test is then finished. If there are large numbers of emissions during filling, the test may be terminated to avoid the possibility of a failure. Tanks fall into three categories generally:

- Those that do not emit, indicating that they are structurally sound.
- Those that give large numbers of emissions, indicating serious problems (the area of the problem is known from the AE result).
- Those that do emit, but not 'significantly'. Care has to be exercised here, since AE usually means damage is propagating, so the tank is not up to standard, but it could still last for a year or more. Periodic AE surveys will indicate if the problem is getting worse. The CARP recommended practice would 'fail' tanks in this category since the basic premise of the code is that no damage propagation in normal service is acceptable. (Finding problems at this early stage may allow economic repair.)

The entire test can be carried out by one engineer (plus a plant operator to control filling, etc.), and a continuous record is kept of each stage so the results are known on completion. Sometimes the problems identified and their cause are obvious – lack of support for a pipe or strain corrosion where the corrosion barrier has failed, for example. If damage is general, perhaps due to overheating, it may not be visually obvious.

The cost of the test is low compared with using AE on metallic structures, since the instrumentation is simple and one person can carry out the test. Small tanks can be tested very rapidly when filling is organised well; four tests per day is possible with tanks about two metres in size.

Details of setting up and conducting tests on FRP vessels are in reference [6] published by the Society of the Plastics Industry.

4. IN-SERVICE INSPECTION OF METALLIC STRUCTURES USING ACOUSTIC EMISSION

There are a number of reasons why our AE testing service has been used on plant in-service:

- Prior to shutdown to give a 'general survey' to steer conventional inspection to any critical areas during the shutdown period.
- To establish if newly found defects are growing in service or if they are just the result of the continuous improvement in conventional inspection methods.

- To find out if and *when* defects are growing (i.e. during start-up, shut-down or continuous operation).
- To find out if known changes in operating conditions have led to any continuing defect growth.

I would like to make two clear-cut distinctions here, since the confidence in the result is different for each. One is the situation where a known problem exists and it can be monitored locally. The sensitivity of a test monitoring a local area can be so high that even yielding or corrosion (active 'rusting') can be heard. The other situation is the one in which an entire structure is monitored with widely spaced sensors. Here only high amplitude signals (from energetic or brittle sources) will be detected and located. The understanding of emission 'amplitude distributions' [7, 8] and how to use them to calculate the effect of sensor spacing on the number of located emissions from a particular source type is a fundamental technique, which if ignored will result in poor interpretation of test results or no results at all!

The feasibility stage of an on-line monitoring exercise has a number of stages:

- Define the objective of the exercise.
- Accumulate the information about construction and operation of the structure. This will include:
  engineering drawings and materials,
  temperature and pressure ranges,
  contents and flow rates,
  plus any other relevant history.
- If this type of structure has been monitored before, make use of the relevant information.
- Assess the validity of materials characterisation data or generate new data.

It may be necessary to measure the operating noise level at different frequency ranges (a simple task using a portable instrument). It may also be necessary to measure signal attenuation in the structure. Usually this is predictable from past experience and is checked on-site rather than at the feasibility stage, but if there is doubt, then it is better to measure it before plans are too far advanced.

Discussion will also centre around the type of test which is most relevant to the structure. In the USA, for example, where many tests have been carried out on hydrocrackers, defects have been found to grow during the thermal stresses of start-up. These would not necessarily be found during a continuous operation test (different defects may be found then).

Monitoring during start-up or shut-down will take place during the most highly stressed part of the operation. Stress engineers have to define this for the particular structure, but it is usually the periods of maximum temperature and pressure change. If process problems are encountered during start-up, AE gives immediate information about whether the structure is being damaged as a result. It is usual to feed as many process parameters as possible (at the least pressure and temperature) into the AE system computer for correlation purposes.

One point not mentioned so far is the effect of temperature on the AE test. The most significant effect in common tests occurs when the contact mounted sensors have to be replaced by waveguides owing to the high temperatures of some plant. Waveguides have an insertion loss of about 20 dB, which immediately loses a factor of ten in absolute sensitivity. Whether this affects the *test* sensitivity depends upon the operating noise level. On 'noisy' plant (ammonia converters, hydrocrackers, powerformers, etc.), the background noise will be the limiting factor and waveguides will lose this as well, so there is no *net* loss in sensitivity. On certain other types of reactor and vessel, the loss in sensitivity will be real and has to be accepted or more sensors used to compensate (at greater cost).

Some 'on-line' tests are a compromise in that the process is temporarily interrupted for the duration of the test in order to reduce the noise level. This interruption may last up to a few hours, but is a method of achieving 'hydrotest' noise levels and hence a high sensitivity on otherwise operating plant.

Monitoring during continuous operation for an extended period is often too costly, so a 'stimulus' is given to encourage emission to occur during the monitoring period. This usually involves raising the pressure 10% above the normal operating pressure (this obviously has to remain below design pressure, so is not always possible). Storage tanks are generally used less than 100% full, so the test involves simply filling them up.

5. ON-LINE EXAMPLES OF ACOUSTIC EMISSION MONITORING OF METALLIC STRUCTURES

5.1 Jacketed chemical plant reactors (1501 carbon steel, 28 ton)

Problems were suspected between the reactor shell and jacket, very difficult to inspect conventionally since there is no access. The process itself involved alternately heating and cooling the reactor by passing steam and water through the jacket, an operation leading to wide variations in background noise and so a challenge for on-line monitoring. Four sets of waveguides were mounted through the insulation to try to locate any acoustic emission sources and identify when they were occurring.

A Dunegan/Endevco DART testing system was used, since this has the ability to 'float' each detection channel automatically above the noise level and to detect AE transients above this. There was no interruption or modification to normal process operation. Fig. 2 shows the AE sources, located next to the steam inlet. These were active only during the period immediately after the steam valve was opened (for up to 20 minutes). The high thermal stresses are the cause of the problem, coupled with corrosion. This test was carried out several times and illustrates a number of important points:

- Successful on-line operation of an AE test in a high (and varying) noise environment is possible (see Fig. 3).
- Identification of the location and cause of damage in an otherwise uninspectable area (it was not as widespread in the jacket as had first been suspected).
- AE source location in a 'low strength' material (the welding and environment change the material).

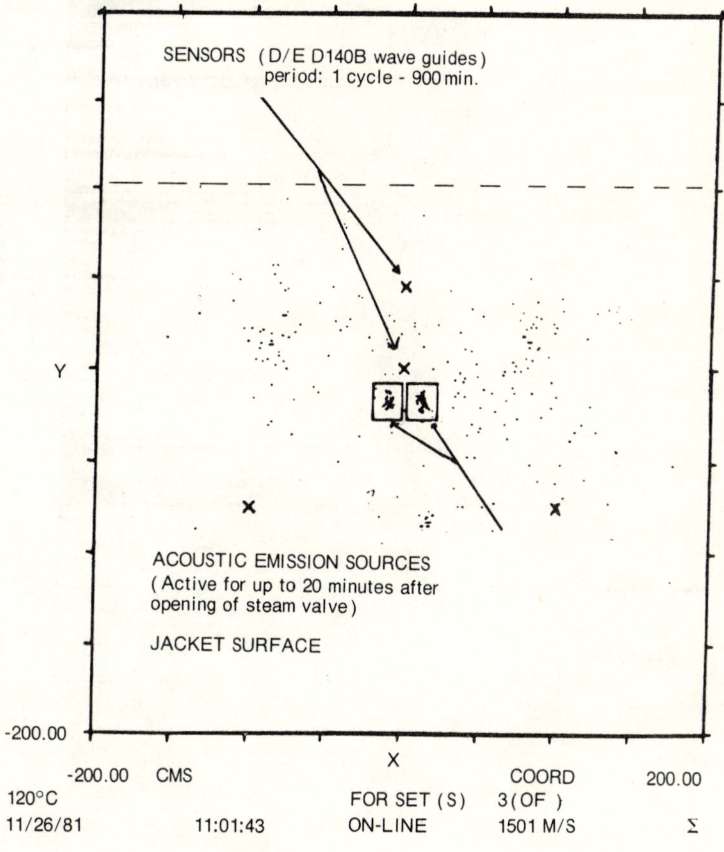

Fig. 2 – On-line location of thermally-induced cracks in reactor jacket.

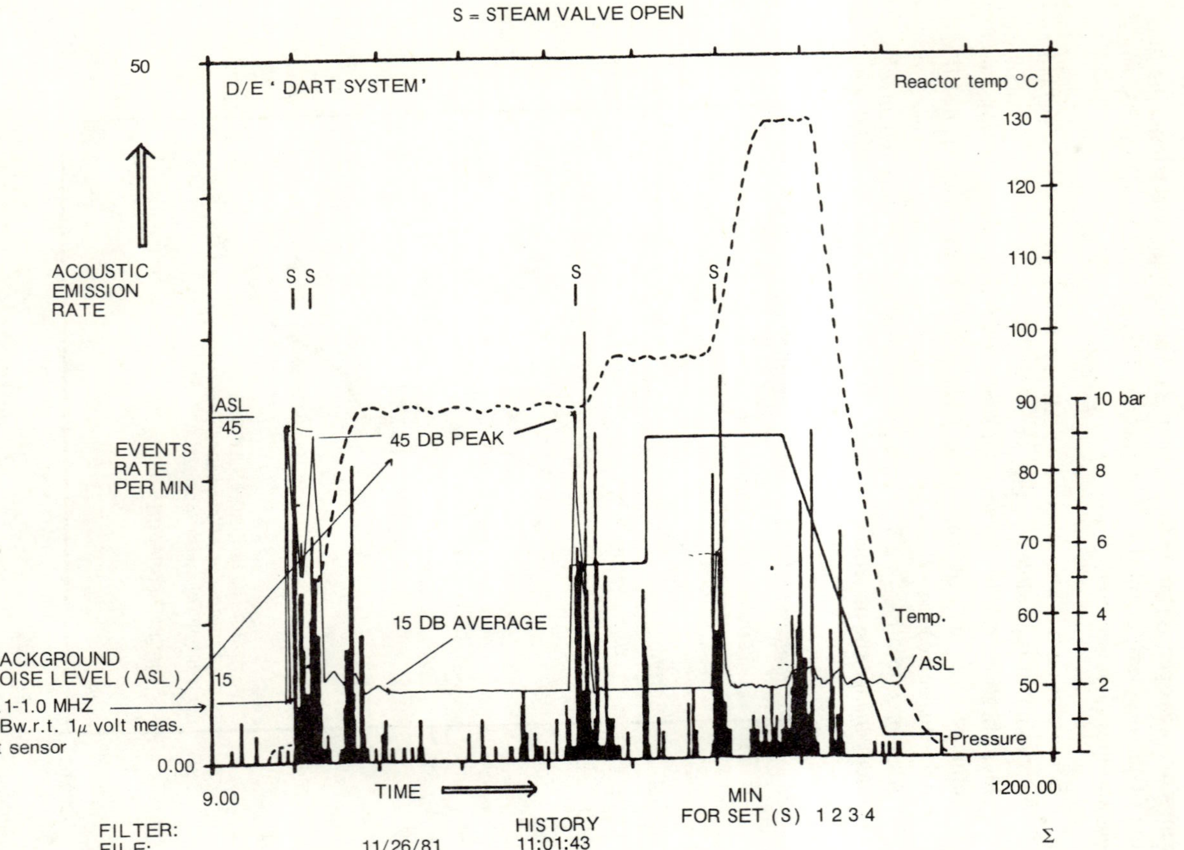

Fig. 3 – Acoustic emission and background noise (one cycle) from reactor jacket in Fig. 2.

5.2 A process chain of reactors with toxic contents (stainless steel)

A minor process abnormality gave reason to suspect that problems of corrosion and stress corrosion cracking (SCC) may have initiated. The AE test was carried out to assess whether the problems were present or serious in order to plan corrective action efficiently.

To detect the very early stages of SCC, it is necessary to have a high test sensitivity, since the AE signals are small. For this reason, the test was not quite 'on-line', the process being temporarily halted.

Prior to arrival on site of the AE testing crew, the customers had bonded to the reactors at each designated sensor location two small mild steel plates using a special bonding agent (maximum temperature was less than 125°C, insulation was removed where necessary). This was to allow magnetic sensor clamps to be used (no problem on ferritic structures, of course). Sensor installation (64 in total) took two days. On day 3 the process was stopped for the test, which consisted of 30 minutes' background noise check, followed by pneumatic pressurisation using nitrogen (the reactor contents were left in during this temporary production halt). The process was then re-started, the shutdown being only a matter of hours.

This was an interesting test, since many small emissions were occurring at below operating pressure from most areas and only one or two areas showed pressure dependence of activity. Thermal stresses (the reactors started to cool) no doubt played some part in this, but most of the emissions were caused by corrosion occurring internally.

No areas of the reactors showed activity characteristic of serious structural problems, so each 'zone' was graded accordingly to overall activity and its dependence upon pressure (i.e. does activity increase with pressure and how much). At the following shutdown, two cracks were found in active areas (on the outside), on a wall crack, and one from an external lug. In one active area nothing could be found. The inside of the reactors showed variations from little damage to severe corrosion (very rough surface with fissuring). Correlation of AE activity with depth of corrosion does not produce very good results, since the low amplitude AE is only telling you what areas are actually corroding *during* the test period.

5.3 Multi-layer reactor

Many high pressure vessels have multi-layer shells which are very difficult to inspect conventionally. Fig. 4 shows a view of one of these, together with sensor locations as 'seen' by an acoustic emission system. Monitoring during start-up is the usual method of test. At low pressure, the layers rub together (crushing of oxides creates AE at low pressures); at higher pressures, the layers are held tightly together and signal transmission between layers becomes acceptable (high contact pressure is all that is used to get acoustic signals to transmit into waveguides, no couplant being used at high temperatures).

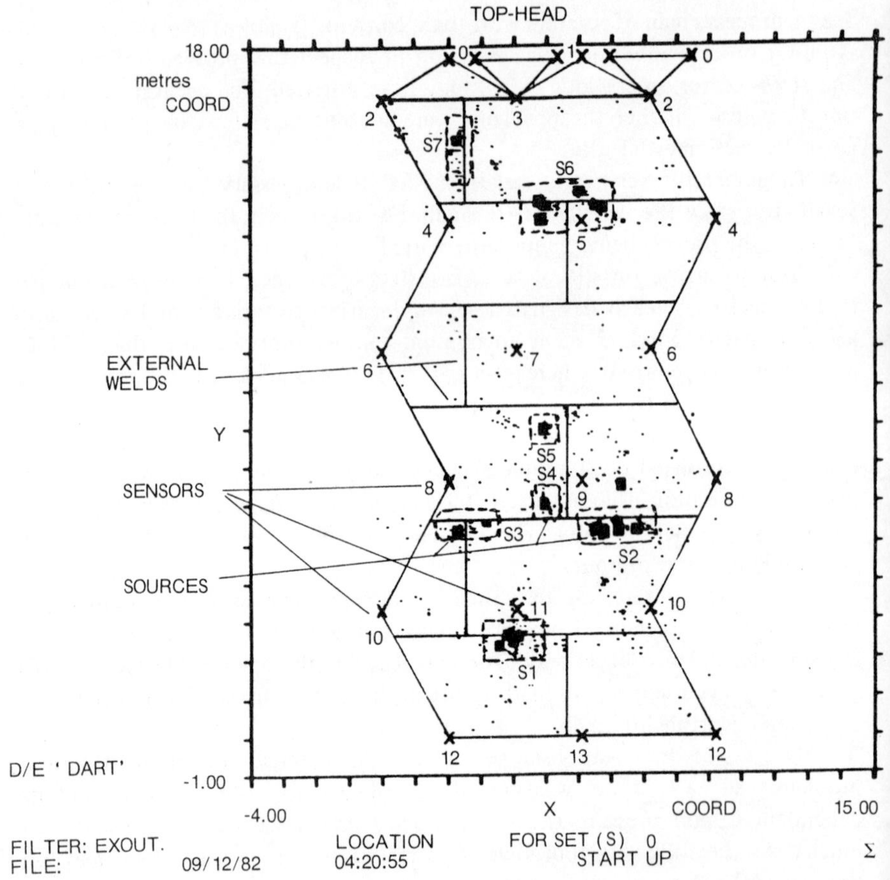

Fig. 4 – Unwrapped view of reactor (AE system O/P) AE sources during start-up.

5.4 Storage tank

Fig. 5 illustrates the capability of AE to monitor large areas, in this case a storage tank. High amplitude AE sources from structures such as this can often be located to an individual weld junction, allowing use of conventional techniques to size and identify the source (it may be decided on the basis of the AE results alone not to do this). Tests on large storage tanks may take a week or more of monitoring due to the time taken to fill them, but once the test is set up, the AE system computer needs little skilled attention since it will automatically bring to the operator's attention any areas of high activity and all data is stored on floppy disk for analysis at any time during or after the test.

The source in Fig. 6 is due to slag inclusion in a repair weld of the tank (correlated with X-ray).

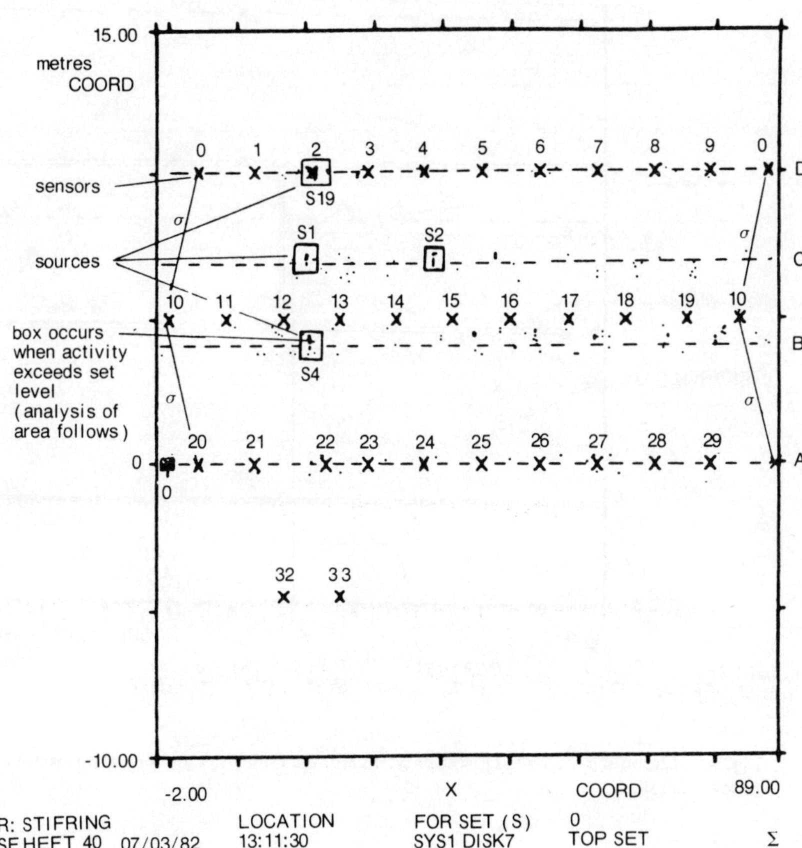

Fig. 5 – Unwrapped view of 90 metre circ. storage tank (top half) AE sources during filling.

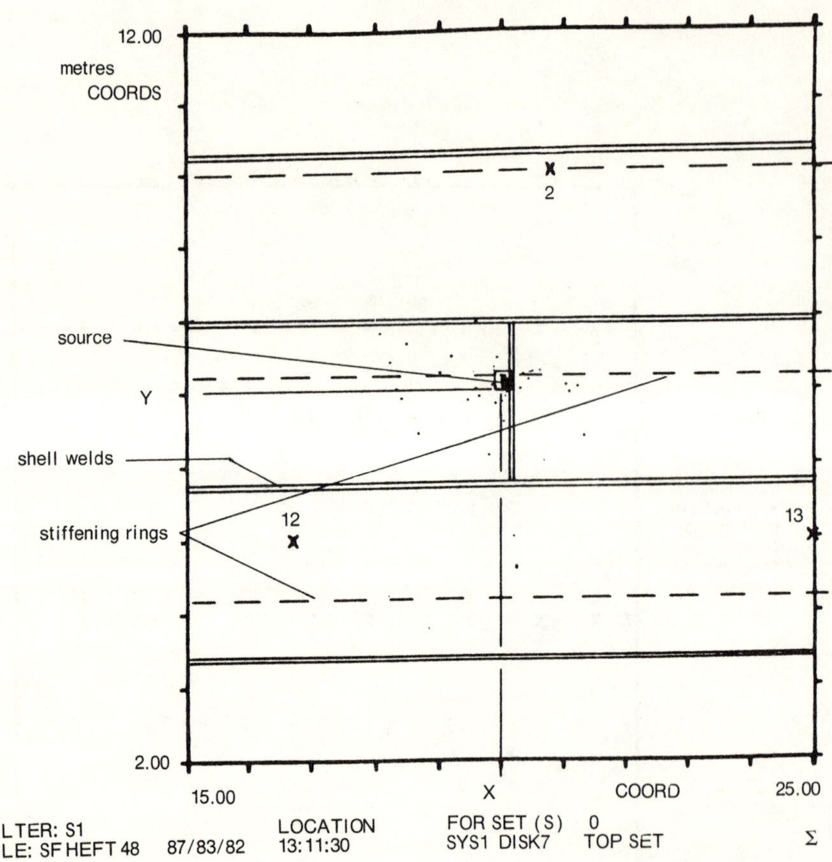

Fig. 6 – Expanded view of AE source S1 in storage tank showing location relative to local detail.

6. CONCLUDING REMARKS

Acoustic emission monitoring offers many exciting new possibilities, only a few of which have been mentioned here. Already high-frequency monitoring of rolling-element bearings [9], acoustic leak detection and several other new techniques have been proven feasible and just await commercial development.

The author would like to acknowledge the help and assistance of Ian Simms, Malcolm Schofield, Brian Schofield of ICI, Cliff Carr of Shell Chemicals, Bob Williams of Monsanto, Ron Dye of UKF, Les Wilkinson of Esso Petroleum, and Martin Peacock and Mike Balmer who carried out much of the practical work.

REFERENCES

[1] L. M. Rogers, Acoustic emission system for early detection of stress corrosion cracking in steel, Proceedings ASNT 1980 Fall Conference, 5–9 October 1980, Houston, Texas.

[2] L. M. Rogers, Monitoring fatigue and stress corrosion cracking in process installations by acoustic emission, Proceedings of Eurotest Conference on new trends in NDT, 24–26 March 1982, Brussels.

[3] H. L. Dunegan and A. T. Green, Factors affecting acoustic emission response from materials, *Materials Research and Standards, MTRSA*, **11**, no. 3, p. 21.

[4] M. A. Hamstead, E. M. Leon, A. K. Mukherjee, Acoustic emission under biaxial stress in unflawed 21-6-9 and 304 stainless steel, Elastic Waves and Microstructures Conference, 16–17 December 1980, University of Oxford, England.

[5] A. Nozue and T. Kishi, An acoustic emission study of the interangular cracking of AISI 4340 steel, *Journal of Acoustic Emission,* **1**, no. 1.

[6] CARP/C. Howard Adams, Recommended practice for acoustic emission testing of fibreglass reinforced plastic tanks/vessels, 37th Annual Conference, Reinforced Plastics/Composites Institute, The Society of the Plastics Industry, Inc. 11–15 January 1982.

[7] A. A. Pollock, Acoustic emission – 2. Acoustic emission amplitudes, *Non-destructive Testing,* October 1973.

[8] A. Pollock, Structural calibration technique for quantitative application of acoustic emission, *Acustica,* S. Hirzel Verlag, Stuttgart, **38**, no. 5, 1977.

[9] P. Bloch, Predict problems with acoustic incipient failure detection systems, *Hydrocarbon Processing,* October 1977.

CHAPTER 3

Acoustic emission signal processing for the characterisation of crack growth in steels

J. L. Roux and D. Bernard

Framatome has implemented a R & D program in order to obtain further knowledge of acoustic emission signals produced by crack growth. The main objectives of this program were to develop a wave pattern recognition method which could discriminate between crack growth and interference signals and to establish a rapid, reliable method of monitoring critical zones in pressurised water reactors.

The objectives of the study were to:

- compile a 'library' of crack signals,
- select digital methods of crack signal recognition,
- correlate crack development and acoustic emission.

1. TRADITIONAL METHODS OF ACOUSTIC EMISSION MONITORING

The principal traditional methods of acoustic emission monitoring involve:

- Analysis at count rates, using one or more thresholds (see Fig. 1) (this technique does not enable source location and is highly sensitive to system gain).
- Analysis of the characteristic parameters of the signal such as signal length and peak value.
- Source location using delayed TTL pulses, generated each time the threshold is crossed. Fig. 2 illustrates the difficulty in ascertaining signal arrival times.

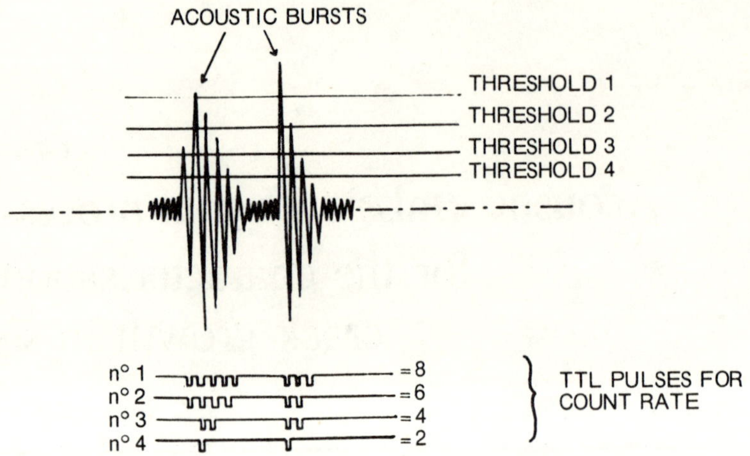

Fig. 1 – Count rate with different bias levels.

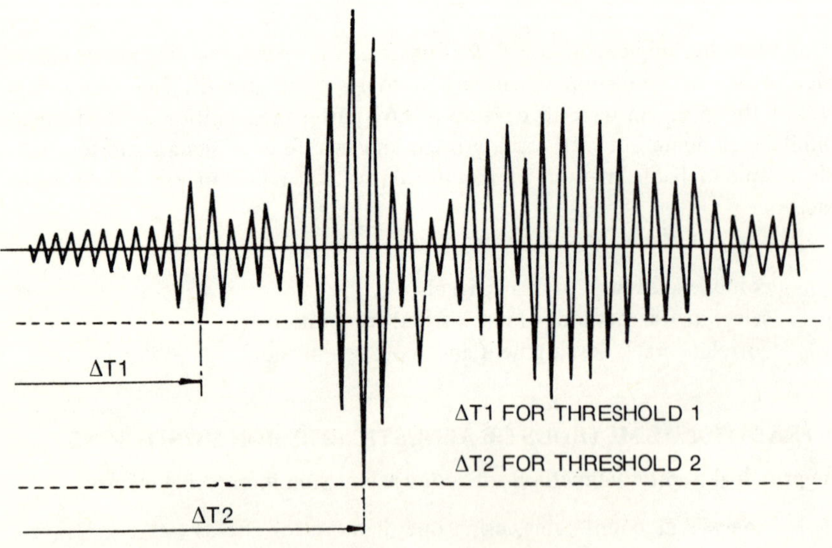

Fig. 2 – Uncertainty in determination of ΔT.

The main disadvantage of the above methods is that they do not take the signal pattern into account and consequently process all signals detected by the instrumentation channels in exactly the same way.

TENSILE TEST MACHINE
WIDE BAND TRANSDUCERS
HYDROLIC CYLINDER
AMPLI
AMPLI
HIGH PASS FILTER
HIGH PASS FILTER
DIGITALIZER
IEEE
HP 8845 CONTROLLER
IEEE
IEEE
DIGITAL VOLTMETER
DIGITAL COUNTER
FORCE
TIME
INSTANTANEOUS FORCE
NUMBER OF CYCLES
MACHINE CONTROLLER

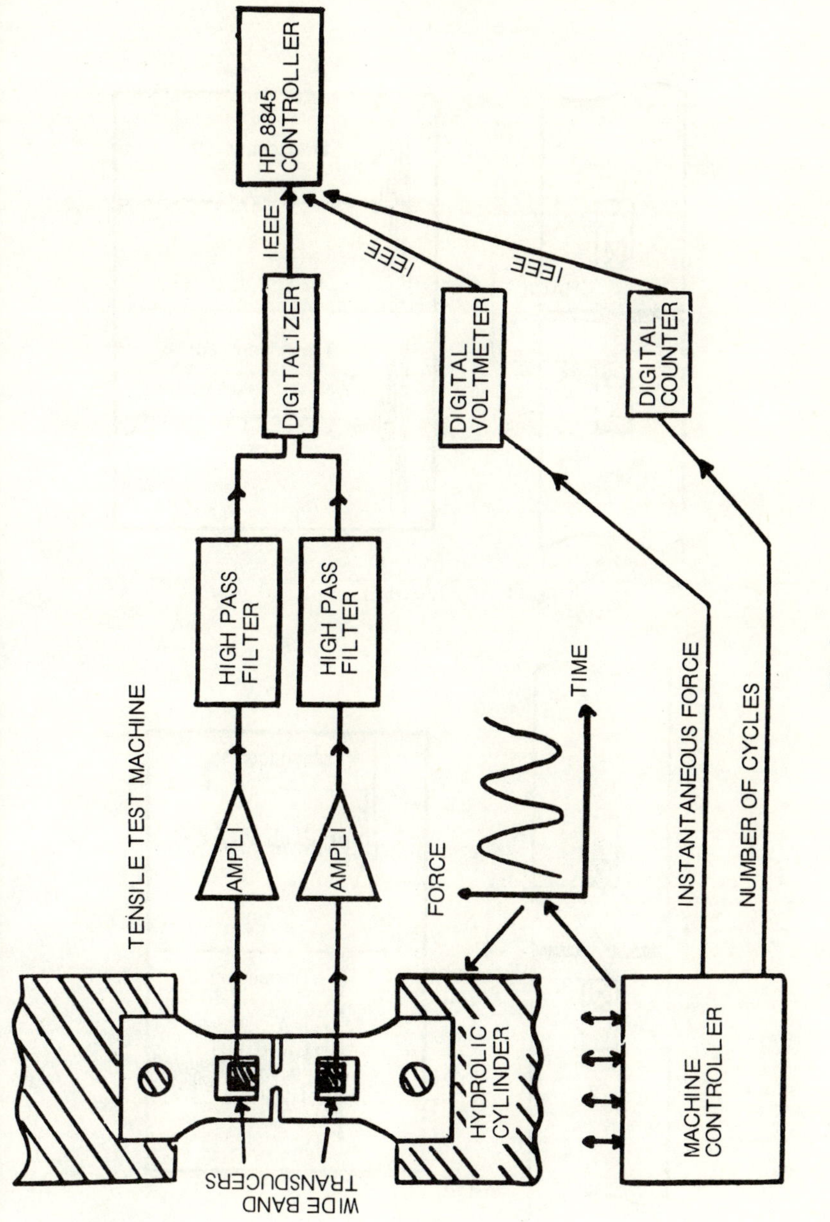

Fig. 3

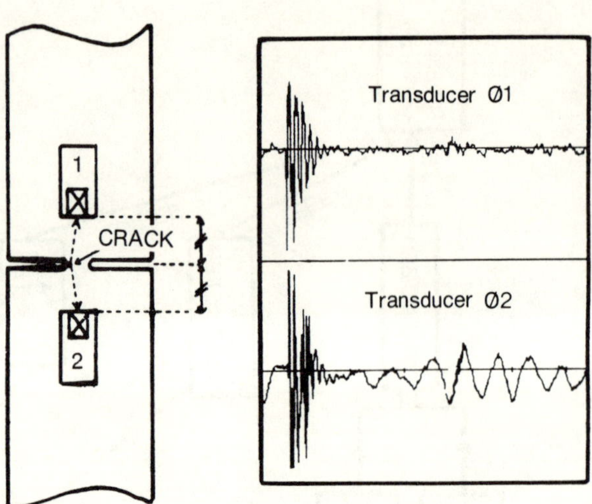

Fig. 4 – Symmetrical sample.

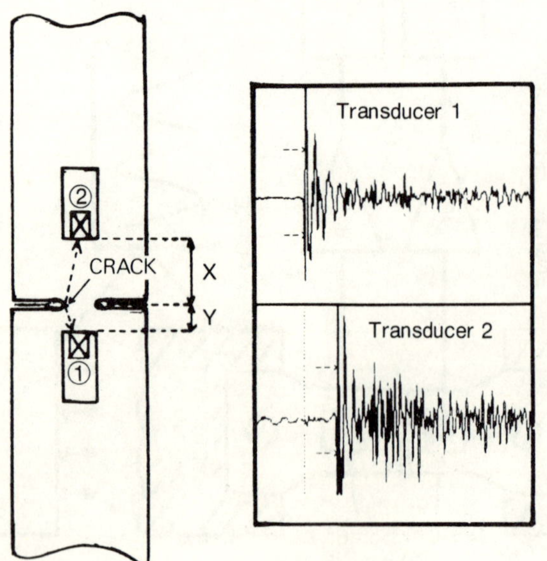

Fig. 5 – Asymmetrical sample.

2. SIGNAL LIBRARY

Our first task, therefore, was to build up a library of crack signals. With this aim in view, specimens pre-cracked by electron discharge machining were subjected to fatigue cycle tests using appropriate instrumentation, (see Fig. 3). Symmetric and asymmetric specimens were used (see Figs. 4 and 5). For the latter, the theoretical Δt of the emission signals was known, signals due to electrical interference and gripper noise could therefore be eliminated (see Figs. 6 and 7).

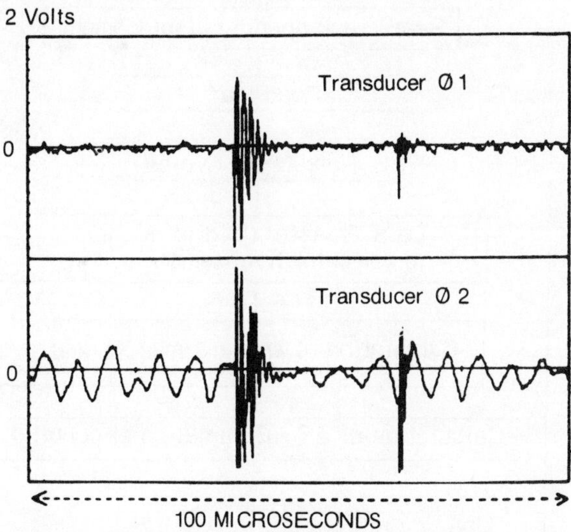

Fig. 6 – Electrical sturbs.

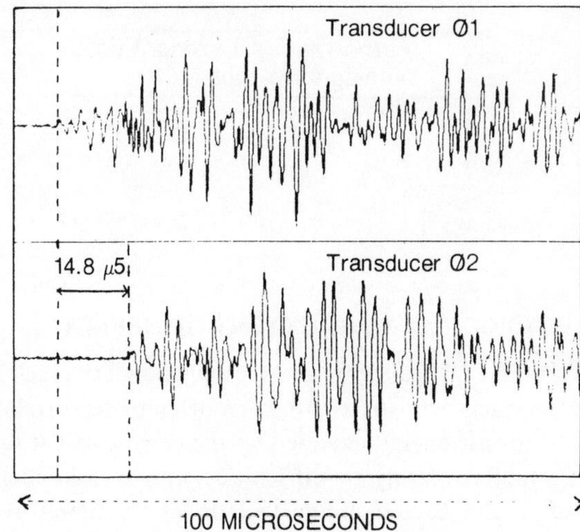

Fig. 7 – Grip noise.

The tests were conducted using a simple program, shown in Fig. 8. Figs. 9 and 10 show two bursts recorded at an interval of a few cycles. Note the similarities.

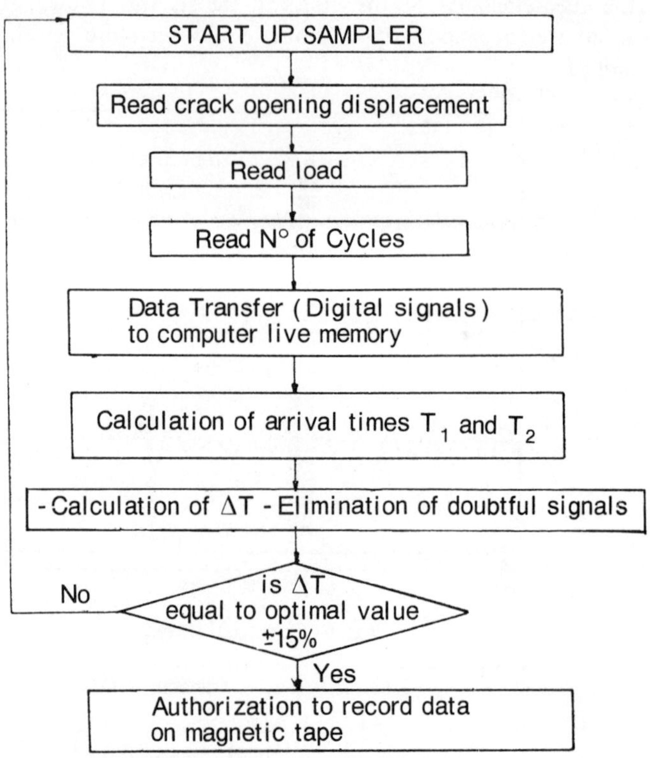

Fig. 8

3. CORRELATION OF AE WITH EXTENT OF DEFECT

Using an electronic COD gauge of a size proportional to crack depth, recordings were made after each acoustic event in an attempt to correlate crack development with the acoustic energy detected by the sensor. As yet results have proved inconclusive, doubtless owing to difficulties with measurement techniques. In the future we intend to use in-line techniques for monitoring crack growth, either by ultrasonic or X-ray examination.

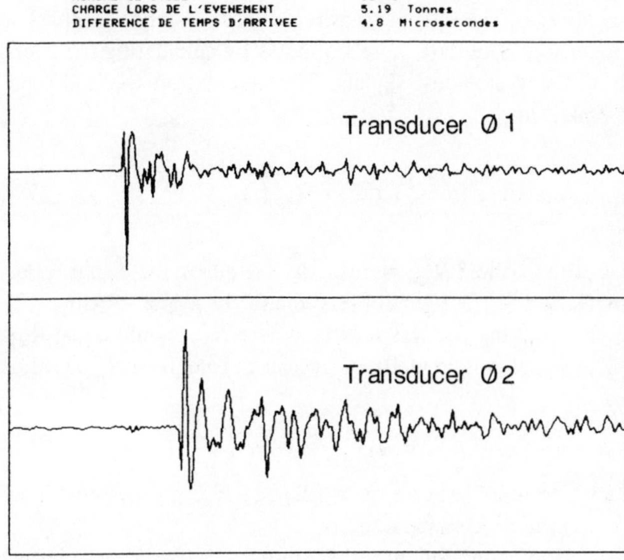

Fig. 9 – Crack acoustic signals.

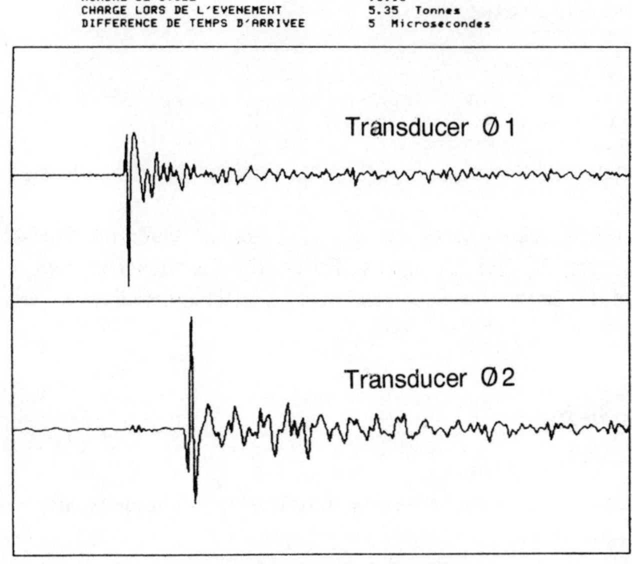

Fig. 10 – Crack acoustic signals.

4. SIGNAL PATTERN RECOGNITION

A statistical analysis of signal types was required in order to establish a method of pattern recognition. The data were obtained by calculating the coefficient of the digital filters using standard signals. The calculation method used was the Werner–Hopf equation:

$$\int_0^\infty h_{op}(Z) R_{xx}(t-Z)\,dZ = R_{xy}(t)$$

where h_{op} = optimal filter; R_{xx} = autocorrelation of the signal observed; and R_{xy} = intercorrelation of the signal observed and the signal sought.

What we were looking for was a filter where R_{xx} would equal R_{xy} delayed by one step. The signal was modelled using an autoregressive procedure (sampling) as follows:

$$X_n + a_1 x_{n-1} \ldots + a_p x_n - p = U_n$$

where U_n is a white noise process, $X = (X_n \ldots X_{n-p})$ the remaining samples and $a = (a_1 \ldots a_p)$ the process parameters.

The Wiener–Hopf equation now becomes: $R_{xx}a = R_{xy}$ – known as the Yule Walker set of equations. The equations may be resolved using the Levinson or Durbin algorithm.

To come back to the problem of shape recognition, or rather, the problem of discriminating between valid signals and extraneous noise: two types of signals are assumed to exist,

$$X^1 \;—\; a^1$$

$$X^2 \;—\; a^2$$

a = statistical breakdown, i.e. coefficient obtained by the autoregressive procedure.

Let R be the signal to be classified, a^1 and a^2 represent filters F_1 and F_2 adapted for type X_1 and X_2 signals. The method is shown in Figs. 11 and 12. Figs. 13 and 14 show the classification of signals emitted by a 316L stainless steel specimen.

5. CONCLUSION

We conclude that:

- Traditional acoustic emission monitoring techniques are not reliable enough;
- digital methods of processing broadband signals represent a valuable addition to acoustic emission techniques.

μ_1
X
F_1
$|\,|^2$
Σ
TEST
R
F_2
$|\,|^2$
Σ
TEST
Y
μ_2
Y
μ_2
X
μ_1
μ_1 and μ_2 : Thresholds

Fig. 11

←BURST THRESHOLD
Signals recognised as burst but not as interference
Signals recognised neither as burst nor interference
INTERFERENCE THRESHOLD
μ_1
Signals recognised as interference and burst
Signals recognised as interference but not as burst
μ_2
0 1 2 3 4 5 6 7 8 9 10
1 2 3 4 5 6 7 8 9

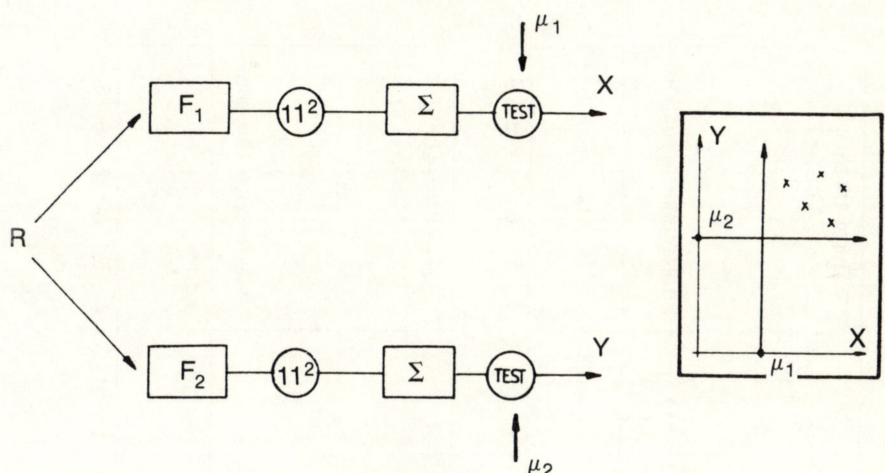

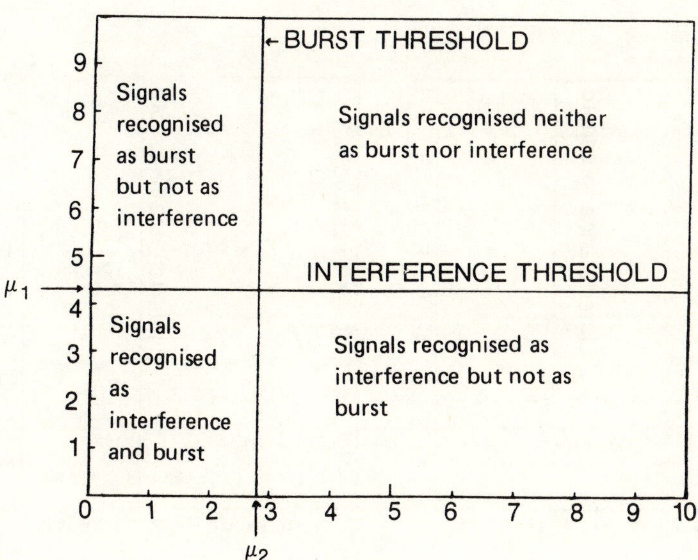

Fig. 12

Burst threshold
(Order 11)
Interference threshold
Burst threshold
(Order 11)
Interference threshold
0 1 2 3 4 5 6 7 8 9 10
1 2 3 4 5 6 7 8 9

Fig. 14

Burst threshold 3.31
(Order 9)
Interference threshold
Burst threshold 3.31
(Order 9)
Interference threshold
0 1 2 3 4 5 6 7 8 9 10
1 2 3 4 5 6 7 8 9

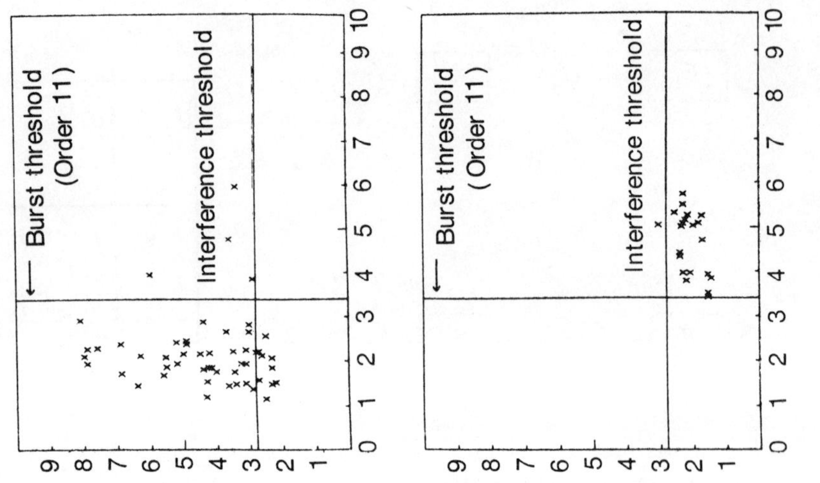

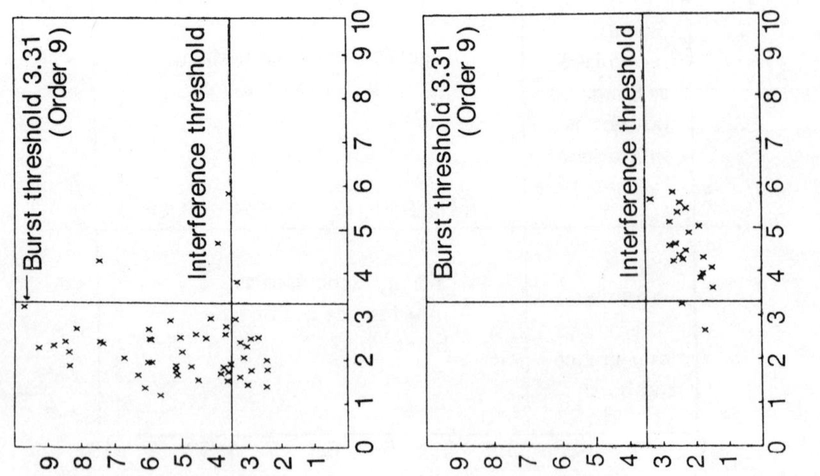

Fig. 13

CHAPTER 4

Condition monitoring using mechanised ultrasonics

W. Browne and D. A. Constantinis

1. INTRODUCTION

The accurate condition monitoring of the general and localised corrosion of the internal walls of pipeline systems – both process and service – has an ever-increasing importance. The risk of actual leakage with its concurrent hazards and costly loss of production must be minimised.

There are four commonly used corrosion monitoring techniques for assessing internal corrosion conditions.

(1) Corrosion coupons.
(2) Pipe spool insert pieces.
(3) Corrosion probes and monitors.
(4) Iron counts.

With the exception of (2) all these systems measure the general corrosion potential of the fluid or gas in the pipeline and give no warning of any localised attack.

The use of manual ultrasonics to establish first a base line datum of wall thickness, at pre-selected position on lines, followed by regular re-survey of the same points is a well established approach, and is worthwhile in systems where high rates of erosion are occurring locally. It is nevertheless slow and, being a manual system, operator sensitive.

The accurate condition monitoring of plant is increasingly important because of risk and cost of leakage and production loss.

Corrosion coupons, insert pieces and so on only measure general corrosion potential.

Manual ultrasonics can give base line and regular re-survey data but the technique is slow and operator-sensitive.

Mechanised ultrasonics has the advantage of being more rapid and gives more reliable results, especially in heavily corroded areas, and a hard copy result.

The approach using mechanised ultrasonics offers several possible advantages.

(1) Accurate data base repeatability.
(2) Hard print-out facility – good comparability.
(3) Increased speed of operation.
(4) Greater area of coverage.

2. MECHANISED INSPECTION – DESIGN PHILOSOPHY

2.1 General

Radiography, although a useful comparator giving a presentable record, is technically limited.

Using traditional approaches, manual ultrasonic examination can suffer from subjectivity in operation and interpretation – the latter being compounded by lack of hard copy evidence.

The inspection coverage afforded in manual operation can be more accurately and controllably reproduced mechanically, and the basic data available from ultrasonic inspection can be readily digested and processed electronically for recording in a variety of formats giving visually descriptive and consistent presentation of inspection results.

This philosophy is referred to as *mechanised* (as opposed to automated) inspection and the criterion is to eliminate the variables of manual inspection by providing ultrasonic test systems based on simple design principles, and using the 'screening' technique.

The requirement for condition monitoring thus points to mechanised UT.

Alternative techniques, for instance radiography, are costly and less site-practicable, giving only an indication of actual wall thickness.

2.2 Mechanical principles

To enable repeatable scanning with comprehensive inspection coverage the mechanical system should be designed to:

2.2.1 Accurately locate probes in a fixed positional relationship (stand-off) to the weld and maintain probe orientation.
2.2.2 Feedback positional data to instrument and have a controlled scan rate.
2.2.3 Maintain all probe units in regular contact with inspection surface to ensure good coupling conditions. Maintain good couplant condition.
2.2.4 Be simple to set up and convenient to operate and versatile.

To satisfy these objectives, Sonomatic manipulators incorporate the following design concepts (see Fig. 1):

2.2.5 Non-motorisation unless absolutely necessary (i.e. inaccessible or repetitive task).
2.2.6 Light weight, durable construction.

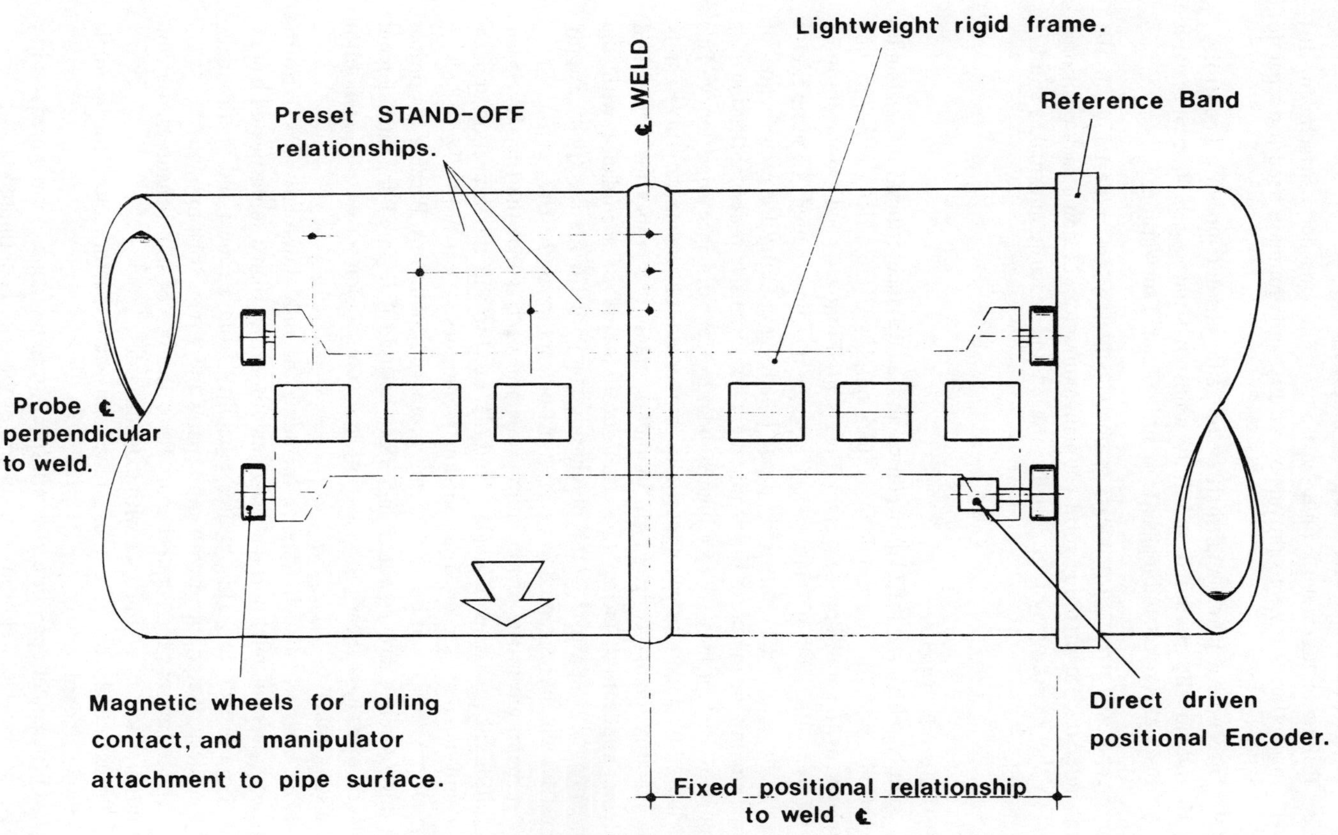

Fig. 1 – Manipulator design concepts.

2.2.7 Ergonomic design with accent on simplicity.
2.2.8 Rolling contact with (and where possible, magnetic attachment to) inspection surface to provide versatility and eliminate need for separate rails etc.
2.2.9 Direct drive incremental shaft encoders for absolute positional control.
2.2.10 Individually mounted and fully adjustable purpose built probes with bi-axial gimbals and magnetic or spring-loaded suspension.

The mechanism used must locate probes accurately and repeatably on the item under test, maintaining good couplant contact and feeding back positional data to the recording system. A light, rugged and versatile unit makes site use practicable.

2.3 Ultrasonic principles

To comply with established procedures and acceptance criteria, multi-channel operation aside, the ultrasonic principles employed are strictly conventional. However, with a multi-probe system it is essential that each and every transducer performs to the same standard – as if a single probe were used. To some extent this can be achieved by providing probe balancing circuitry for each channel of operation but to minimise the requirement for extensive pre-amplification it is important that probes are to a high standard of performance consistency. This is achieved by design, providing accurate component tolerancing, careful assembly and exacting quality control. All Sonomatic miniature probes for mechanised systems application incorporate transducer inserts in accordance with the company's ESI approval – the purchasing standard for the Electricity Supply Industries which is accepted to be the most demanding in the UK (see Fig. 2).

A major advantage of the multi-probe approach is that the transducers can be optimised (in terms of angle, stand-off, etc.) to suit the inspection application and tandem/through transmission techniques – not normally feasible in manual practice – can be used to suit geometric conditions. Because the probes are purpose-built, non-standard angles can be used if required and dual element units comprising a shear probe and integral compression wave coupling monitor are also possible.

The basis of all such systems, including the fully automatic ones, is a proprietary single channel portable flaw detector rather than a sophisticated laboratory type instrument. This ensures reliability from a robust and operationally proven piece of equipment with good spares and service back-up.

The ultrasonic techniques, apart from the use of multi-channels, are conventional, allowing compliance with existing codes.

Probes used therefore must have consistent performance, for example ESI 98-2 standard.

Multi-channel operation allows inspection to be focused, if required, on key areas, for example, weld root erosion and probe criteria optimised to suit.

Signal-Noise Ratio in Steel
Curves plotted from 1.5 ø cross drilled holes at half FSH (high gain) on USM2
Outer curve representative of test block quality steel.
Inner curve representative of coarse grain weld material.
GAIN IN HAND dB
80
70
60
50
40
30
20
10
INTERNAL NOISE
EFFECTIVE WORKING ENVELOPE
NOISE
SIGNAL
5
10
15
20
60
100
140
180
220
260
300
INTERNAL NOISE
RANGE mm
GRAIN NOISE
Sonomatic
B.T.I. Probe
SGS
NOMINAL DATA
Angle
60
Degrees
Single/Twin
S
Single
Xtal size
10
mm ø
Xtal material
5a
PZT
Frequency
2.5
M Hz
PERFORMANCE CHARACTERISTICS
PULSE SHAPE
See Fig. 1
CYCLE SPACE
< ± 15%
FREQUENCY
< ± 10% See Fig. 2
PULSE LENGTH
2.5 μ sec
INDEX POINT
NOT MARKED
BEAM ANGLE
± 2 °
SQUINT
< 1.5 °
SIDE LOBES
> 30dB weaker than main beam
LONGIT^L WAVES
Normal Beam not detectable
Angled Beam > 40dB weaker
SURFACE WAVES
> 34dB weaker
REVERSE SHEAR
> 40dB weaker
SUBSIDIARY MAXIMA
Not detectable outside main beam
SENSITIVITY
> 65dB in hand (½FSH off 100mm)
NOISE
See S/N Curves
Unrectified Display of Probe Pulse
Frequency Spectrum
AMPLITUDE
1
2
3
4
FREQUENCY MHz
Fig. 2
AMPLITUDE
1
2
3
4
TIME MICROSECONDS
Fig. 1

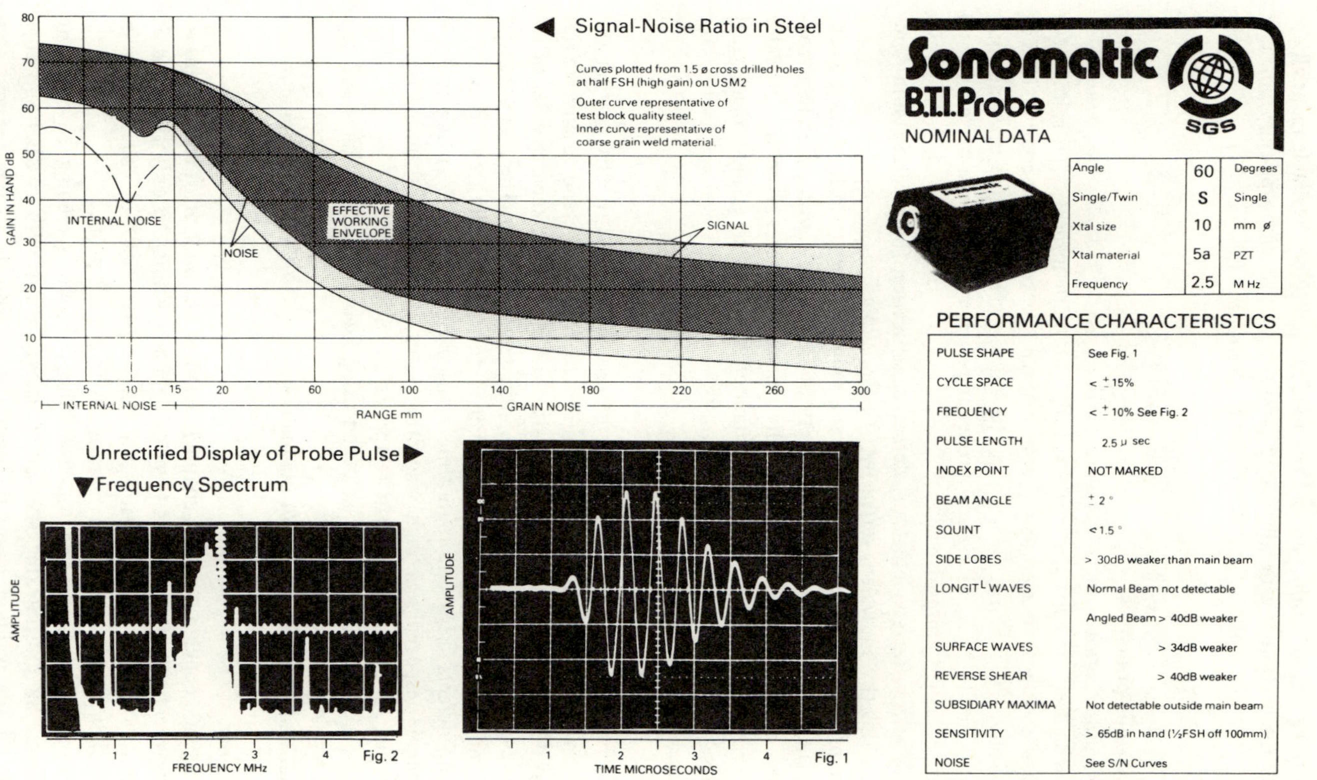

Fig. 2

Coupling can be monitored and recorded, a facility not available manually.

Instrumentation is generally based on a robust, proven proprietary portable flaw detector which allows spares and service back-up.

2.4 Electronic principles

2.4.1 Weld inspection using shear wave techniques

With mechanised, and for that matter, fully automatic inspection, it is impracticable to size defects in accordance with manual procedure. The complexities of mechanically reproducing intricate probe movements for signal maximisation and the difficulty in acquiring and processing such a vast amount of data in real time remain, as yet, unresolved.

However, go/no go detection, defect classification (by position) and comparative sizing (by equivalent reflectivity) are possible.

One of the difficulties confronting the manual operator as he scans a workpiece is the sheer volume, rate and variety of information with which he is continuously confronted. This is especially the case where geometry provides natural reflectors which must be discriminated from defects giving similar signal responses.

Gating

The multi-probe approach itself overcomes some of the difficulty in that the probe is in a fixed positional relationship to the workpeice, and to any regular geometry likely to generate 'ghost' signals. This means that each channel of inspection can be individually 'gated' to eliminate such effects, The true potential of this simple technique can best be explained diagrammatically (see Fig. 3).

Sizing defects using automated/mechanised equipment to existing codes is impracticable, especially at high inspection rates, through XY scanning frames for example, AXYS equipment can be applied to critical areas.

Go/no go screening can be achieved using mechanised techniques and can be commercially attractive.

Final sizing can be carried out manually or using AXYS type equipment.

Multiplexing

Extending this concept to multi-channel coverage not only provides a means of confining the field of vision (or level of data acquisition) to only the areas of inspection worthy of consideration but also enables a rate of coverage far in excess of that possible in manual operation with a single probe.

The following example demonstrates that adequate 'screening' coverage can be achieved by multiplexing six 'gated' shear probes with 6 dB beam edge coverage of the weld at a circumferential scan rate of 50 mm/sec and scan resolution of 0.16 mm. This means that a 10″ butt weld could be fully evaluated

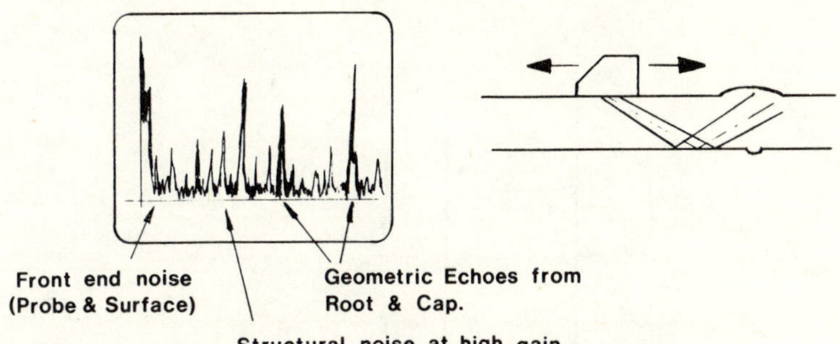

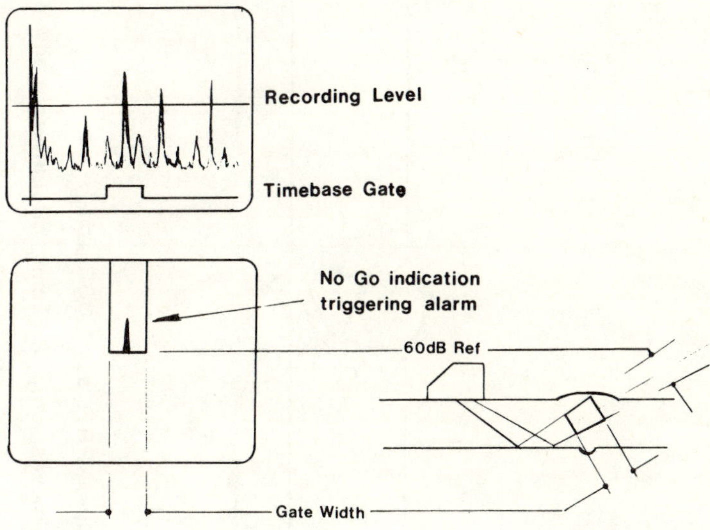

Fig. 3

for the presence of non-acceptable defects in 16 seconds, so affording the operator much more time to assess in detail those areas identified as defective (see Fig. 4).

Taken a stage further, multiple gate and quantised amplitude techniques can be employed to locate defects more accurately in terms of time base (and through wall) position and provide echo dynamic information for equivalent sizing (see Fig. 5).

Amplitude recording

In the majority of inspection situations acceptance levels are specified for reporting and/or sentencing. This means that amplitude recording can be limited to one or two levels. Electronically, a preset video threshold can be set up such that only signals breaking the preset level(s) are processed.

MULTIPLEXING

Circ. Scan
50mm/sec

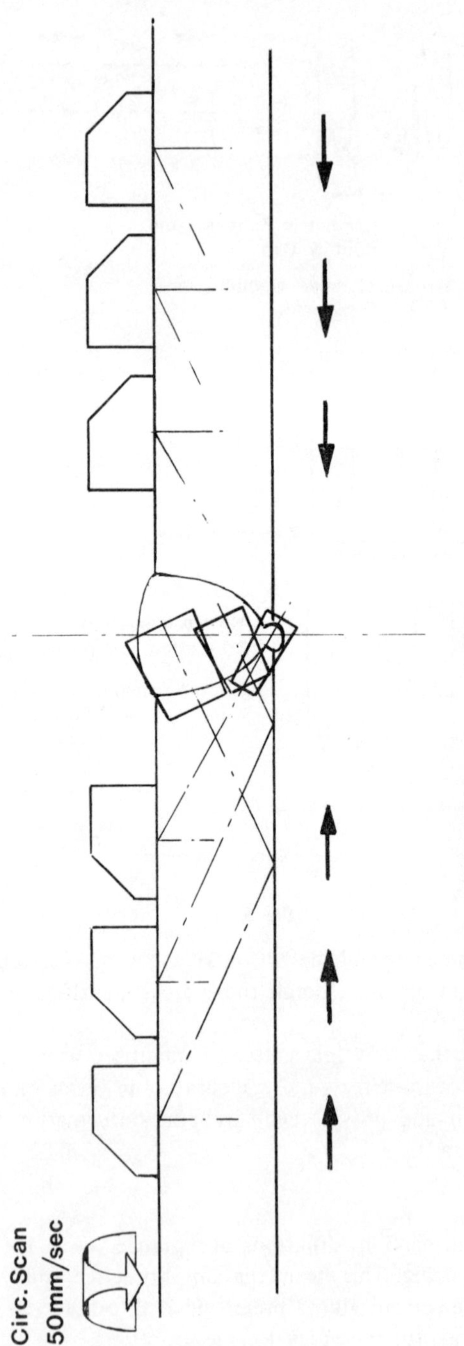

Probe switching PRF say 1800 p.p.s.

ie Manipulator moves only 0.16mm circumferentially for each complete cycle with a 6-probe array.

Fig. 4

The principle of screening, using fixed probes, can be used effectively by 'gating' areas of interest and using the beam width for coverage.

If necessary a number of probes can be used and switched sequentially (multiplexed) to cover more volume of material.

Thus, indications within a 'gate' and above the reference amplitude are recorded for further investigation.

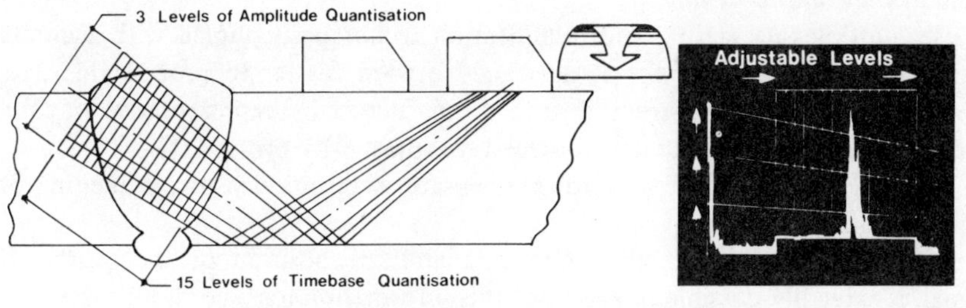

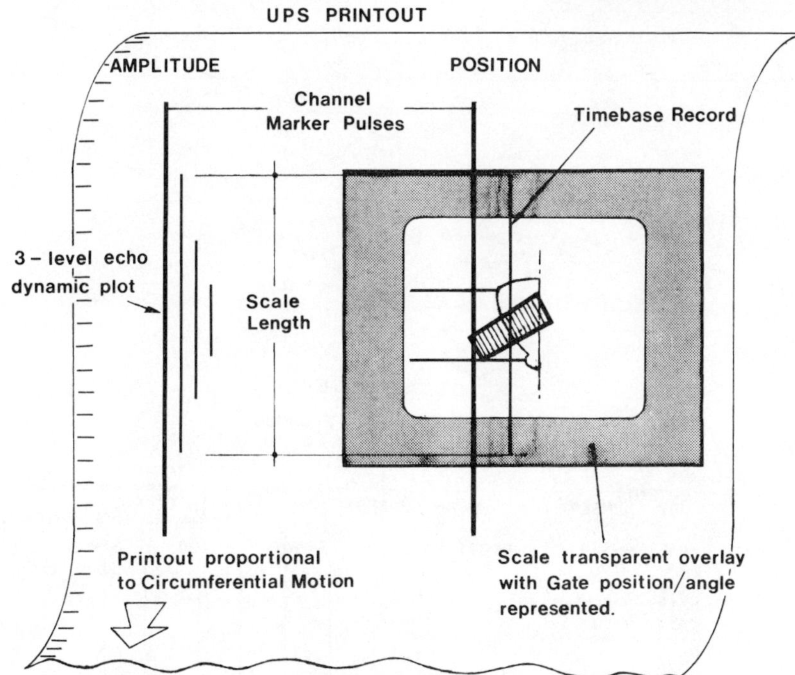

N.B Where timebase record intersects overlay gate, defect position (relative to weld), and approximate depth can be determined.

Fig. 5

To account for beam spread and material attenuation these recording levels can be arranged to represent the DAC characteristics of the test situation. The recording levels and inspection sensitivity can be adjusted to compensate for beam spread, material attenuation, surface condition and so forth. This can be achieved by simply 'sloping' the threshold or adjusting the gain of each probe individually or by more complex and accurate digital means.

Data display and recording

These multiplexing, gating and quantisation techniques, coupled with accurate positional information make it a straightforward matter to process this data electronically and reconstruct it in a form pictorially representative of, and proportional to, the area under test (see Figs. 6(a), 6(b), 6(c), 6(d)).

It is sometimes necessary to compensate electronically for weakening of signals coming from longer ranges.

Having obtained the ultrasonic data and the position of the source of signal, it is readily possible to print out this information for record purposes.

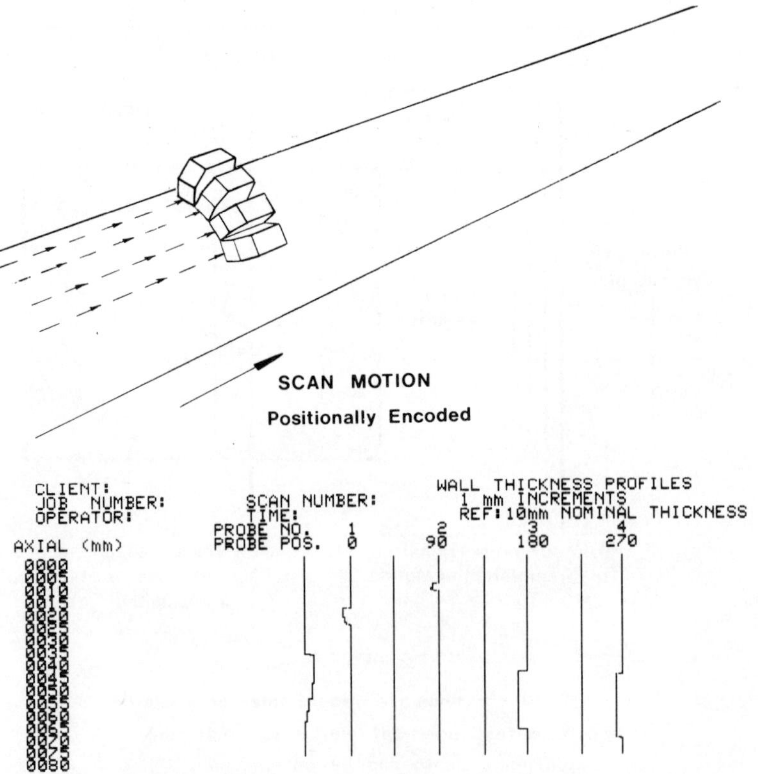

Fig. 6a – Corrolog. Shear or compression wave corrosion detection using timebase gated multi-probe array scanning axially along pipe.

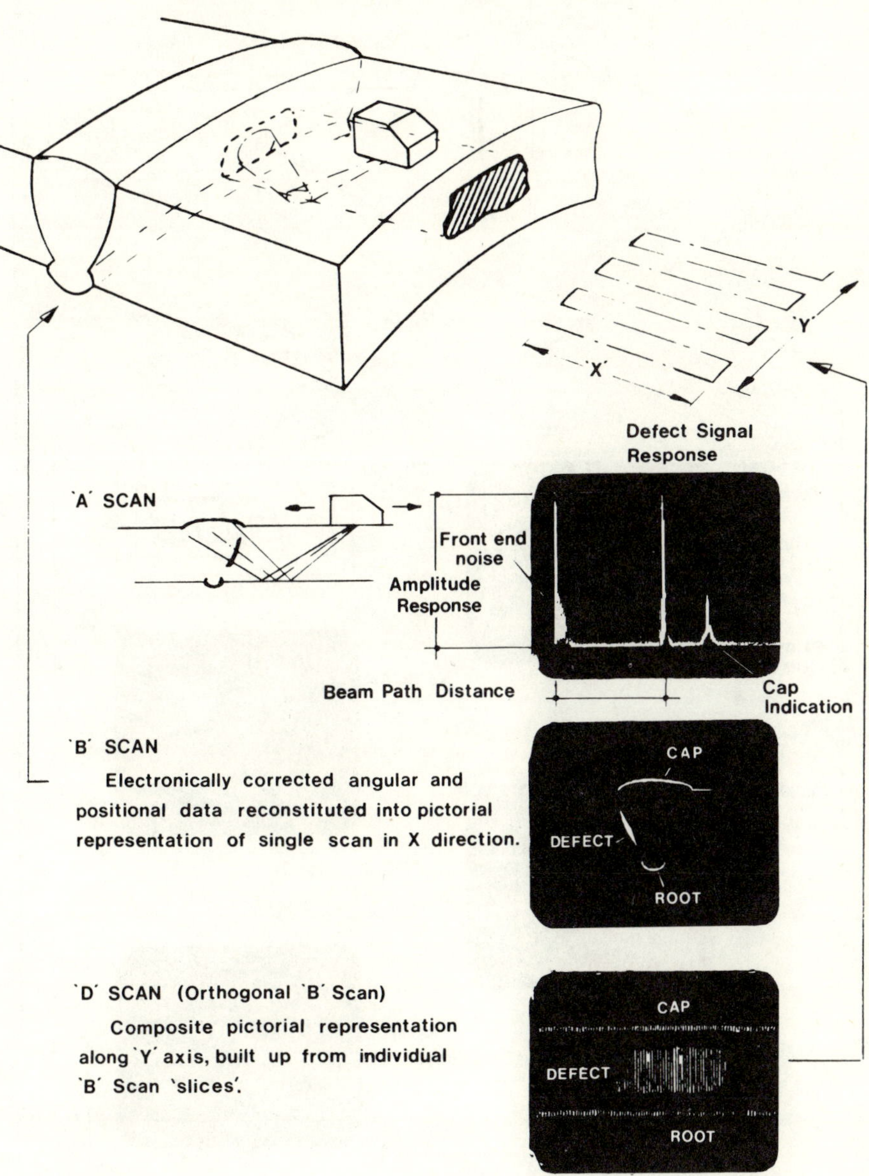

Fig. 6b – Shear wave testing. On-line display techniques for defect visualisation.

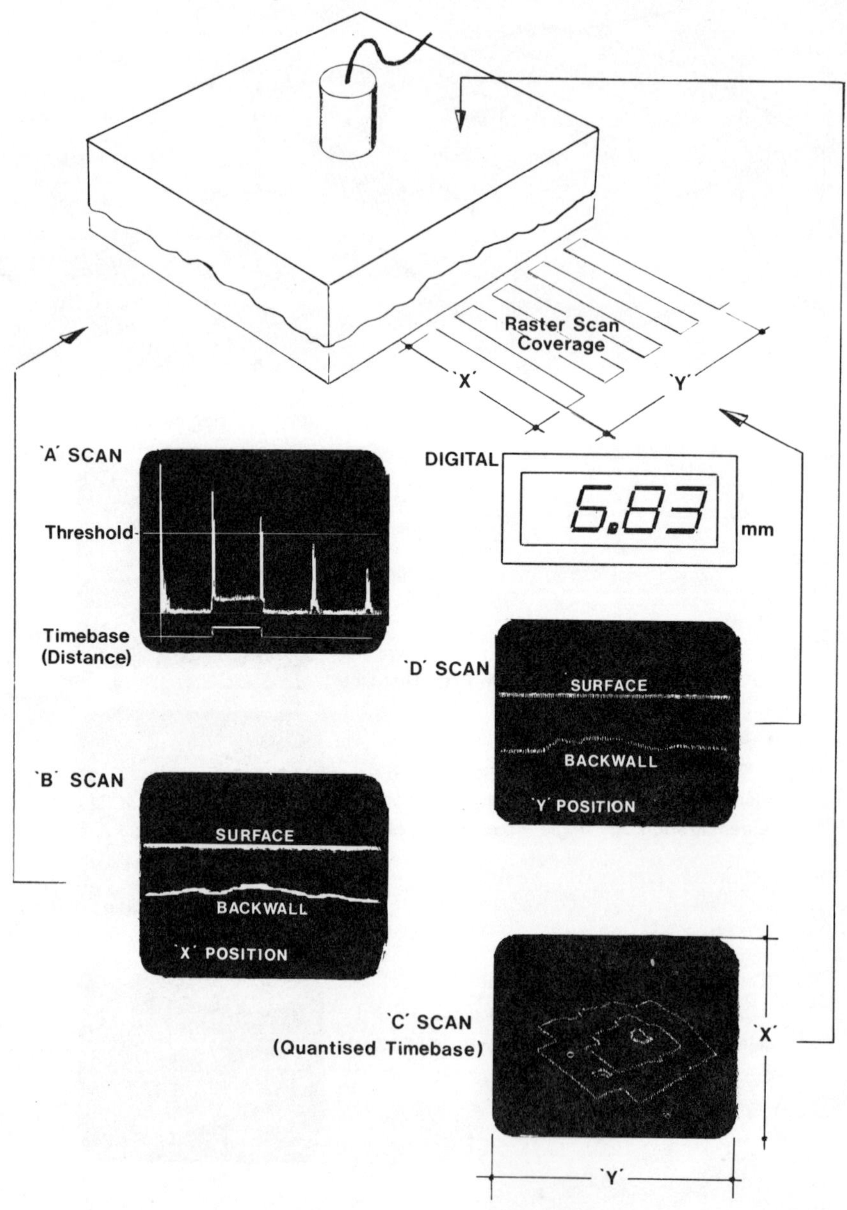

Fig. 6c – Compression testing. On-line display techniques for corrosion assessment.

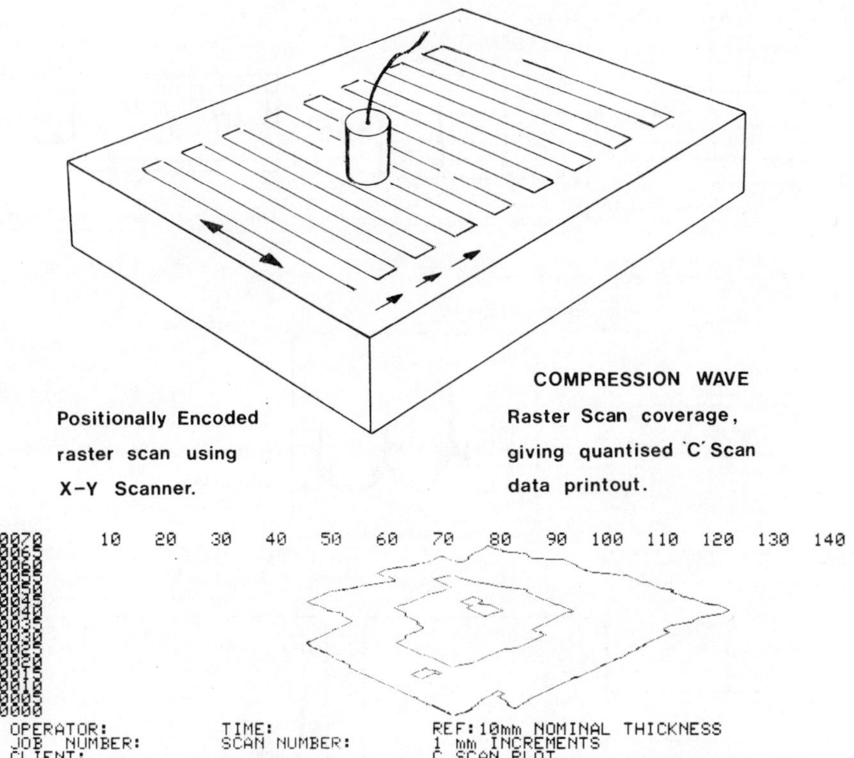

Fig. 6d – AXYS. Isophonic plot.

2.4.2 Thickness (corrosion) testing using compression wave techniques

The traditional approach to thickness testing as a corrosion assessment technique is to use either the calibrated A-scan of a portable flaw detector (usually in conjunction with a twin compression probe) or a proprietary digital thickness meter.

The former is limited by the low resolution offered by the majority of such instrument/probe combinations, and the general inaccuracy of taking readings of a conventional CRT.

The portable thickness meter, with its data display giving actual digital representation of thickness, is useful for the wall thickness measurement of clean flat plate. Fig. 7(a)). Problems can arise, however, on corroded components, in that for a given beam width, various beam path distances are apparent – having the effect of widening the pulse or causing multiple echoes (see Fig. 7(b)).

Thickness testing is often carried out using flaw detectors or digital meters.

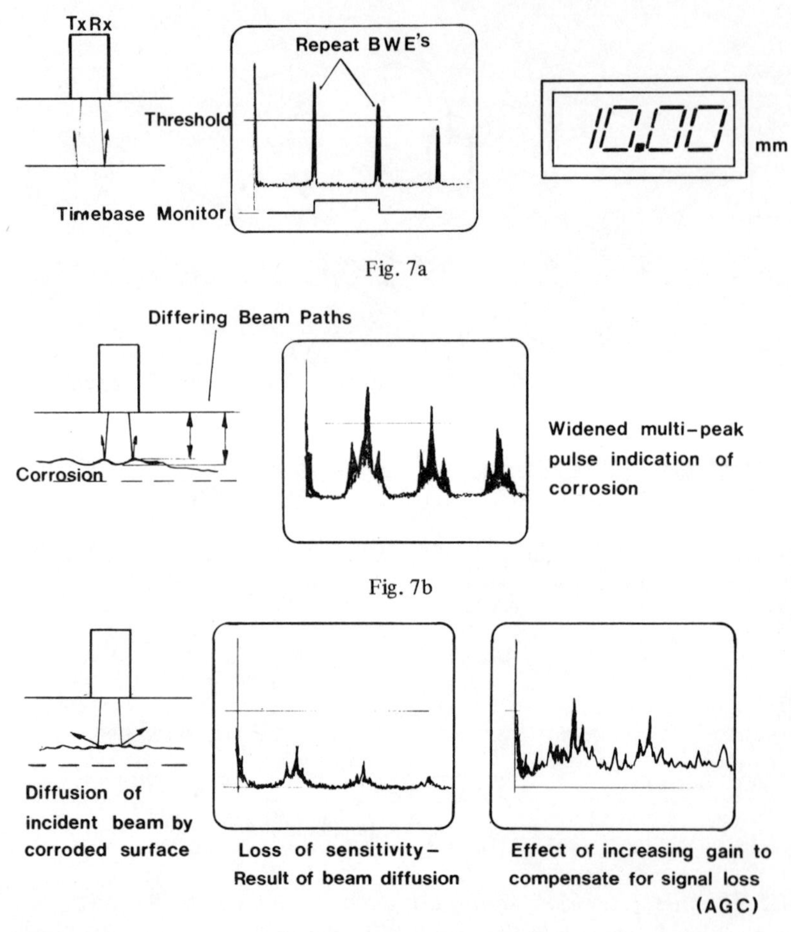

Fig. 7a

Fig. 7b

Fig. 7c

The application of these units must be carefully controlled and is limited by the sort of corrosion levels expected.

The beam is dispersed by the geometry of the reflector and the resultant received signals are of lower amplitude. This normally means that more 'gain' has to be used (AGC) to reinstate the signals to an amplitude that can be monitored but this increase in sensitivity introduces structural 'noise' from the material under test and, in some cases electronic noise from the instrument itself (see Fig. 7(c)). The effect of this is that the instruments electronic monitoring circuitry is incapable of discriminating between actual back wall indications and other spurious signals. This could cause mistriggering, giving a random display reading (see Fig. 7(d)).

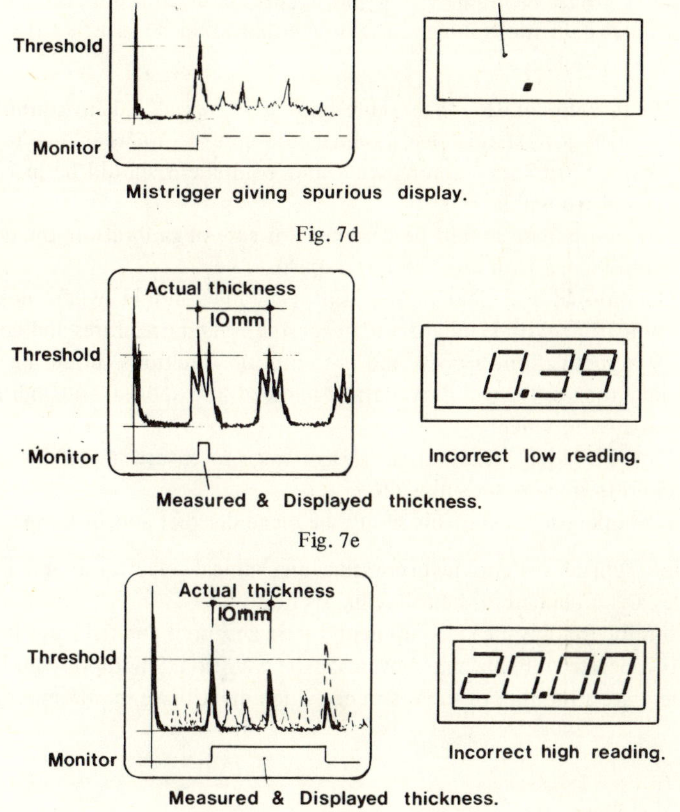

Fig. 7d

Fig. 7e

Fig. 7f

The instrument could also measure between successive 'noise spikes' giving the impression of extreme thinness (see Fig. 7(e)).

If background noise coincides with repeat B.W.E.s further down the time-base, then the signals become superimposed causing a 'double-up' of thickness measurement so that if the original nominal wall was 20 mm, a display could be registered indicating that no loss of metal has occurred – when in fact the material under test has corroded down to 50% its nominal thickness! (See Fig. 7(f)).

Beam dispersion on corroded areas gives distorted and reduced signals, difficult to interpret on a flaw detector and liable to give misleading readouts on digital thickness meters. Paint and other coatings can aggravate the problem.

2.5 Site practicability

In addition to these techniques of acquiring, processing and presenting ultrasonic and positional data the following requirements should be satisfied if site practicability is to be achieved.

2.5.1 The instrumentation 'packaging' should be rugged and environmentalised.

2.5.2 Construction should be modular for ease of maintenance/repair and proven proprietary components and equipment should be incorporated wherever possible.

2.5.3 The equipment should be designed for ease of calibration, operation and maintenance without specialist expertise.

2.5.4 Operational and data presentation principles should, where possible, be compatible with standard practice, established procedures and codes.

2.5.5 Data acquisition, process and presentation functions should be designed such that only necessary data is handled and spontaneous hard copy is not a requirement.

2.5.6 Care should be taken at the design stage to ensure 'noise' immunity by radio frequency screening etc.

2.5.7 Non-operational controls should be made discreet and/or tamper proof.

Some typical systems incorporating mechanical, electronic and ultrasonic principles are explained in the following section.

Naturally for advanced equipment to be accepted for field use it must be simple to operate, reliable and proven against traditional methods over extended trials. Here are a number of such systems which meet these requirements.

3. TYPICAL SYSTEMS

3.1 UCM (Universal Corrosion Monitor)

UCM is a semi-automatic ultrasonic tool with single channel facility enabling the inspection of badly corroded components in difficult access situations which are normally impossible to inspect (see Fig. 8).

3.2 SPIE/SPUPS

The SPIE (Small Pipe Inspection Equipment) and SPUPS (Small Pipe Ultrasonic Printout System) (see Figs. 9 and 10), are both portable multi-channel flaw detection systems differing only in the fact that the latter is equipped with on-line hard copy facility for go/no go recording of defect and related positional information.

Both instruments can be used with a variety of manipulators for butt weld defect detection and weld-related corrosion assessment using pulse echo shear wave techniques.

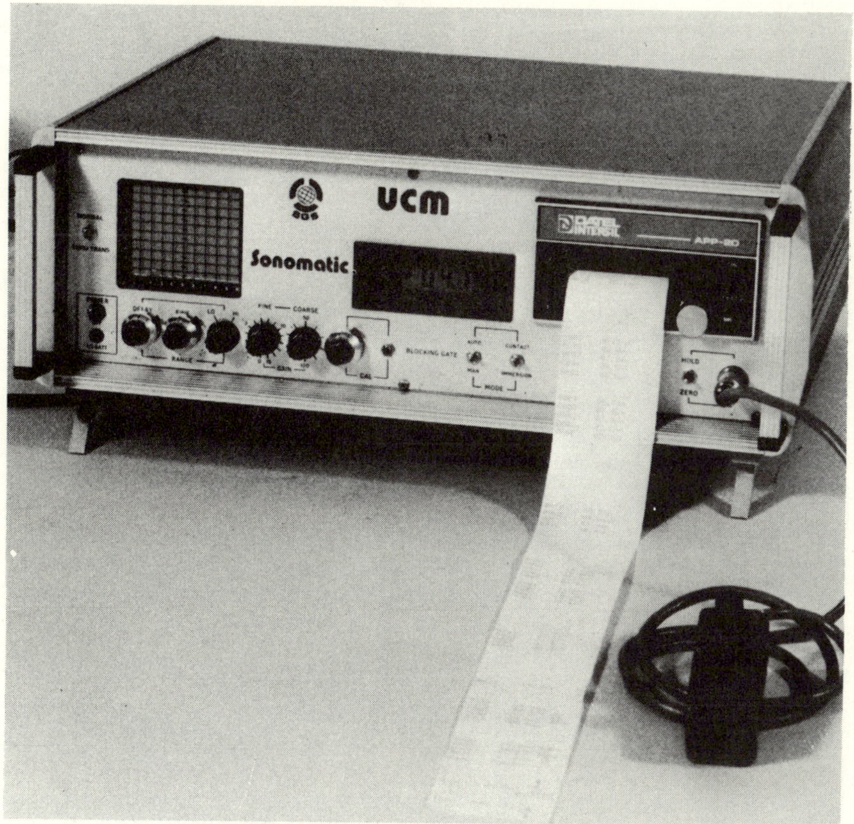

Fig. 8 – UCM (Universal Corrosion Monitor).

3.3 AXYS (Automatic XY Scanner)

This scanner is used for evaluating and sizing critical areas and giving a variety of information displays to help the Engineer assess the condition of the component (see Fig. 11).

3.4 UPS (Universal Pipe Scanner)

UPS is a fully automatic weld inspection which has been successfully used in the field for two years (see Fig. 12).

At the heart of the system is a lightweight (20 kg), robust, waterproof tractor using a unique tracking system based on a toothed, reinforced rubber belt threaded through its drive system and enveloping the pipe. This belt can be cut to size on site and it is joined using a simple demountable pin linkage.

Being inexpensive, light and durable the belt does not damage the pipe or any surface coatings and enables the transition from one pipe size to another without the need to fabricate and fit special precision machined rails.

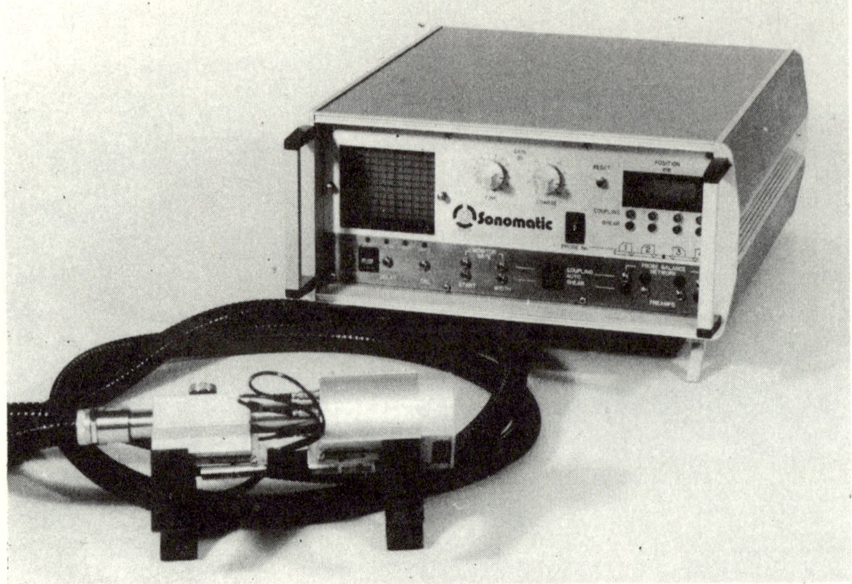

(a)

(b)

Fig. 9 – (a) Small Pipe Inspection Equipment (SPIE). (b) Small Pipe Ultrasonic Printout System (SPUPS).

REPORTING
PROBE ANGLE :65 degrees
REF SENSITIVITY :56dB
TEST SENSITIVITY:66dB
POSITION
WELD CENTRE
CHANNEL
0000
0015 2,3
0030
0045
0060
0075
0090
0105
0120
0135 0,5
0150
0165 1,4
0180
0195
0210
0225
0240 0,1,2,3,4
0255 2,3,4
0270 1,2,3,4
0285 0,1,2,3,4
0300 1,2,3
0315 2,3,4
0330 2,3
0345
0360 1,2
OPERATOR:
DATE:
2
3
Lack of root penetration
0
5
Isolated defects in weld body
1
4
Intermittant indications at various positions (Ch. 0,1 & 4)
Continuous lack of root penetration (Ch. 2 & 3)
Partially penetrated root (Ch.2)
Isolated indication 4/5mm below cap (Ch.1)

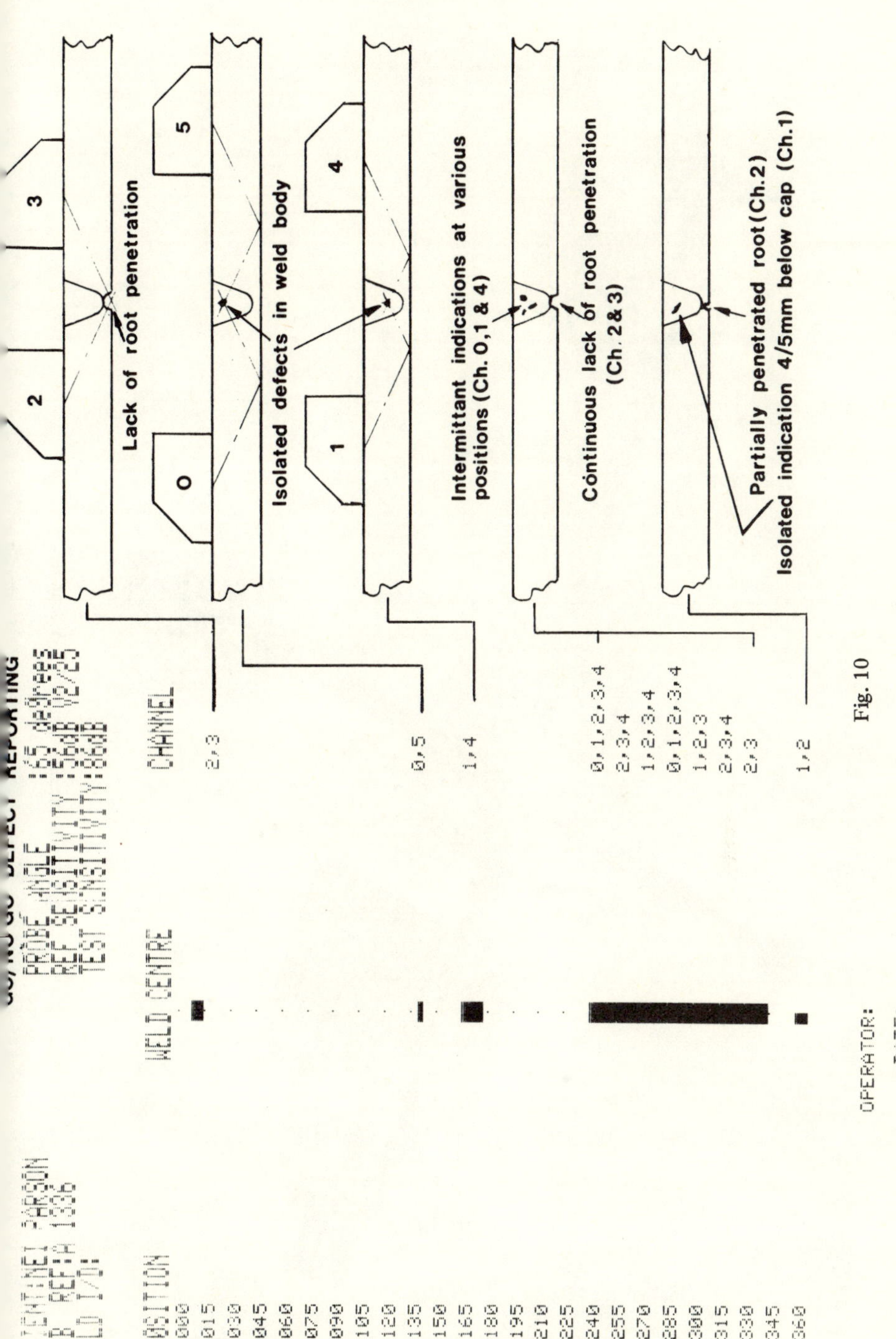

Fig. 10

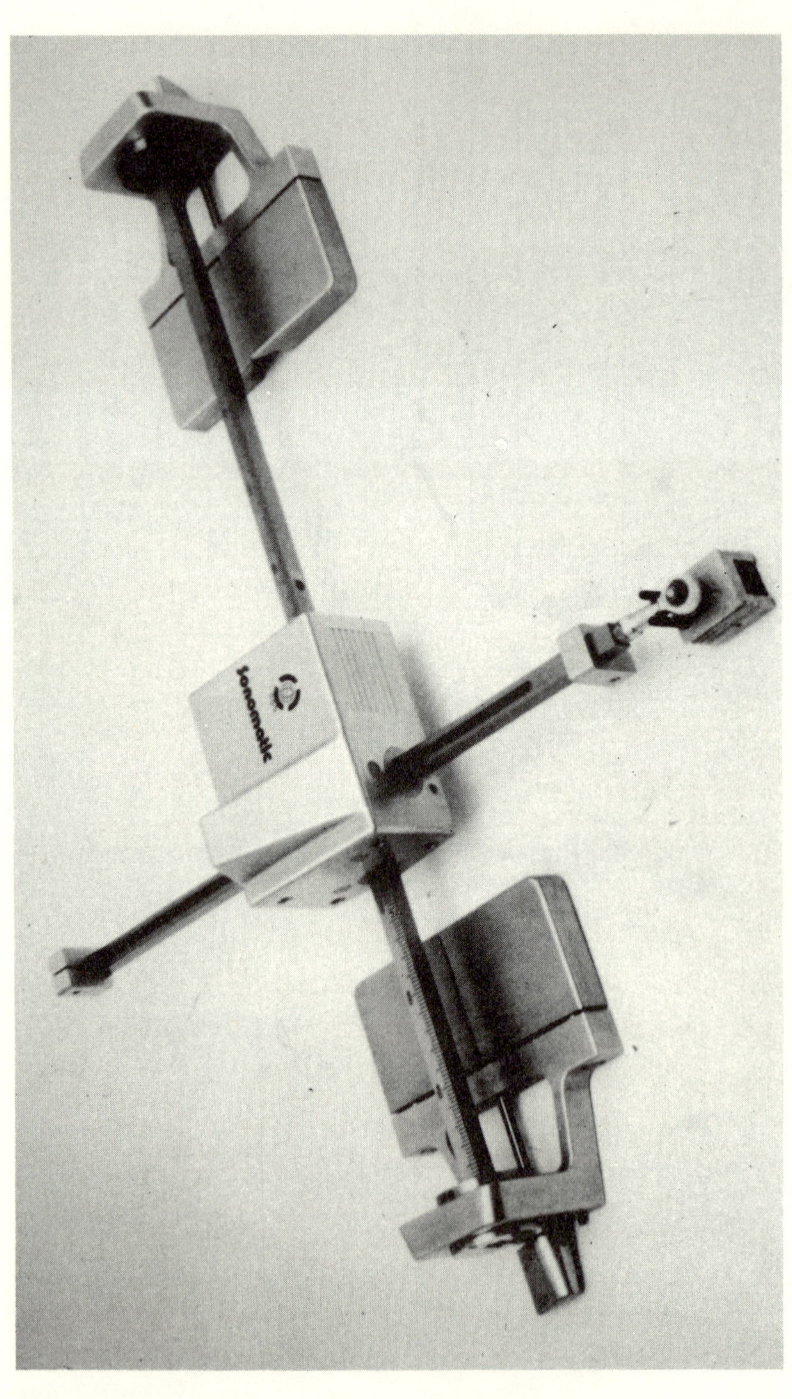

Fig. 11 – Manually-operated X-Y scanner.

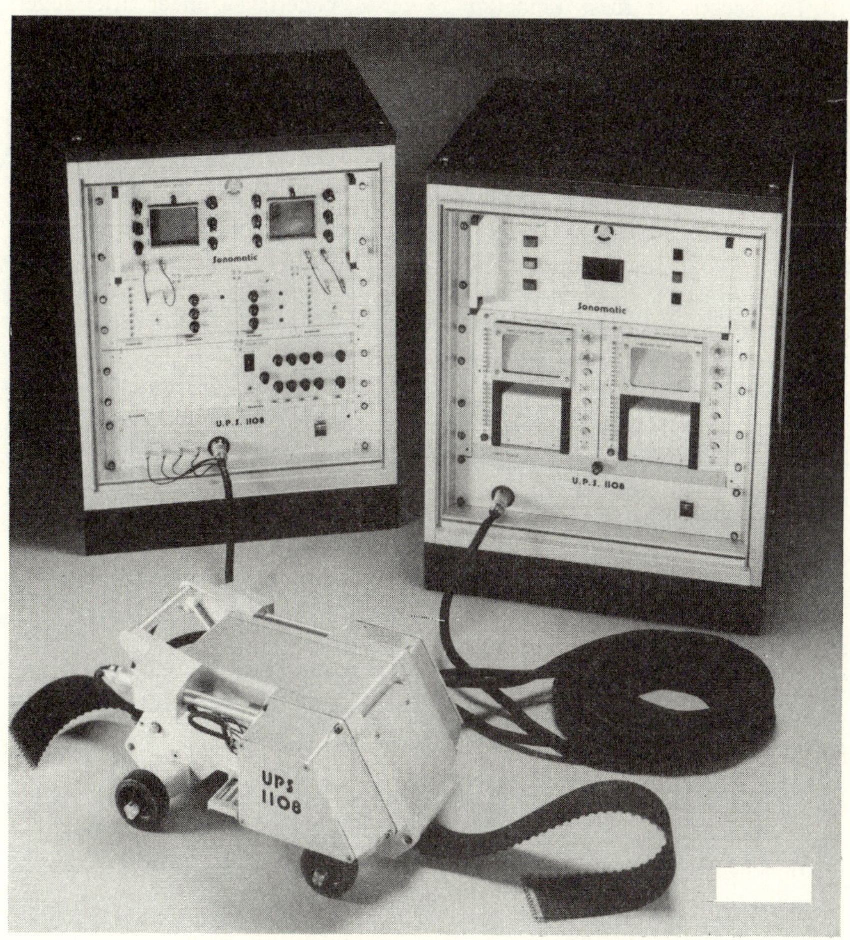

Fig. 12 – UPS (Universal Pipe Scanner).

4. OFFSHORE CONDITION MONITORING

In June 1982, SGS Sonomatic were commissioned by Shell Expro Aberdeen to carry out offshore corrosion surveys. Techniques and procedures were devised specifically for the task and those using the above systems were comprehensively validated to satisfy Shell's requirements, prior to mobilising offshore.

4.1 Validation

Shell prepared test coupons containing representative 'known' defects and these were used to validate the SPIE and SPUPS systems and inspection procedures.

The success of these validation exercises, undertaken under strict supervision by Shell Expro, were sufficient to demonstrate that the UCM, SPIE and SPUPS equipment, and related procedures and techniques, were fit for purpose as a means of offshore condition monitoring on their platform in the Brent Field.

4.2 Scope of work

The object of the corrosion survey was to demonstrate the technical and commercial viability of using such techniques as a field 'practicable' means of providing a data base for condition monitoring purposes.

A three-man team comprehensively examined more than 80 welds in various locations on the three platforms. Inspection coverage was geared to provide both weld defect inspection and to monitor corrosion in the weld root, HAZ and adjacent parent pipe.

Over 2,800 thickness readings were recorded together with exact positional information as a reference for future re-inspection. All areas between these points were monitored but not recorded.

The area of tubular components covered by shear wave testing is estimated at 170 sq. ft. on pipe sizes ranging from 6″ NB to 8″ NB. All indications exceeding threshold were evaluated in detail and reported.

4.3 Techniques

Because of the volume of coverage required and the uncertaintly of the nature and extent of defective material, a 'screening' approach was used to detect and locate defects followed by back-up detailed evaluation.

Ultrasonic compression wave technique of thickness testing for internal corrosion
The UCM equipment was used as a systematic means of measuring and recording the thickness of metal adjacent to welds.

4.4 Reporting

Since the objective of the exercise was to provide meaningful data, the question of accurately referring positional information to a fixed datum was carefully considered.

4.5 Results

The validation and offshore inspection exercise proved, to the client's satisfaction, that mechanised ultrasonic examination is technically capable of satisfying condition monitoring requirements.

The techniques and procedures used with the SPIE/SPUPS and UCM systems cost effectively provided accurate and comprehensive hard copy records of corrosion data, which can serve as a basis for subsequent in-service assessment with consistency and reproducibility.

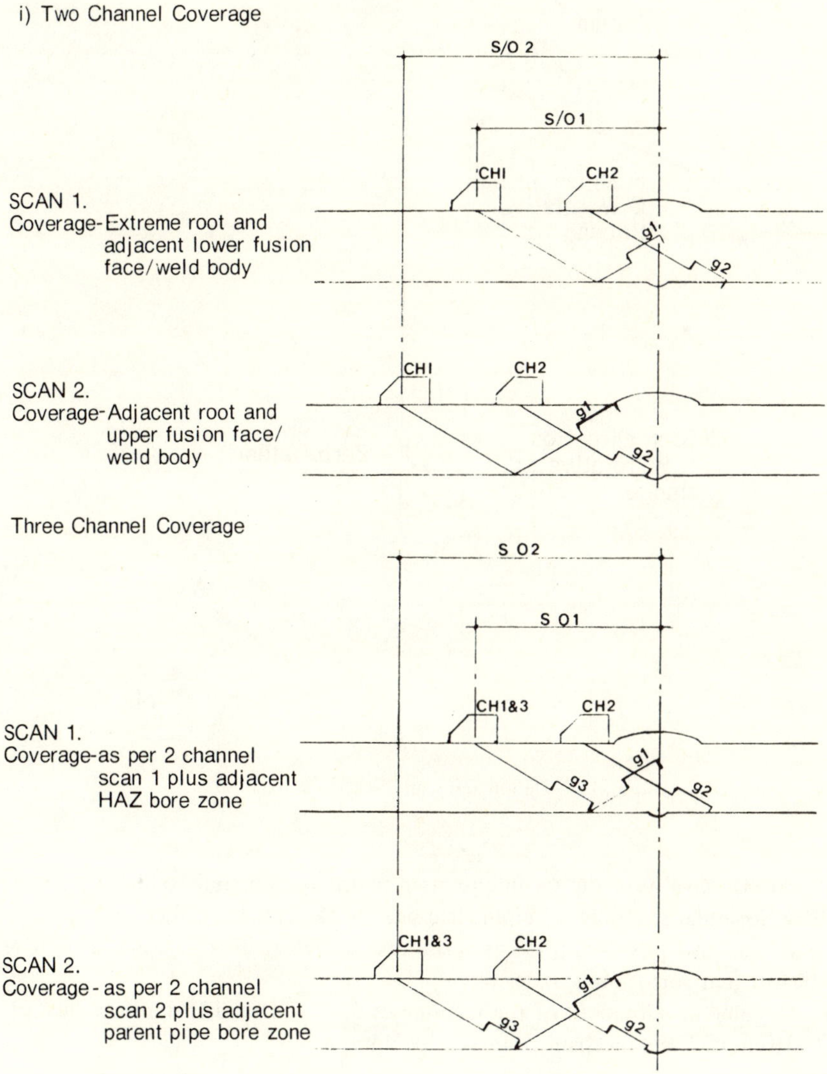

i) Two Channel Coverage
S/O 2
S/O 1
CHI
CH2
SCAN 1.
Coverage-Extreme root and adjacent lower fusion face/weld body
g1
g2
CHI
CH2
SCAN 2.
Coverage-Adjacent root and upper fusion face/ weld body
g1
g2
Three Channel Coverage
S O2
S O1
CH1&3
CH2
SCAN 1.
Coverage-as per 2 channel scan 1 plus adjacent HAZ bore zone
g1
g3
g2
CH1&3
CH2
SCAN 2.
Coverage - as per 2 channel scan 2 plus adjacent parent pipe bore zone
g1
g3
g2

Fig. 13 – Two and three channel coverage. Shear waves.

6/8mm
6 5 4 3 2 1 0
7-13
0-6
ZERO DATUM
14-20
21-27

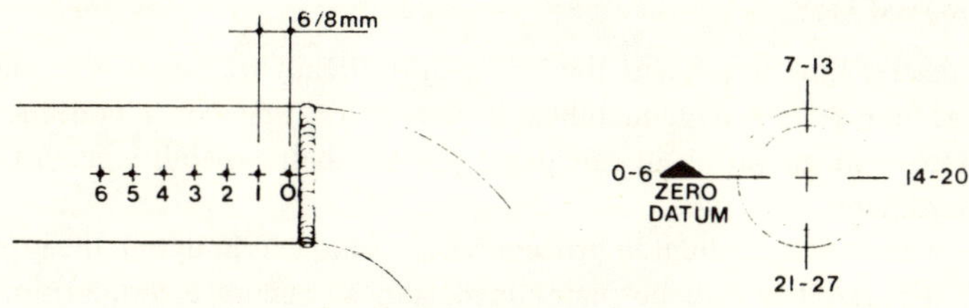

Fig. 14 – Grid coverage. Compression waves.

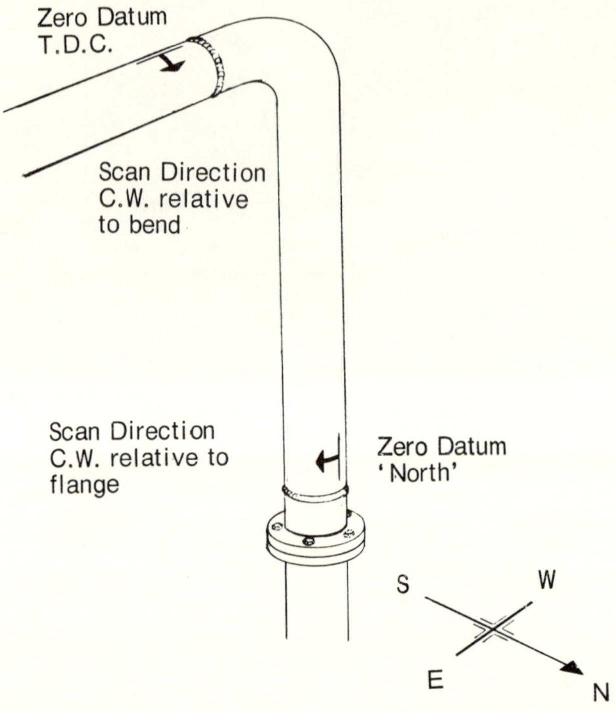

Fig. 15 – Datum and scan direction convention.

So effective were the techniques used that it was possible to identify a joint where dissimilar schedules of piping had been butt welded together.

It was also possible, in some instances, to detect liner disbond, a positive indicator that corrosion is under way.

Ongoing development of the techniques used will further improve the cost-effectiveness of the mechanised inspection approach.

5. SUMMARY

No matter how important the information being presented, the quality in terms of accuracy and dependability is the over-riding consideration. The system described in operation in this paper proved their capability against the specified requirements.

If previously validated procedures are followed in detail, the proven accuracy and repeatability will be maintained, and an accurate comparison of like with like will result from subsequent inspection.

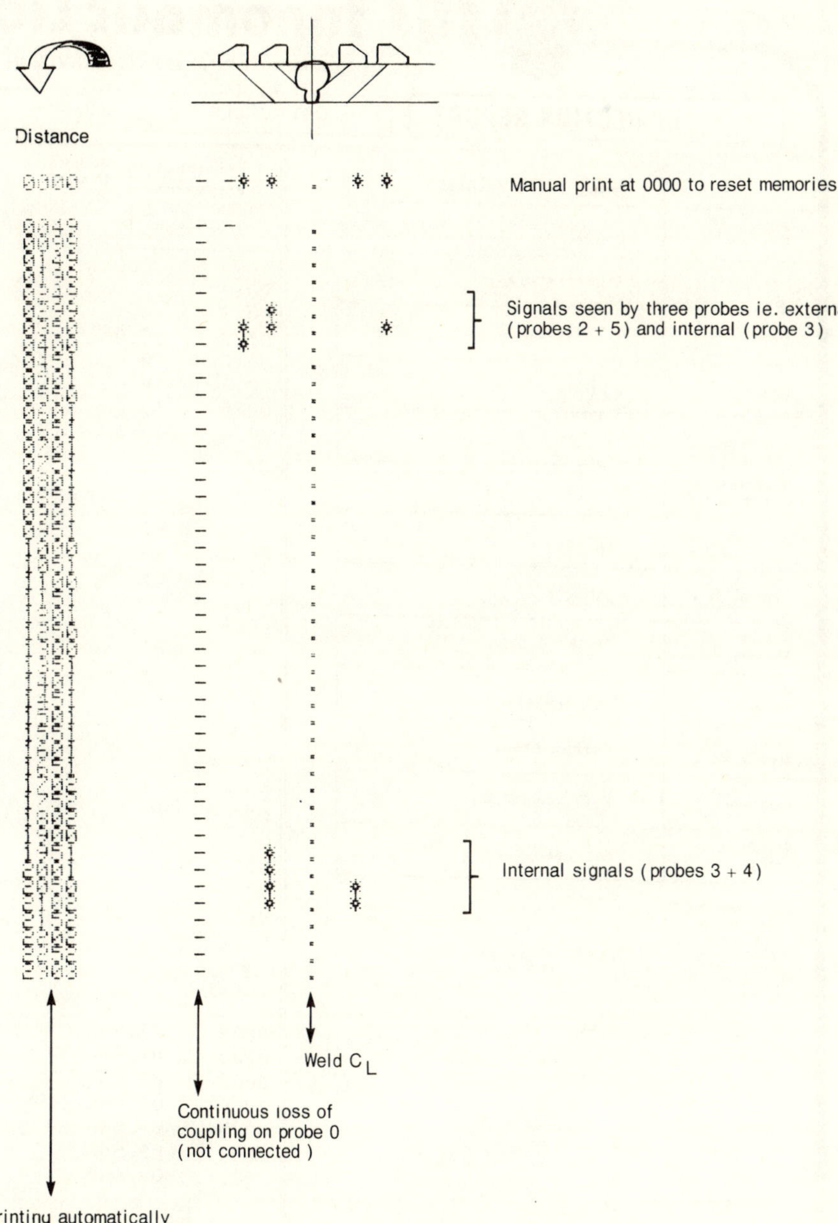

Fig. 16 – Typical SUPS printout.

SGS Sonomatic Ltd.

CORROSION MONITORING SERVICES DIVISION

INSPECTION REPORT

REPORT NUMBER 5331/024

Client	SHELL EXPRO ABERDEEN
Job Reference	C5331
Weld Ref.	A Deariators 24
Date	4.7.82
Engineers	[signature]
Technicians	
Inspection Code/Mode	UCM/UTP - 2
Nominal O.D.	14.625"
Nominal I.D.	14" N.B.
Nominal Thickness	9.38/10.72mm
Critical Thickness	Not Specified
Tube Material	Carbon Steel
Increment	See Convention
Acceptance Criteria	Not Specified

Other Observations

Senior Operator

Positional Reference (See Conversion)	Measured Thickness
0031	10.2mm
0030	10.1mm
0029	10.1mm
0028	10.0mm
0027	10.0mm
0026	10.0mm
0025	10.0mm
0024	10.1mm
0023	09.2mm
0022	09.0mm
0021	09.0mm
0020	09.0mm
0019	09.0mm
0018	09.1mm
0017	09.1mm
0016	08.5mm
0015	08.5mm
0014	08.5mm
0013	08.6mm
0012	08.6mm
0011	08.6mm
0010	08.8mm
0009	09.9mm
0008	09.6mm
0007	09.6mm
0006	09.6mm
0005	09.5mm
0004	09.5mm
0003	09.6mm
0002	12.0mm
0001	10.0mm
0000	08.0mm

20 Rivington Court, Hardwick Grange, Woolston, Warrington WA1 4R.
Tel.: Padgate (0925) 810511 Telex: Sonmat 627539

Fig. 17

This improved reporting provides corrosion engineers with the ability of comparing like with like, which facilitates the establishment of more accurate corrosion rates and subsequently the prediction of life to failure of systems.

It can be confidently affirmed that continuing emphasis will be placed on the full range of factors influencing the credibility of information produced by these and other similar systems.

Part 2
CHEMICAL CONDITION MONITORING

CHAPTER 5

The assessment and prediction of plant corrosion using an on-line monitoring system

M. J. Robinson and J. E. Strutt

1. INTRODUCTION

The incentives for on-line corrosion monitoring are high in terms of improved plant reliability and reduced downtime. If the corrosion rate can also be regarded as a process variable then the plant operating conditions may be controlled in such a way as to optimise the quantity and the quality of the product produced while at the same time maintaining the corrosion rate at an acceptable level. In addition, the possibility exists for using lower grade and thus less costly constructional materials provided that their integrity can be reliably assessed during their service life.

This chapter discusses an on-line corrosion monitoring system based on the electrochemical a.c. impedance technique. Initial development studies were conducted on a stainless steel flow loop containing glacial acetic acid with various catalyst additions. The corrosion rate measurements were made with an annular type flush mounted probe and the results were stored together with the operating temperature and flow rate in a logging microcomputer. The software was developed to carry out correlations of the data and to present graphical displays at predetermined intervals [1]. The results of both flow loop experiments and plant trials are described and techniques are discussed by which a corrosion monitoring system may be employed to give predictive information concerning the condition of plant in susceptible yet inaccessible locations.

2. THE A.C. IMPEDANCE TECHNIQUE

2.1 The equivalent circuit

The processes occurring at a corroding metal surface can be considered to behave electrically in the same way as a resistor, capacitor and inductor network. A generalised Randles-type equivalent circuit which describes a range of corrosion impedance processes is shown in Fig. 1.

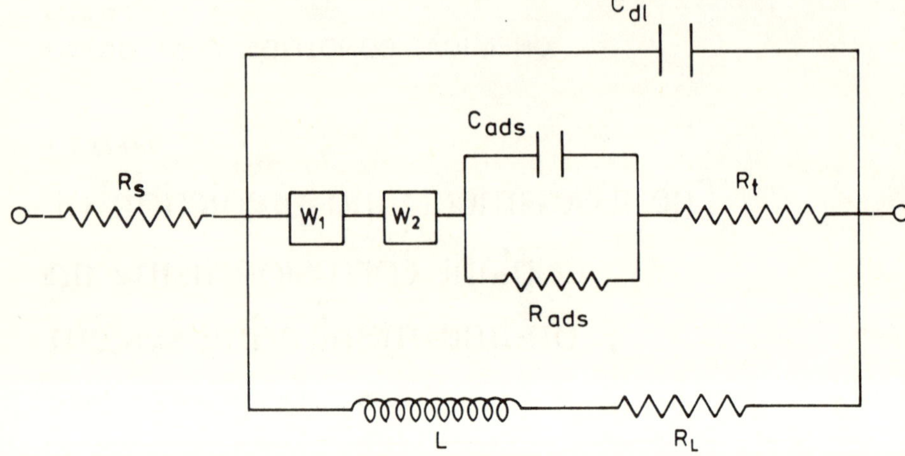

Fig. 1 – A Randles-type equivalent circuit representing the reactions occurring in a corrosion process.

The electrochemical significance of each of the components is as follows [2]. The resistor, R_s, represents the ohmic resistance of the solution and of any surface film that may be present. Clearly in a low conductivity electrolyte such as acetic acid, the solution resistance can be relatively large. Transfer of electrical charge is an inherent part of the anodic and cathodic corrosion reactions and R_t is the resistance to electron transfer across the surface. R_t is of prime importance being, inversely related to the corrosion rate and the value from which the rate is calculated. For an activation controlled corrosion process, provided that the solution resistance is small and complex adsorption – desorption processes are not occurring, then R_t is equivalent to the linear polarisation resistance measured by some commercially available d.c. corrosion rate instruments.

In either case the corrosion current density, I_{corr}, may be obtained from the Stern–Geary equation [3].

$$I_{corr} = \frac{ba\,.\,bc}{2.3\,(ba + bc)} \cdot \frac{1}{R_t} \tag{1}$$

$$= \frac{B}{R_t}$$

where *ba* and *bc* are Tafel constants.

In commercial instruments the value of the constant, *B*, for iron is generally taken to be 20 millivolts [4]. A current density of 1 microamp/sq. cm corresponds to a corrosion metal loss for iron of 0.0116 mm/year [5]. It follows, therefore, that for practical purposes;

$$\text{corrosion rate} = \frac{232}{R_t . A} \tag{2}$$

where A is the electrode surface area.

In addition to the solution and charge transfer resistances, in Fig. 1, ions and water molecules adsorbed on the surface introduce a double layer capacitance, C_{dl}. In some corrosion processes the diffusion of electroactive species towards or away from the interface may have a controlling effect on the corrosion rate. The diffusion of dissolved oxygen to cathodic sites, for example, is generally rate limiting for the corrosion of steels in aerated neutral aqueous environments. W_1 and W_2 are Warburg terms which represent the effects of concentration and diffusion of species within the electrolyte on the a.c. response of the circuit. Finally, the inductor, L, and its associated resistance, R_L, relate to adsorption and desorption of species on the surface and appear to have most significance at potentials where passivating alloys break down, exhibiting pitting, intergranular corrosion and other localised corrosion phenomena.

2.2 Impedance measurements

The impedance of a corrosion process can be measured using a Transfer Function Analyser (TFA) over a range of frequencies from 1000 Hz to 0.001 Hz. A sinusoidal a.c. signal, $V_0 \sin wt$, of 10 mV amplitude is applied, at each frequency, w, in turn, between a working (specimen) electrode and an inert auxiliary electrode. The amplitude and phase response of the current, $I_0 \sin (wt + \theta)$, is measured and the impedance, Z, calculated from the relationship:

$$|Z| = V_0/I_0 \tag{3}$$

and

$$Z' = |Z| \cos \theta, \quad Z'' = |Z| \sin \theta$$

where Z', Z'' are the real an imaginary components of impedance respectively and θ is the phase angle between the current and voltage signals.

2.3 Interpretation of impedance information

To aid the interpretation of impedance data it is usual to plot a graph of the real versus the imaginary components at each frequency. This generates a Nyquist figure from which the required corrosion parameters can be determined. The semicircular shape, shown in Fig. 2(a), is characteristic of the results commonly found in studies with stainless steel in acetic acid. The same shape results from measuring the impedance of a dummy network representing solution resistance, R_s, and the parallel components of charge transfer resistance, R_t, and double layer capacitance, C_{dl}. As impedance is frequency dependent, the points near the origin correspond to the high-frequency measurements. It can be seen that R_t may readily be obtained from such a plot, being the diameter of the semicircle. For practical on-line monitoring at times when the corrosion rate alone is required then, in this situation, it is only necessary to take two measurements, one at high and one at low frequency. There are advantages, however, in performing a complete frequency scan at perhaps daily intervals. Firstly, the frequencies used in the two point measurements can be checked to ensure that they correspond to data points at the intersection of the semicircle and the real axis. Secondly, the tendency to form corrosion pits, for example, can be readily detected by observing whether an inductive loop is present below the axis of the plot as shown in Fig. 2(b).

Other corrosion systems, including steel in aerated sea water, display a Nyquist plot similar to Fig. 2(c). The straight line region results from the Warburg diffusion term and as the influence of diffusion increases, the plot approaches more nearly a straight line with a gradient of 45 degrees. Clearly obtaining the value of R_t is more complex in such cases and depends on the construction of the semicircular form from the available data. This may not be possible if the process is entirely diffusion controlled and in such circumstances an alternative monitoring technique must be employed.

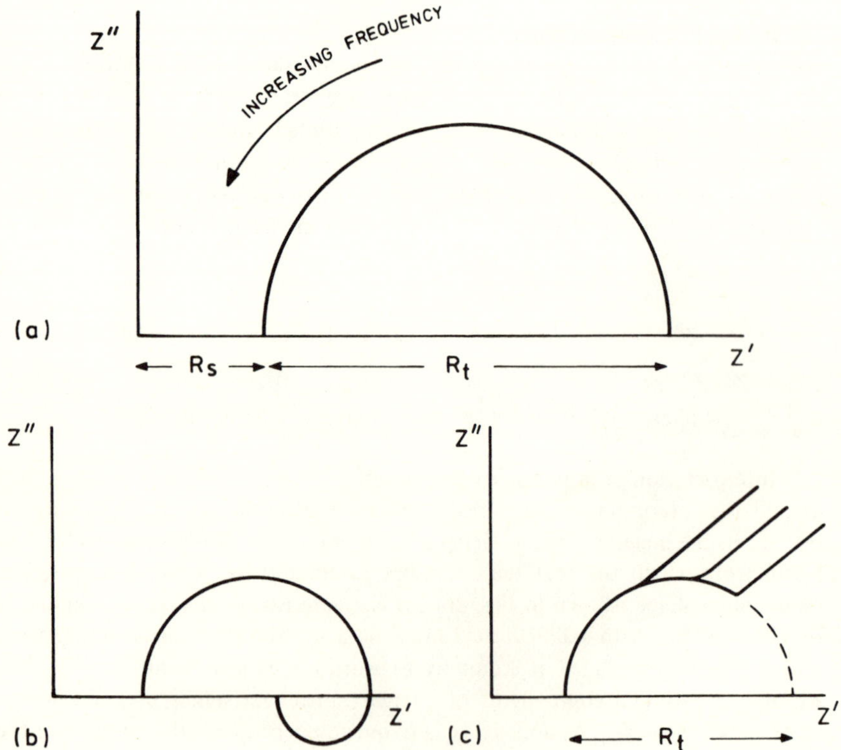

Fig. 2 – Schematic Nyquist plots showing the influence of: (a) Charge transfer resistance; (b) Inductive components; (c) Warburg diffusion terms.

The effect of increasing temperature on the corrosion rate and upon the Nyquist plot is illustrated in Figs. 3(a) and 3(b) from the laboratory testing of 316L stainless steel exposed to a glacial acetic acid/bromide mixture. In each case, a reduction in the dimensions of the plot accompanied an increase in temperature. A similar reduction in R_t occurred when the specimens were anodically polarised to artificially increase the corrosion rate. At potentials of 600 mV and 900 mV (SCE) the stainless steel was in its pitting region, the Nyquist plot showing an inductive loop at low frequencies.

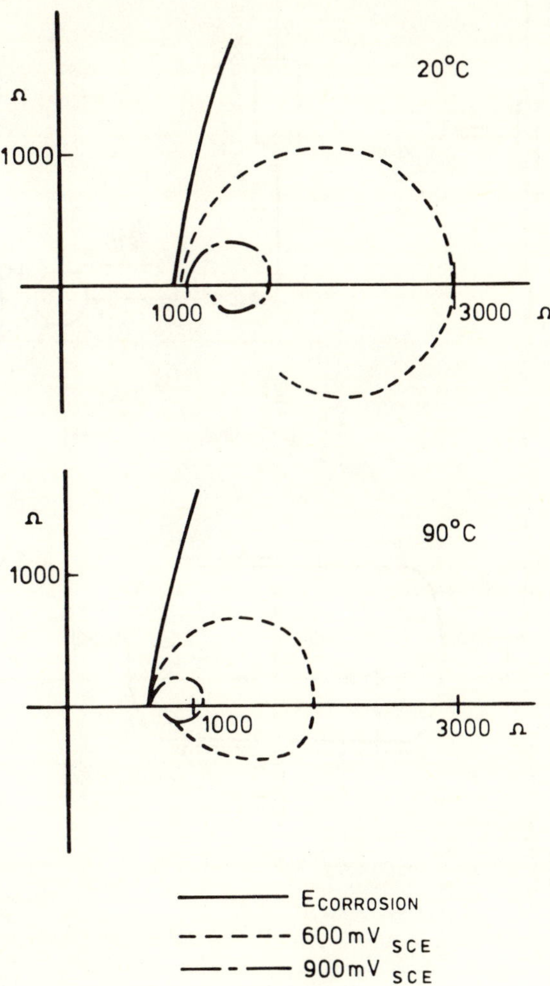

Fig. 3 – Graphs showing the effects of temperature and applied potential on the Nyquist plots for Type 316L exposed to glacial acetic acid/bromide mixtures.

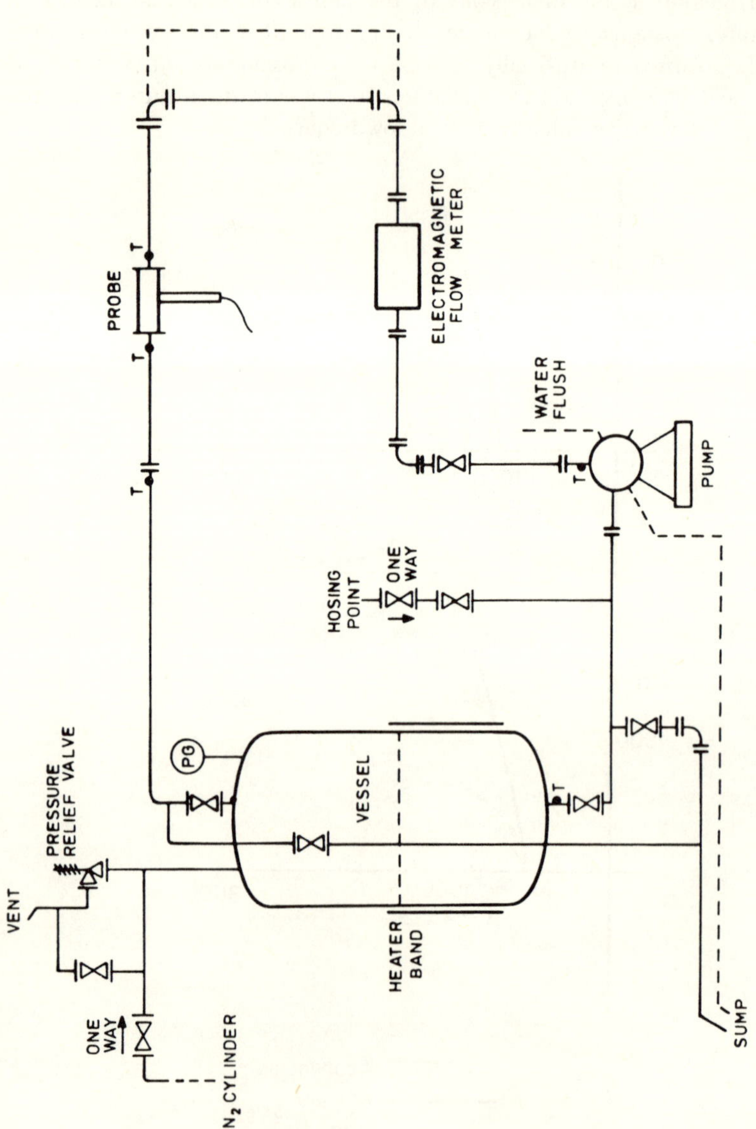

Fig. 4 – Schematic diagram of the flow loop.

3. FLOW LOOP STUDIES

3.1 General design

A schematic diagram of the 316L stainless steel flow loop is shown in Fig. 4. Acetic acid was pumped from the 90-litre heated pressure vessel through 20 mm I.D. pipework containing an electromagnetic flowmeter. A corrosion probe was situated in the upper self-draining limb. The loop was assembled from flanged sections of pipe such that its size and shape could be altered and the probe repositioned. In addition, lengths of pipe constructed from other materials could be inserted in order to assess their corrosion behaviour. The operating limits of the loop were set at 150°C and a pressure of 10 bar.

The electrical signals from the thermocouples and the measuring instruments were lead to a North Star microcomputer (56K RAM) housed in an adjacent laboratory. The TFA function described above was provided by a purpose-built board in the computer which was designed to operate up to four probes simultaneously.

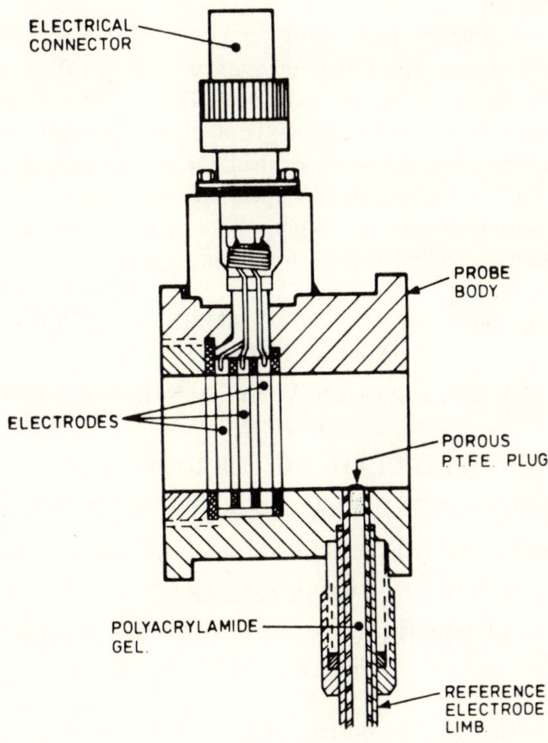

Fig. 5 – Diagram of the corrosion probe used for a.c. impedance measurements.

3.2 The corrosion probe

The probe, which was constructed at Cranfield, is shown in Fig. 5. It was designed to be easily inserted between the flange faces of the pipework. The probe body contained two identical annular 316L electrodes located on either side of a silver auxiliary electrode. The stainless steel rings were machined from a section of the pipe to ensure that they would behave chemically in an identical manner to the rest of the loop. The electrodes were separated from each other by a washer of FEP polymer and from the body of the probe by a backfilling of chemically resistant epoxy resin. While one of the stainless steel rings was used as the specimen (working electrode), the other acted as a reference electrode, establishing a stable potential when exposed to the acetic acid. In the tests described, the reference ring was short-circuited to the pipework thereby ensuring a reference potential characteristic of the rest of the loop. In all the loop investigations the free corrosion potentials of the steel electrodes differed by less than 5 mV indicating that this approach was valid. The electrode potential was calibrated on a thermodynamic scale against a standard sulphate electrode held in the limb containing a polyacrylamide gel salt bridge [6].

3.3 Liquor testing

In addition to testing the probe design and developing monitoring software, the flow loop was used to study the corrosion of Type 316L in a range of commercial liquors based on glacial acetic acid. The experiments were designed to determine the effect of different levels of process catalysts and other liquor additions on the corrosion rate of the steel, the object being to establish suitable catalyst levels and the maximum safe operating temperature.

The evaluation of one liquor composition, tested at a single flow rate, took five hours and was conducted in the following sequence:

FULL IMPEDANCE SCAN AT AMBIENT TEMPERATURE
DISPLAY NYQUIST PLOT

HIGH & LOW FREQUENCY IMPEDANCE MEASUREMENTS
AT 64 Hz & 0.01 Hz

CALCULATE THE CORROSION RATE

RECORD THE TEMPERATURE AT EACH THERMOCOUPLE

INCREASE THE TEMPERATURE & RETURN TO STEP 2

END THE SEQUENCE WHEN TEMPERATURE REACHES 120°C

FULL IMPEDANCE SCAN AT HIGH TEMPERATURE
DISPLAY NYQUIST PLOT

Scatter in the solution resistance values was minimised by carrying out the 64 Hz measurement five times and averaging the results.

3.4 Results

Graphs of the real components of impedance, (Z'), at the two test frequencies are shown in Fig. 6. The charge transfer resistance, R_t, was obtained at each temperature, from the difference between the real components at 64 Hz and 0.01 Hz. The corrosion rates were then calculated from the Stern–Geary relationship, equation (2). As the 64 Hz reading represents the electrolyte resistance it will be appreciated that this resistance was a significant proportion of the total low frequency impedance $(Z' = R_t + R_s)$. Simply calculating the corrosion rate from a low frequency measurement or a d.c. linear polarisation reading, in this case, would lead to a serious underestimate of the corrosion rate. It should be understood that the value of R_s is dependent upon the separation of the working and reference electrodes and the corrosion probe should be designed to keep this distance small.

From the wide range of liquor compositions tested, results are discussed here for a liquor containing 15% water. Graphs of corrosion rate versus temperature are shown in Fig. 7. Above 90°C, (curve A), the rate increased considerably and this temperature marked the upper operating limit.

An Arrhenius plot of the data produced a straight line graph indicating further that the corrosion was controlled by a single activation process (i.e. charge transfer), Fig. 8. The activation energy for the process was calculated to be 28 KJ/mol.

It was also found that repeating the test with the same liquor gave increased rates of corrosion on each successive occasion. Subsequent analysis of the liquors indicated that the dissolved chromium, nickel and iron from the 316L were reaching accumulated levels of 260, 285 and 900 ppm respectively. The time-dependent effects, shown in Fig. 7, appeared to result from modifications to the complex solution chemistry. While this occurrence would not be expected in a once-through or a partially recycled system it serves to indicate the sensitivity of the measurements to process conditions. Replacement of the liquor caused re-establishment of passivity and a return to the former corrosion rates.

4. PLANT TRIAL

The technique described above has been successfully applied to corrosion monitoring on plant. The objective was to allow the process conditions to be optimised while ensuring that the corrosion rate was kept at an acceptably low level. While weight loss coupons and other electrochemical methods are frequently used, it is believed that this trial represented the first successful plant experience with the a.c. impedance technique. Unlike the weight loss coupon, however, which is a long-term integrating method of corrosion rate measurement, the two point impedance method, depending on the frequencies used, gives a rate reading at approximately five-minute intervals. The effects of short-term fluctuations in the process conditions can be observed therefore. The

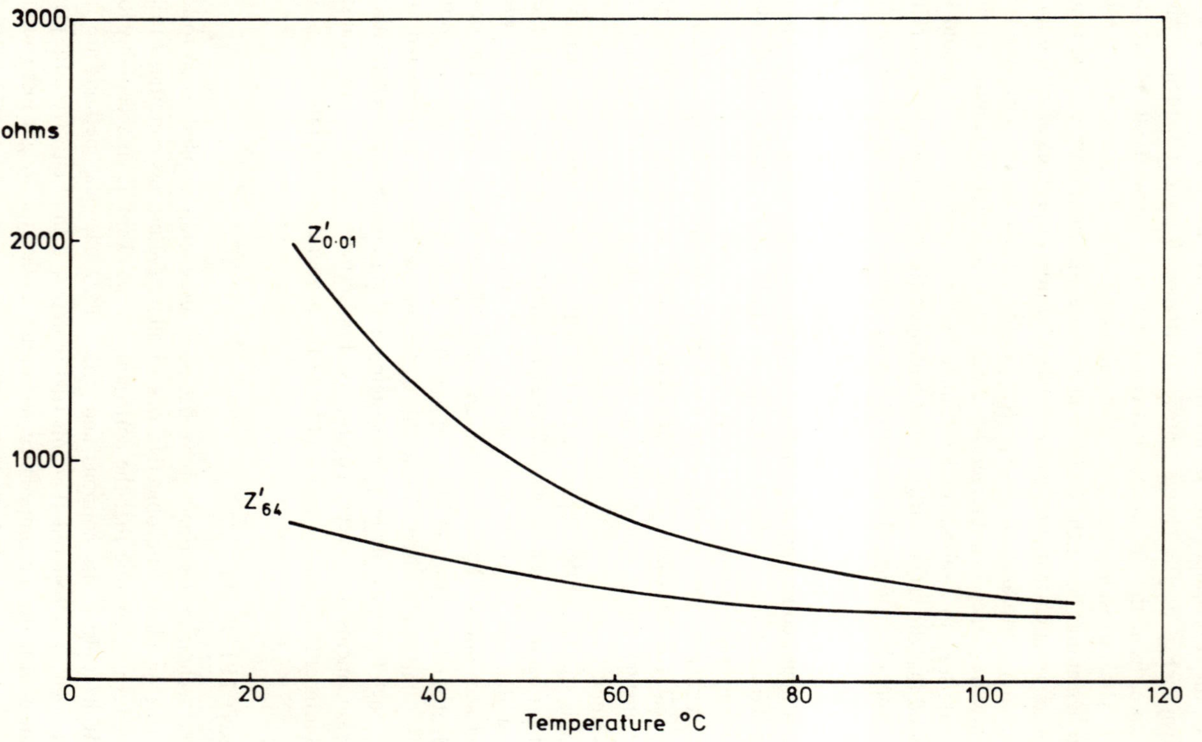

Fig. 6 – Graph showing the effect of temperature on the real component of impedance for Type 316L exposed to glacial acetic acid/15% water/catalyst liquor.

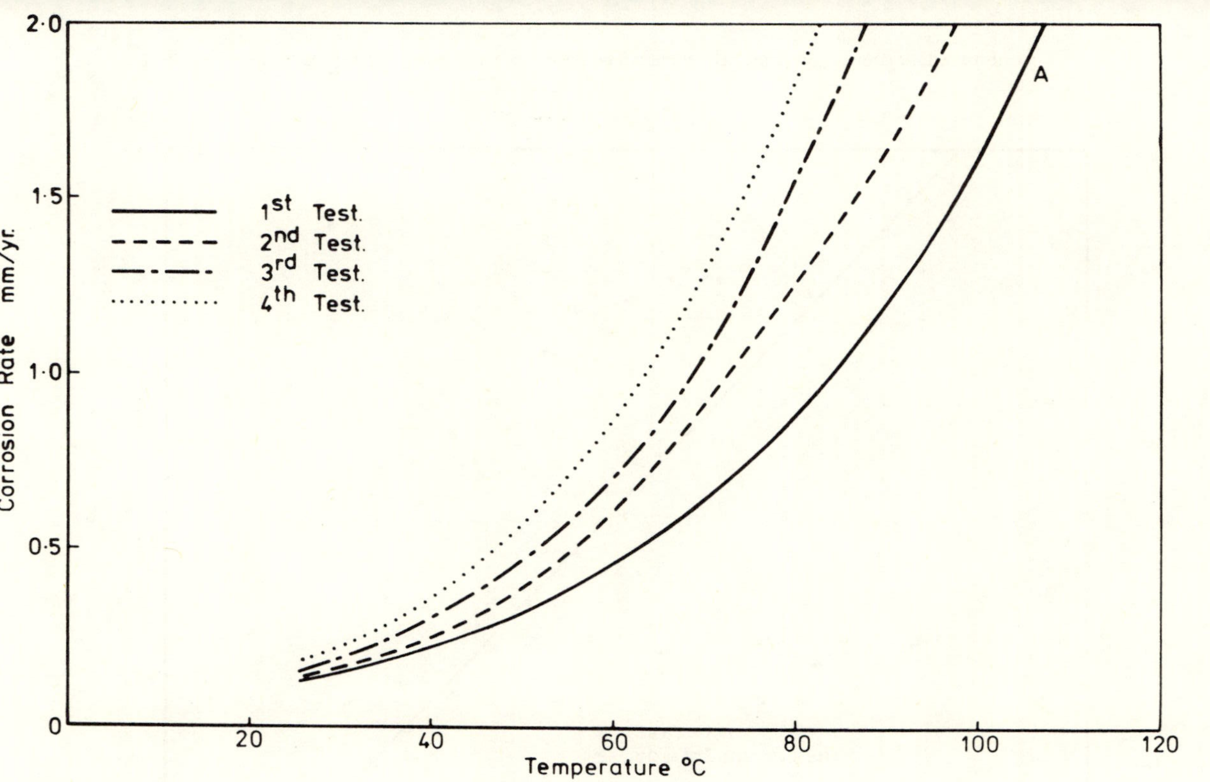

Fig. 7 – Graph showing the effect of temperature on the corrosion rate of Type 316L exposed to glacial acetic acid/15% water/catalyst liquor.

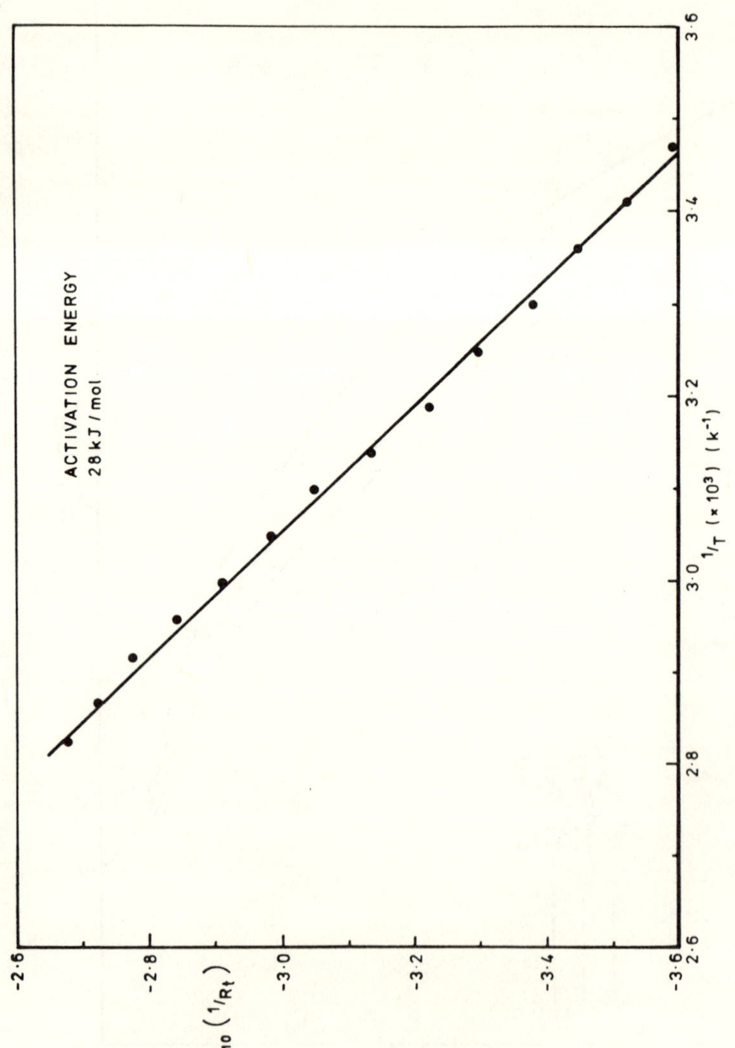

Fig. 8 – Arrhenius plot for Type 316L exposed to glacial acetic acid/15% water/catalyst liquor.

Nyquist plot shown in Fig. 9, for example, corresponded to steady state operating conditions (full semicircle), indicating a very low corrosion rate of 33 micrometres per year. The plot became distorted at low frequencies (dotted line) by a transient temperature increase on the plant of 14°C. The displacement of points towards the origin was consistent with an increase in corrosion rate and a reduction in charge transfer resistance.

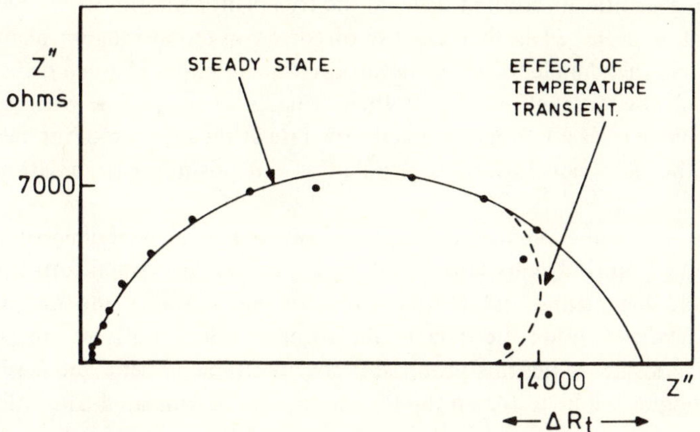

Fig. 9 – Nyquist plot from the plant trial showing the effect of a temperature transient of 14°C.

Some inductive behaviour was apparent in the Nyquist plots although this had not been detected in the flow loop studies or the laboratory tests conducted at the free corrosion potential. Corrosion problems in stainless steels are generally associated with localised breakdown of passivity and may show a marked time-dependence. For this reason the plant trials are currently continuing over an extended period.

5. PREDICTIVE ANALYSIS

The form and the extent of corrosion damage on plant may be deduced satisfactorily by the methods described, at all points where a probe can be sited. In many cases, however, locating a probe in a particularly critical location may not be possible. An alternative approach, worthy of investigation, is to determine whether the corrosion information can be inferred from the measurements made on probes at remote locations.

The detailed relationships between corrosion rate, flow rate, temperature and potential, for the particular metal and environment, are required for such an analysis. To this end, the results of long-term plant corrosion monitoring would form a data base.

In its simplest form, the temperature versus corrosion rate graphs in Fig. 7, from the flow loop studies, enable predictions to be made of the corrosion rate expected in that specific liquor and in that temperature range. This type of information is used, of course, by the corrosion engineer in selecting suitable plant materials and in calculating corrosion allowances. Where the concept differs, however, is that the data used by the engineer generally arises from laboratory pot tests, often performed in advance of the plant being constructed and having significant differences from the true plant conditions.

What is proposed is that the use of corrosion monitoring on plant should provide all the detailed corrosion information required characteristic of that metal and environment system. It then remains to obtain the basic operating conditions of at least temperature and flow rate at the location of interest and to extract the corrosion rate by interpolation and possibly extrapolation of the data base.

Two techniques appear appropriate. Mathematical modelling of the data to provide empirical equations could be attempted. As several forms of corrosion, including pitting and crevice corrosion are unstable and discontinuous effects, however, using the data in the form of a look up table appears to be more practicable. While this proposal is speculative at present, the feasibility of the technique will be tested on the flow loop. The accumulated data will be used to indicate the corrosion rate at a point on the loop and the predictions will be validated by placing a further probe at that point.

ACKNOWLEDGEMENTS

The authors are indepted to ICI, Agricultural Division, Billingham, Cleveland, for the support of this work and for permission to include the experimental results.

REFERENCES

[1] J. E. Strutt, M. J. Robinson and W. H. Turner, Recent developments in electrochemical corrosion monitoring techniques, *The Chemical Engineer*, No. 375, Dec. 1981, p. 567.

[2] K. Hladky, L. M. Callow and J. L. Dawson, Corrosion rates from impedance measurements, *British Corrosion Journal*, 1980, **15**, No. 1.

[3] D. D. Macdonald and M. C. H. McKubre, Electrochemical impedance techniques in corrosion science. In *Electrochemical Corrosion Testing* (editors F. Mansfield and U. Bertocci), ASTM STP 727, 1981, p. 110.

[4] L. L. Shrier, ed. *Corrosion* 2nd edn., Newnes-Butterworth (1976), Section 20:41.

[5] Reference [4], Section 21:64. Table 21.29.

[6] I. R. Mcgill and B. McEnanly, A novel reference electrode arrangement for high temperature polarisation studies, *Corrosion Science,* **18**, No. 3, 1978, p. 257.

CHAPTER 6

Thin-layer activation – a new plant corrosion-monitoring technique

J. Asher, T. W. Conlon, B. C. Tofield and N. J. M. Wilkins

1. INTRODUCTION

Corrosion is a degradation mechanism of particular concern for a wide range of industrial plant and plant components. The theme of the Conference reflected the increasing requirement to be able to perform plant condition monitoring on-line. On-line corrosion monitoring is, therefore, clearly an important component of such a capability. As well as being a monitor of plant condition, corrosion monitoring can often supply valuable process control information.

Some techniques widely applied to monitor corrosion such as visual inspection, the use of corrosion coupons or ultrasonic thickness measurement are either inapplicable to on-line monitoring or frequently not sufficiently straightforward to apply or sufficiently sensitive.

Among the requirements of a corrosion monitoring technique, if it is to be widely applicable to on-line plant monitoring, are:

- it should be straightforward to apply
- it should be rugged and compatible with normal plant operation and be capable of being made intrinsically safe where necessary
- it should ideally be non-invasive
- it should be capable of continuous operation with the possibility of remote data-logging
- it should not require frequent replacement
- the technique should be sensitive with a rapid response to changes in corrosion rate
- the results should be unambiguous, independent of process conditions, and present no difficulties in interpretation
- the results should be characteristic of the plant or component of interest
- the technique should be cost effective to apply.

This chapter presents the first results from plant-scale demonstrations of a new corrosion-monitoring technique, Thin Layer Activation (TLA), which meets many of the requirements above. The basis of the technique is to introduce a small quantity of radioisotope tracer into the surface of a coupon or plant component of interest. Material lost by corrosion or other processes can then be monitored remotely and with high sensitivity. TLA is already well established as a technique possessing unique advantages in the measurement of wear and erosion, particularly in oil and automobile industry research laboratories. The industrial-scale demonstrations reported here bear witness to its utility as a technique for on-line corrosion monitoring. There is little doubt that TLA will find wide application in situations where the monitoring of corrosion or other material loss processes will be of benefit for on-line plant surveillance or process control.

2. THIN LAYER ACTIVATION

The principle of TLA is that trace quantities (typically 1 in 10^{10}) of a radioisotope are generated in a thin surface layer of the component under study by an incident high energy ion beam. The activity levels are modest and only elementary handling precautions are necessary. Because of the trace quantities of material affected, the surface structure of metals is not changed significantly and there are no detectable effects of the activation on corrosion or other chemical behaviour. The technique and its applications have recently been reviewed [1].

Loss of material from the surface of the component can readily be detected using a simple γ-ray monitor. The penetrability of γ-radiation may be exploited to enable measurements to be made non-invasively, for example from the outside of a pipe wall. The reduction in activity is converted to give a depth of corrosion loss directly and, provided that the corrosion is not highly localised, this gives a reliable measurement of the average loss of material over the surface. The depth distribution of the activity is known accurately so that the depth of loss can be determined to a precision of about 1% of the layer depth on the basis of a single measurement or to a higher precision if, as is normally the case, measurements are made at regular intervals. Much higher sensitivity may be achieved in circumstances where corrosion of wear debris can be collected, as in the filter of a circulating fluid system, although this latter option is not normally available in plant corrosion monitoring because of the rapid dilution of debris in the fluid environment.

There is a clear distinction between TLA and the process of ion implantation which involves the deliberate introduction of large concentrations of foreign atoms into a surface, often with the express purpose of modifying the properties of the surface. Indeed, TLA has been used to evaluate in a particularly sensitive manner, improvements to wear characteristics of materials which may be brought about by ion implantation [2].

TLA may also be distinguished from neutron activation, where bulk samples are irradiated in a nuclear reactor inevitably resulting in much higher activity levels than result from TLA. Because TLA is restricted to the near surface, it allows a much higher sensitivity to material loss than is possible with neutron activation and it can be applied to a much wider range of materials.

On entering a material, the charged particles in an ion beam rapidly lose energy through collision processes with electrons and come to rest as a well-defined penetration depth. However, a small fraction of particles above a certain threshold energy interact with the nuclei of the material to induce a nuclear reaction transmuting the host atom to a radioactive form. For iron-based metals, ^{56}Co may be produced from ^{56}Fe using a proton beam (Fig. 1) or ^{57}Co using a deuteron beam. The cobalt isotopes have half lives of 79 days and 271 days respectively and main γ-ray energies of 0.85 MeV and 0.12 MeV. Useful monitoring periods may extend over several half lives so that monitoring may be continued at least for one year using ^{56}Co and for several years using ^{57}Co. γ-Ray penetration is a function of γ-ray energy. ^{57}Co may be remotely monitored through pipe walls in many circumstances and ^{56}Co may be observed through material layers up to a few inches of iron equivalent in thickness. No absolute reference is required; the initial and final levels of activity yield the material loss directly.

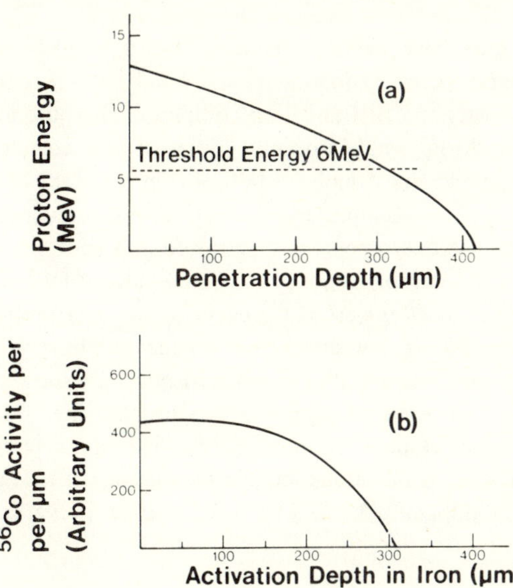

Fig. 1 – A comparison of the penetration depth for 13 MeV protons in iron (a) with the depth profile for activated ^{56}Co (b). The ^{56}Fe(p, n) ^{56}Co reaction has a threshold energy of 6 MeV so that no activity is produced beyond ~ 300 μm although most protons penetrate to ~ 400 μm.

The depth distribution of active species depends on the incident beam energy and the orientation of the beam to the surface. In the applications described below, ^{57}Co was produced to depths of 100 μm and 50 μm using 9.2 MeV deuterons at 90° and 30° to the surface respectively. The penetration depth may be increased by increasing the incident beam energy and for high beam energies sub-surface 'sentinel' active layers may be produced. For example, a sub-surface layer of ^{56}Co, about 500 μm thick may be produced at a depth of about 2.5 millimetres by activation with a 40 MeV proton beam. Thinner active layers may be produced by bombardment with a heavier ion such as Li.

3. CURRENT APPLICATIONS OF TLA

TLA is already applied on an extensive scale to industrial problems involving wear and erosion processes in ferrous and non-ferrous metals and alloys. In these applications, the component of interest is itself activated at Harwell, generally using the Tandem van der Graaf generator. A wide range of component sizes are handled and items up to several hundred kilograms in weight have been studied. The ion beam can be focused to an area of a few square millimetres as it emerges from the beam line for direct irradiation. Alternatively, a larger area may be covered by rotation or oscillation of the component or by scanning the beam across a stationary component. The only restriction is that a line of sight to the area of activation must be available.

Wear measurements have been particularly useful in test-bed studies of engines, lubricants and fuels. Components activated include cam followers, cylinder liners, valve seats and fuel injection equipment. Because measurements can be made *in situ* during engine running, very precise correlations of the effects of lubricants, fuels and running conditions on the wear of components can be determined [3]. For example, two types of engine were investigated [4] by British Petroleum. Cylinder liners were activated at a spot where maximum wear was expected to a depth of 75 μm of ^{56}Co and material loss was monitored through the liner. The wear rates were very dependent on the running conditions. Running-in of a Ruston 7X-HR low-speed truck engine on a base oil was followed by a more extended run with a fully formulated oil. TLA showed clearly the severe wear associated with the running-in procedure and the effects of subsequent engine faults, which resulted in poor combustion and increased wear, were also apparent. Other wear applications include measurements on machine tools and on artificial hip-joint components in which wear on polyethylene bearing material was measured.

4. TLA AS A CORROSION-MONITORING TOOL

These applications have been conducted under laboratory or test-bed conditions and in some cases the associated equipment for γ-counting has been designed to cope with specific hostile environments. Successful studies have included measurements of corrosion and corrosion/erosion, for example in boiler-tube simulations [5] and the potential for plant application of TLA as a corrosion monitoring tool was evident [6].

The technique offers several advantages in comparison with other corrosion monitoring methods. The actual component of interest may itself be activated to provide a direct measurement of plant corrosion, or, if more convenient, a coupon in the same metallurgical form as the plant component can be irradiated and flush mounted. Flush mounting of other corrosion monitoring probes is often less straightforward. Conventional corrosion coupons can, in some circumstances, provide misleading results because of accelerated corrosion at the edges of the coupons; such effects can be completely eliminated in TLA through activation of only a selected area of a coupon away from the edges. The material loss can be remotely monitored through the plant wall, so avoiding the need for electrical connections to the coupon or probe. In addition, the results provide an unambiguous measurement of material loss, independent of the process conditions; the technique is equally applicable to high resistance aqueous or non-aqueous media or even to erosion in gas/particulate streams. Because TLA is non-invasive, the test item is very robust and the interpretation does not depend on any property of the process fluid such as electrical conductivity.

5. PLANT DEMONSTRATIONS OF TLA FOR CORROSION MONITORING

The success of TLA in laboratory and test-bed studies of wear and corrosion, together with the potentially attractive features of TLA as a plant corrosion-monitoring technique, has led to a programme to evaluate TLA under realistic plant conditions and to compare it with other corrosion-monitoring techniques. Two successful demonstrations are described in this paper. As a result of these, attention is now being given to designing equipment specifically for plant monitoring. This will require, in many instances, certification of intrinsic safety for use in flammable environments. Systems for pipeline application using high pressure access and retrieval tools with monitoring of flush mounted coupons are also under development.

In one demonstration, a plant component was activated and corrosion loss was monitored using an externally mounted detector. Apart from removal of the component for activation, no penetration of the plant was required. The second demonstration used TLA to monitor the effectiveness of an inhibitor treatment programme for a chiller plant where blockages had occurred through the deposition of corrosion product sludge. In this case, a test piece was activated and the results were compared with weight loss measurements from conventional mild steel corrosion coupons. It was notable that the weight-loss result for this inhibited system greatly exceeded the TLA result because of preferential attack at the edges and corners of the coupons so that the coupon results seriously underestimated the effectiveness of the new inhibitor on the steel parts of the system.

The first demonstration was set up on an evaporative water cooling system at Harwell [7]. Under normal operating conditions, the evaporative losses from the system are about 7500 gal/hr with a blowdown of about 840 gal/hr. The system uses acid-treated mains water make-up with continuous blowdown to maintain impurity conditions within predetermined limits. The TLA monitoring point was established in the blowdown line which carries cooling water through a 1 in mild steel pipe.

A deuteron beam was used to produce an approximately uniform distribution of ^{57}Co to a depth of 50 μm in the pipe wall (Fig. 2). The pipe was installed during one of the regular shutdown periods when the water flow was stopped.

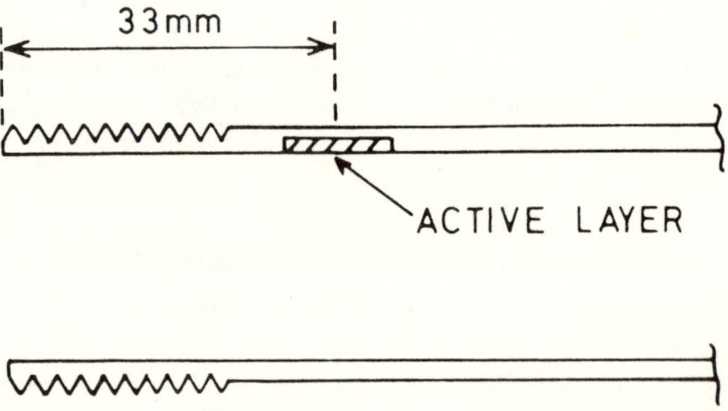

Fig. 2 – Diagram of the test pipe section showing the position of the active layer.

A ruggedised 1 in diameter NaI(Tl) scintillation counter was located in a mount firmly attached to the exterior of the pipe and activity measurements were made over 100 sec intervals every hour and recorded. Measurement of the reduction in activity arising from corrosion loss continued over three cycles of plant operation extending over 84 days, and included two scheduled shutdown periods. The results of the activity measurements condensed from the hourly record are displayed in Fig. 3 in terms of μm of surface loss. The zero loss of corrosion products during shutdown periods when the flow ceased is apparent as is the higher rate of activity loss immediately after shutdown.

The ability to measure corrosion losses non-invasively under plant operating conditions is clearly revealed. Fluctuations in corrosion rate are apparent and can be associated with changes in operating conditions. The results of this demonstration show very clearly the sensitivity of TLA, with incremental material loss being measured at less than 1 μm sensitivity. The total corrosion depth indicated by the TLA results was 26.5 μm prior to removal of the pipe for examination. Subsequent cleaning after termination of the experiment yielded a further loss of 21.5 μm which was associated with corrosion products retained on the surface of the pipe under operating conditions.

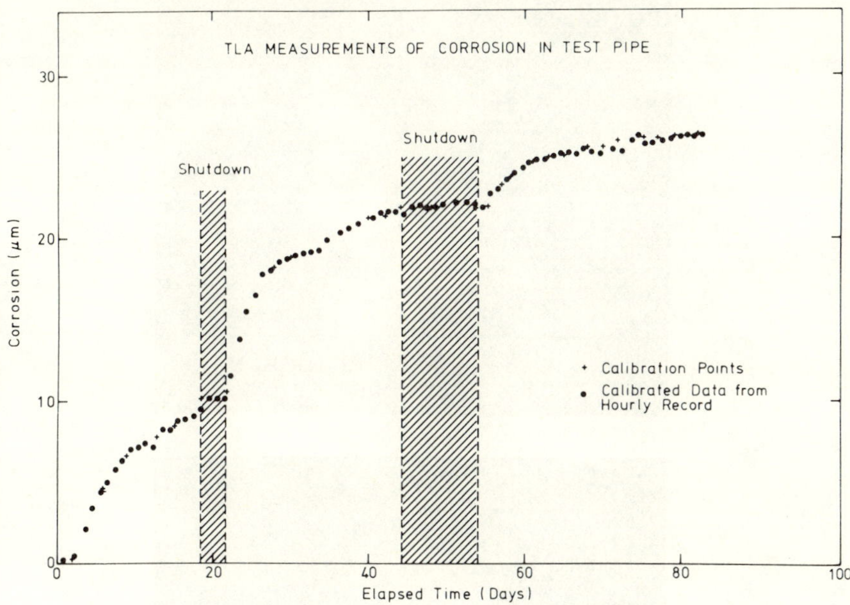

Fig. 3 – TLA measurement of corrosion in a cooling-water blowdown line. Material loss, in micrometres, deduced from the activity measurements, is plotted against time (days).

Although TLA and other corrosion monitoring results are interpreted in terms of uniform corrosion, shallow pitting is frequently observed during general corrosion attack and was seen in this case also. This was revealed by autoradiography taken using a Polaroid film pack in direct contact with the inner pipe surface (Fig. 4(a)). The autoradiograph of a similar pipe which was activated at the same time and used for calibration purposes showed the uniformity of the as-activated area (Fig. 4(b)).

A more detailed study is planned in the same plant to demonstrate the capability of a custom-built scintillation counter, modified to locate within a standard pipeline access and retrieval tool which will be used to insert an activation flush-mounted coupon. In this work, a more detailed evaluation of the effects of operating conditions and water chemistry on corrosion will also be undertaken and a comparison made of TLA results with those obtained from other types of corrosion monitoring probes.

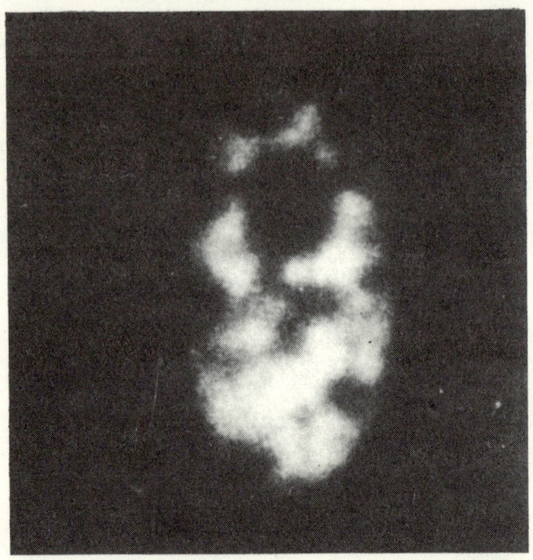

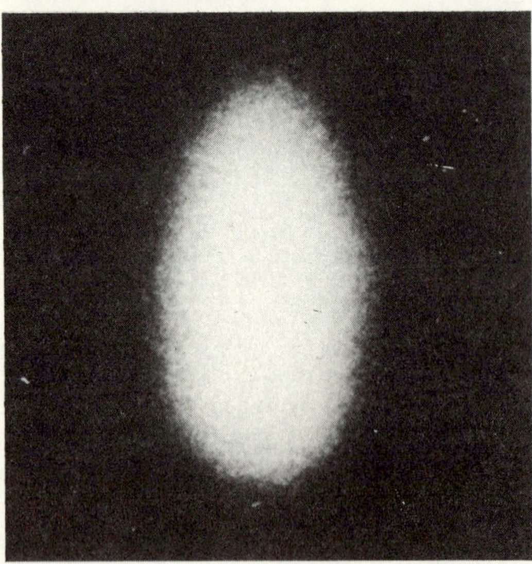

Fig. 4 – Autoradiography showing the active area on the corroded pipe (a) in comparison with that on the pipe used for calibration purposes (b) which was maintained in an uncorroded state. The expected shallow pitting is clearly revealed.

In the second demonstration, TLA was chosen to measure the corrosion rate before and after the addition of inhibitor [8]. The test piece was activated to a depth of 100 μm with ^{57}Co. The activated sample was installed in a by-pass leg of the chiller plant with the flow arranged to impinge upon the activated surface. The standard corrosion coupons were contained in plastic piping downstream from the activated component. Data were again recorded at hourly intervals and the results over the test period are shown in Fig. 5. The non-inhibited system was measured for about 20 days and a loss averaging 0.25 ± 0.05 μm/day of iron established. After addition of the inhibitor, the rate of metal loss dropped by an order of magnitude to about 0.021 ± 0.007 μm/day average.

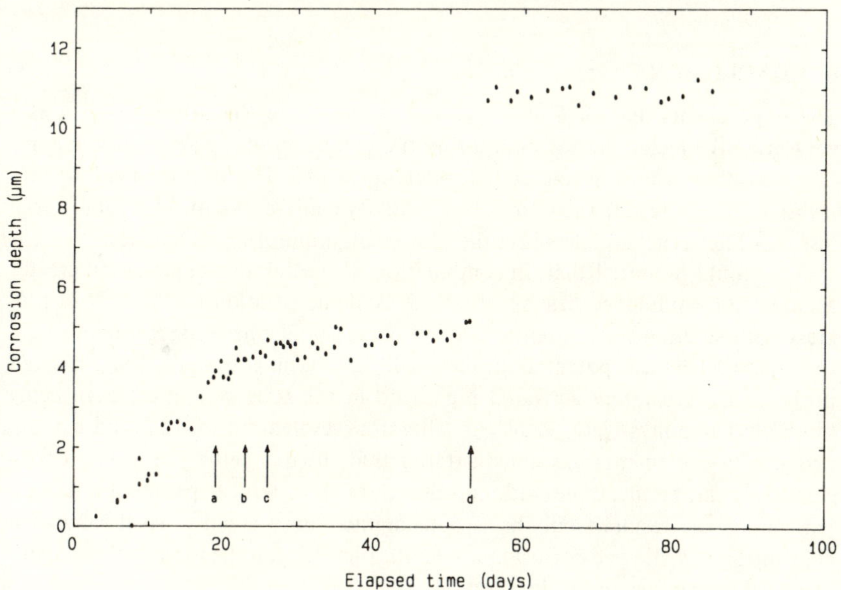

Fig. 5 – The depth of corrosion (μm) with time as deduced from the activity loss of the test component. The correct inhibitor conditions were established between points a and c. At d, the component was removed for inspection and cleaned. After replacement, the apparent corrosion rate was very low. indicating the effectiveness of the inhibitor.

Part-way through the experiment, the sample was removed and chemically cleaned before reinstallation in the system. Activity measurements continued for a further 30 days with a very low apparent corrosion rate of only 0.004 ± 0.005 μm. Chemical cleaning at the end of this period revealed a material loss of only about 1 μm which was probably the passivating film built up by the inhibitor.

In contrast, the weight-loss coupons showed only a reduction of about two in the corrosion rate following the addition of inhibitor. Visual inspection showed considerable corrosion at edges and corners of these coupons both before and after treatment. The weight loss results, therefore, seemed to be dominated by non-typical edge corrosion and to underestimate seriously the effectiveness of the inhibitor.

As with all TLA measurements, only the net material loss is measured. This, however, is the parameter of particular interest where the corrosion problem is associated with the deposition of corrosion products elsewhere in the system. TLA revealed very clearly, by direct non-invasive measurment during plant operation, that the inhibitor treatment was functioning very effectively.

6. CONCLUSIONS

The experiments described above clearly demonstrate the utility of TLA as a very effective technique for monitoring the condition of a plant from the point of view of corrosion during actual plant operation. The measurement is non-invasive and the results show that the sensitivity achievable with TLA for material loss is as high as achievable with other corrosion monitoring techniques.

It should be noted that, in combination with other techniques, even greater accuracy or confidence can be obtained. A demonstration is currently in progress with simultaneous potential and TLA measurements being recorded where the electrochemical potential probe itself has been activated. The activated probe and a separately activated pipe bend in the same system are both being monitored non-invasively. A recent laboratory evaluation of TLA and a.c. impedance measurements has demonstrated that, used together, very low corrosion rates can be measured with confidence in short-term experiments. A comparison of the results from the two techniques on a stainless steel sample in cold nitric acid allowed corrosion losses of order 0.1 Å to be measured in experiments of only a few hours' duration [9].

The demonstrations reported above showed that, even though corrosion products are often partially retained, TLA nevertheless gives reliable quantitative information on corrosion rates and provides estimates of actual material loss as reliable as achieved by other techniques. Either actual plant components or coupons may be irradiated and the possibility of using flush-mounted coupons is particularly attractive for monitoring corrosion of pipelines.

Clearly, activation by ion beams is an essential component of the TLA technique; this is routinely performed on a commercial basis by Harwell using the Tandem van der Graaf generator or the Variable Energy Cyclotron. The demonstration described above, however, used detectors and counting systems not directly appropriate to plant use except under control of qualified technicians. In addition, counter calibration was required and was effected using, as reference standards, duplicate components activated under similar conditions but not exposed to the process fluid. Ruggedised, self-calibrating, intrinsically safe detectors and display equipment are currently under development for application to many types of plant and simple-to-interpret data displays (for example μm or mils/year loss) will be available.

As the results of more plant demonstrations become known, and as appropriate equipment becomes commercially available, it is expected that the single-layer TLA technique described here will become widely used as a corrosion monitoring tool and also as a routine monitor for other material loss processes such as erosion and wear. In the future, there will be exploitation of more sophisticated methods developed at Harwell [1, 6] such as the study of selective corrosion by the activation of individual atomic components of an alloy, or of localised corrosion by the individual monitoring of double layers of, for example, ^{56}Co and ^{57}Co activated to different depths. Also, the use of a buried layer permits a 'sentinel hole' technique to be adopted without any compromise to plant wall thickness or integrity.

REFERENCES

[1] T. W. Conlon, *Contemporary Physics,* **23,** 353 (1982).

[2] P. D. Goode, A. T. Peacock and J. Asher, Harwell Report AERE-R10696 (1982).

[3] R. Evans, *Wear,* **64,** 31 (1980).

[4] R. F. Pywell, *J. Inst. Mech. Engrs.,* **35,** 59 (1978).

[5] D. J. Finnigan, K. Garbett and I. S. Woolsey, *Corrosion Science,* **22,** 359 (1982).

[6] J. Asher and T. W. Conlon, in *Corrosion Monitoring in the Oil, Petrochemical and Process Industries,* ed. J. Wanklyn, Oyez Scientific and Technical Services Ltd., 1982, p. 91.

[7] J. Asher, J. W. Webb, N. J. M. Wilkins and P. F. Lawrence, Harwell Report AERE-R10391 (1981).

[8] J. Asher, J. W. Webb, N. J. M. Wilkins and P. F. Lawrence, Harwell Report AERE-R10574 (1982).

[9] D. E. Williams and J. Asher, Harwell Report AERE-R10477 (1983).

CHAPTER 7

Corrosion monitoring–probe designs

Roe Strømmen

1. INTRODUCTION

The importance of satisfactory corrosion control in oil and gas producing operations is well recognised. It is also well understood and accepted that a prerequisite for proper corrosion control is adequate monitoring and inspection to get information on the condition of the equipment, to get early warnings of failures, to find out whether corrosion mitigation programs give satisfactory results etc.

There are certain general requirements to the monitoring systems and methods to be used. First and foremost it is required that methods and systems which have been chosen can give a true and reliable picture of corrosion/erosion rates and of the corrosion condition of the plant. Secondly monitoring should be carried out at a minimum of risk, at low costs and at a minimum of interference with production operations.

According to references [1, 2] corrosion detection and monitoring methods in oil and gas producing operations can be classified into four categories (see also reference [3]):

- Direct or indirect physical/visual inspection on shut-down ('after-the-fact inspection').
- NDT (non-destructive testing) of operating equipment.
- Analysis and data interpretation of corrodants and corrosion product samples.
- Use of probes and weight-loss coupons inserted into the operating equipment and into pilot systems and/or laboratory simulations.

Normally an operator cannot rely on one or a couple of these methods alone in that each of them have their characteristic deficiencies but also advantages which can be beneficially utilised depending on type of system, operating conditions etc.

1.1 Physical inspection

There is a lot to learn from visual/physical inspection, on shutdown, in regard to causes and nature of corrosion. Dependence on this method alone, however, would leave the operator without any warning on corrosion damage and would result in expensive failures.

Under this heading can also be listed various types of pigs and calipers, some of which have become 'standard tools' in on-line inspection of pipelines, for inspection of tubing etc. [4].

Again these are expensive methods, e.g. due to well down-time for calipering of tubing, but are still considered necessary in combination with surface monitoring techniques. A useful correlation has been carried out by C. J. Houghton *et al.* [5].

1.2 Non-destructive testing

A range of NDT equipment/methods like radiography, ultrasonics, magnetic particle inspection, eddy currents, electrical and magnetic conduction etc. are being used to a considerable extent, and are often found to offer certain advantages over other methods available. NDT can often be carried out on-line, and may allow for direct inspection of critical sections or critical elements in a plant. However, such methods can be expensive and may not always be as accurate as required.

1.3 Product sampling

Iron counts, or analysis of produced water for dissolved iron is the most commonly used product sampling. Although this method can give an indication of corrosion in a system it leaves the operator completely in the dark as to where corrosion takes place, and whether it is localised or general attack. Furthermore the corrosivity of the product flow can be evaluated by analysis for O_2-content, chlorides, H_2S, CO_2. The value of the method is operator-dependent, i.e. dependent on his ability to sample correctly in his interpretation of the data.

1.4. Probes and coupons

A number of different probes and coupons are used to monitor corrosion. These are inserted into pipes and vessels of operating equipment, into pilot systems and also into laboratory simulations. Various types available comprise:

- weight loss coupons
- electrical resistance probes (ER-probes)
- linear polarization resistance probes (LPR-probes)
- galvanic probes
- hydrogen probes
- erosion probes
- film resistivity probes

The first three probes on this list are the more commonly employed. Except for the coupons which are retrieved at regular intervals for weighing and inspection, the probes are monitored either continuously using automatic monitoring systems, or interrogated at frequent intervals, e.g. by portable instruments.

The various probes each have their characteristic features. While some of them provide information on actual corrosion rates or reductions in pipe/vessel wall thickness, others monitor variations in the corrosivity of the product flow.

Coupons, ER-probes, LPR-probes and erosion probes can all provide information about metal loss due to corrosion and/or erosion. By inspection of coupons information is obtained also on the type of degradation, e.g. on pitting and other forms of localised corrosion. Such information on pitting etc. is less easily obtained from probes. For further details on monitoring methods and recent North Sea experience with various probes and monitoring systems, see references [3, 6].

1.5 Probe designs

Some typical features of probe and coupon designs will be reviewed.

Weight-loss coupons

Coupons are inspected and retrieved e.g. every six or every three months for weighing and inspection. Information on the corrosion activity is therefore delayed until after retrieval. For this reason coupons are normally used in combination with probes for continuous monitoring.

Coupons are manufactured of mild steel or of a material quality which is similar or identical to that of the pipe wall itself.

The most commonly used designs are the strip type, mounted projecting into the product flow, or the disc type. The strip type coupons may require an alignment feature to have the pair of strips inserted into the pipe parallel with the product flow and thus avoid excessive erosion.

The disc coupons which are becoming more and more popular are frequently inserted flush with the internal pipe wall, particularly in pipes with high flowrates.

Typical shapes of coupons and coupon holders are shown in Fig. 1.

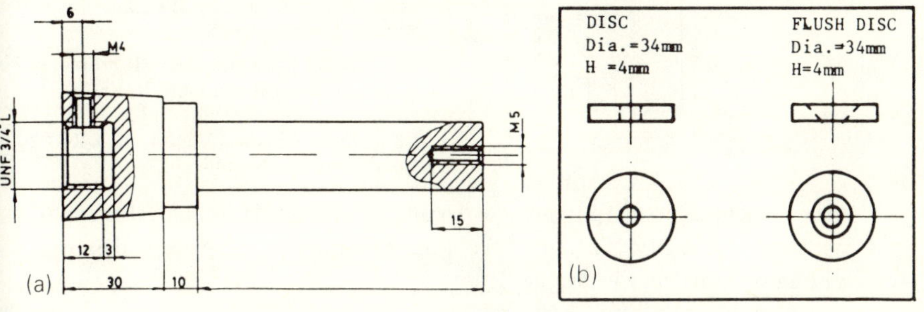

Fig. 1 – (a) Typical coupon holder for disc coupon. (b) Typical disc coupons.

Experience with the use of coupons is normally good, and it is widely used in combination with other methods.

Electrical resistance (ER) probes

This probe provides a means of 'electrical weighing' with weight loss. An element, normally shaped as a wire, a tube or a strip, is exposed to the process environment and corrosion decreases the cross-section with a resulting increase of the electrical resistance. The change with time of the electrical resistance, obtained with an instrument using, e.g., a.c. techniques, is calculated back to a corresponding change in the cross-section (wall thickness) of the ER-element. Temperature compensation is obtained by a reference element inside the probe, made of the same material quality (same temperature coefficient) as the main ER-element.

The wire and tubular type ER-element have been the commonly employed ER-probes over a number of years. Inserted in pipes with high flow-rates, however, such projecting type elements have suffered from mechanical failures like breaks at the base of the tubular element caused by vortex shedding and vibrations. Such weakness together with an increasing popularity for flush-mount designs (see below) has prompted new developments like the strip type, a spiral type developed recently† [7] (see Fig. 2) and a type of slip ring design [8].

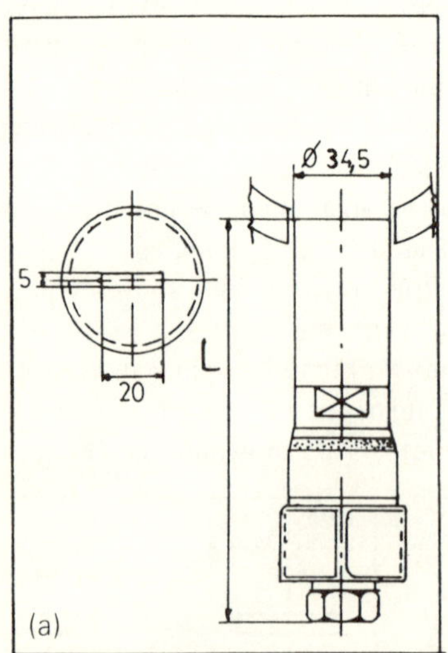

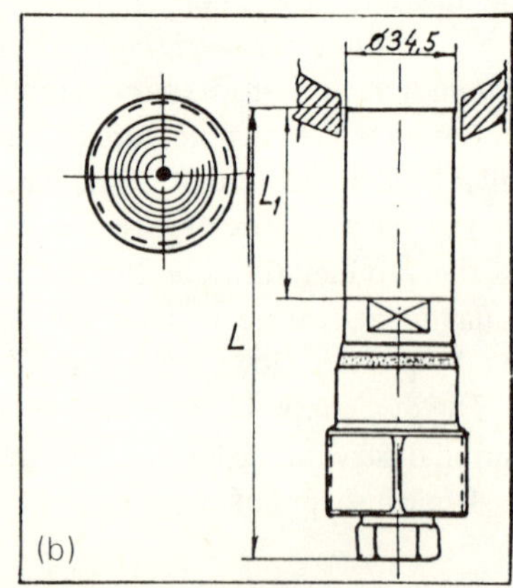

Fig. 2 – (a) Flushmounted ER-probe with striptype ER-element. (b) Flushmounted ER-probe with spiral type ER-element. Typical depth of ER-spiral element is 6 mm.

Major achievements with these new designs are improved mechanical strength and increased durability. Secondly the flush-mounted probe design does not represent an obstruction for pigging.

†Patent pending.

As for coupons the materials for sensing elements should be identical to or similar to that to the pipe wall itself.

LPR-probes

Two-electrode and three-electrode probes are available. The three-electrode probe provides the most accurate alternative and is now the most commonly employed. As for ER-probes, designs with projecting type electrodes have been used more or less exclusively until a few years ago when flush-mount designs were made available on the market. The three electrodes are all made of the same material, similar or identical to the pipe wall. Typical designs are shown in Fig. 3.

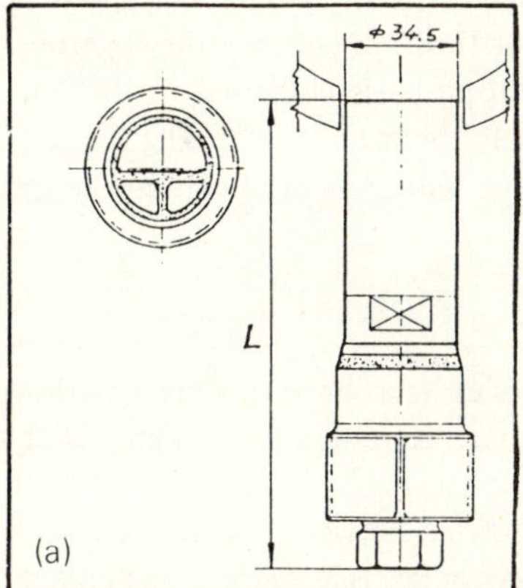

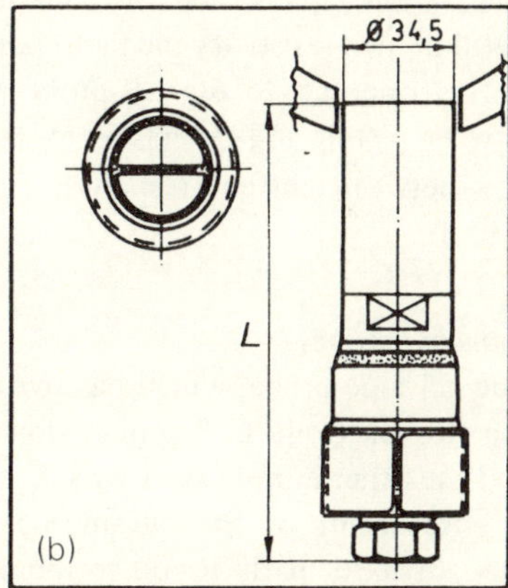

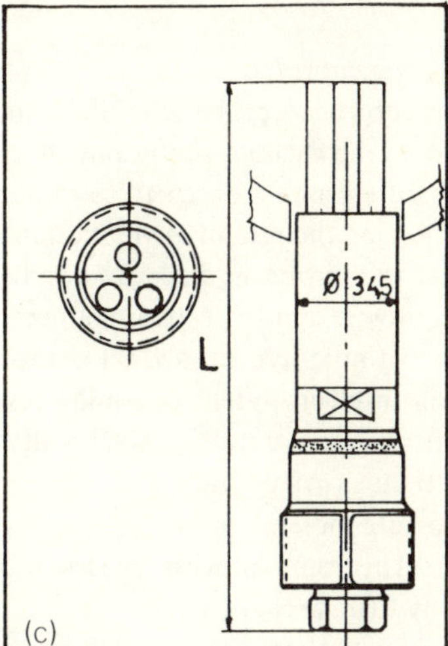

Fig. 3 – (a) Three-electrode LPR-probe for flushmount. (b) Two-electrode LPR-probe for flushmount. (c) Three-electrode LPR-probe with cylindrical projecting electrodes.

While the ER-probe reads accumulated corrosion, the LPR-probe monitors the instantaneous rate of corrosion (mm/year or mpy) and thereby allows monitoring of variations in inhibitor efficiency etc.

The LPR reading is obtained using the well-known Stern–Geary relationship [9]:

$$\begin{aligned} i_{corr} &= B/R_p \\ R_p &= \Delta E/\Delta i = \text{polarization resistance} \qquad (1) \\ \Delta E &= \pm 10 \text{ to } \pm 20 \text{ mV polarization.} \end{aligned}$$

In addition to linear polarization the LPR-probe can be used for recording polarization curves, and for a.c.-impedance tests, the latter being particularly useful in high resistivity media to correct for IR-drops between the electrodes.

As opposed to the ER-probe the LPR-probe is also used to discriminate between pitting and general corrosion. This is obtained by recording the relationship between current responses for anodic and cathodic polarisation (pitting index).

Galvanic probes

The galvanic probe, which has two electrodes, is often very similar in design to the two-electrode LPR probe. However, one electrode is (usually) made of brass and the other of mild steel.

Recording of the galvanic current flowing between the two electrodes is thought particularly useful to monitor oxygen levels in water in the pipe, e.g. to control the efficiency of oxygen scavengers.

Hydrogen probes

The hydrogen probe is available in several versions. These are the high pressure probe, the vacuum probe and an 'electrochemical' version. For example, the high pressure probe is designed to monitor the rate of H_2 pressure build-up in a cavity, reflecting the rate of H^+-reduction in the corrosion process.

The probe is made as a cylindrical shaft with a cavity drilled through it lengthwise and the tube is connected to a pressure gauge and bleeder valve.

Atomic hydrogen (H^+) is formed in the corrosion reaction of steel, particularly in electrolytes containing (even trace amounts of) H_2S. The atomic hydrogen will penetrate the steel wall by diffusion, probably along grain boundaries into the cavity, react and form molecular hydrogen resulting in a volume and pressure increase.

The same process causes blistering and H_2S cracking particularly in low alloy high-strength steels.

Since there are different processes in a corrosion reaction, it is obvious that this probe monitors one element only in a complex process. Based on a number of tests carried out by R. M. Vennet [10] it was concluded that useful qualitative information about the corrosion process rather than corrosion rate data can be obtained from this probe.

Erosion probes
An erosion probe inserted slightly projecting is shown in Fig. 4. The probe is connected to a bleeder valve and a gauge. Erosion and corrosion will eventually penetrate the front section of the probe body and cause a quick rise in the pressure which is monitored by the gauge.

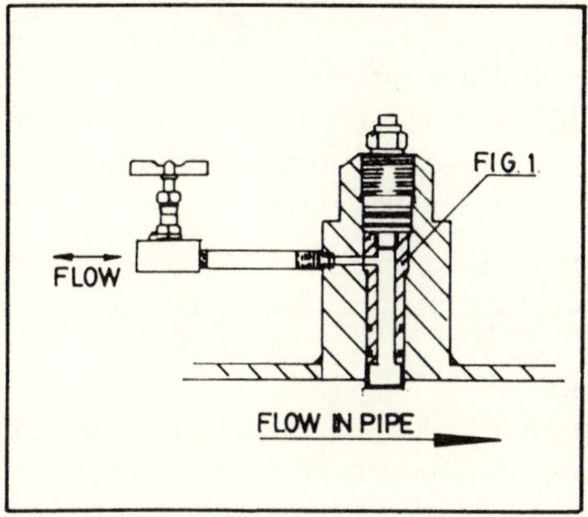

Fig. 4 – Erosion probe connected to bleeder valve, and (not shown) pressure gauge.

A major drawback with this probe is that the corrosion engineer is left in the dark until the day of penetration. This probe is used to a moderate extent only.

An alternative to this approach has been the recent use of ER-probes of flush-mount design but with relatively deep and robust sensing elements. By the use of one probe with carbon steel element and one with a stainless steel element it has been possible to discriminate between corrosion and erosion by comparison of the two.

Film resistivity probes
This is a probe which is seldom used. Utilising an auxiliary electrode to apply an a.c.-voltage, the resistivity of films formed on the surface can be monitored [1]. In this way characteristics of inhibitor films are measured, and their variation with time reflecting speed of film formation, film life, breakdown voltage, heat stability. Films of good quality are reported to have an apparent resistance of 100 times that of a film of poor quality [1].

1.6 Other methods

New methods being studied with great interest are the acoustic emission (for crack detection/monitoring) and the thin layer activation techniques. These are still considered to be laboratory methods.

2. BASIC CONSIDERATIONS IN PROBE DESIGNS

Major requirements of a corrosion-monitoring probe or coupon relate to its capability of monitoring validly and (preferably) instantaneously:

(a) direct metal loss, changes in wall thickness with time and preferably also the distribution of corrosion;
(b) (variation in) process environment or corrosivity, to give feedback on the efficiency of oxygen scavenging treatment, of inhibitor treatment or of other corrosion mitigation efforts.

For selection of a probe or a combination of probes for a specific application, the following parameters are considered important.

2.1 Position of probes

Probes and coupons must be located in relevant positions in the plant. If the intention is to monitor the efficiency of a water separator, a probe of relevant type is obviously required behind the separator and preferably in front of it.

To monitor erosion, probes are located in bends and elbows where turbulence and impingement might occur. For horizontal pipe sections where there is a hazard for deposits of water, the probe should be located at the bottom and so on.

Sometimes flush-mount probes are inserted slightly recessed as this may leave the electrodes in a pool of water and thus simulate 'sort of worst conditions'. However, in a hydrocarbon system with wax deposition on the pipewall, the probe may be left inactive.

To select proper monitoring locations requires knowledge of corrosion as well as of the process and on the process plant.

2.2 Probe type

The principle on which the probe(s) is based must be suitable for the relevant application. For example, the LPR-probe can be used in conducting media only and for this reason not (for instance) in a dry gas pipe.

2.3 Probe material quality

It is vital that probe elements validly represent the components of the system which it is monitoring. Of major importance is therefore material quality of the sensing elements of any probe. That is the probe material should be identical to that of the pipewall itself in terms of composition, thermomechanical treatment, metallurgical condition, surface condition etc. This can be obtained for fairly thick electrodes for LPR and galvanic probes, machined from pipe material, except for the surface condition which will be different from rolled pipe steel.

ER-probes with the wire and strip type utilise very thin elements. The advantage is obviously high resolution of the readings. In regard to material quality such elements pose a problem. Thin cold-drawn wire elements and cold rolled strip elements will have a different metallurgy and grain structure compared to the pipe wall material, owing to differences in the manufacturing process. This will eventually affect the corrosion properties as well.

Also, in regard to insulating compounds between electrodes and elements, different materials have been used. Neoprene and epoxy have suffered from swelling and cracking which can affect local flow conditions; this may cause crevice corrosion etc. Teflon (PTFE) is now the more popular insulating material.

2.4 Geometrical shape – general

Probe design including geometrical shape of electrodes and sensing elements has proved very important in regard to a large number of parameters relating to

- measurement principle
- type of corrosion attack to be monitored
- environmental conditions
- temperature gradients
- species present in the product flow
- hydrodynamics
- operating conditions
- mechanical strength.

Some major considerations have been evaluated.

2.5 Projecting type versus flush-mounted electrodes and elements

The projecting type tubes have been designed to monitor the worst conditions in regard to hydrodynamic effects and erosion which are expected to be found in the midst of the product flow rather than at the pipe wall. Another objective might simply be to reach the bottom of a horizontal pipe section when the probe is inserted from the top.

The flush-mounted probe has gained increasing popularity and the advocates of this design claim that:

- the flush-mounted probe will 'see' corrosive action in the same way as the pipe wall (corrosion should be measured at the pipe wall where it actually takes place);
- the flush design obviously does not restrict pigging;

– hydrodynamic loads are strongly reduced and the probe is furthermore made mechanically very robust.

There is evidence that projecting type electrodes have indeed suffered strong erosion in cases where the pipe wall suffered very little erosion and corrosion, and also vice versa [8, 11].

2.6 Mechanical strength/localised corrosion

As already indicated above, projecting type probes have proved too weak in systems with high flowrates, e.g. tubular elements suffering breaks at the base due to vortex shedding and vibration [6].

Although flush-mount designs are subjected to lower hydrodynamic loads, thin flush elements like the ER-strip have suffered premature erosion damage and breaks at the edge where it is bent into the probe body. Both to increase the strength and useful probe life, and also to allow pitting to occur, the spiral type ER-probe was designed. The depth of the spiral element is typically 6 mm. Apart from strength considerations this allows pitting to occur in a natural manner:

> Pitting rate may be, e.g., 4–10 times the average corrosion rate. I.e. for a thin foil or strip the deepest pit may have penetrated the foil at a stage when only 10%–25% of the average foil thickness has corroded away. Therefore the deepest pit stops corroding at an early stage, as no more material is left at the bottom of the pit. Corrosion will accordingly not proceed in a natural fashion as for the pipe wall itself, and readings tend to become systematically erroneous.

For LPR-probes the surface area of the working electrode should be maximised thereby avoiding one pit alone dominating the corrosion situation.

2.7 LPR-probes and IR-drops

In principle the LPR-probe cannot be used in non-conducting hydrocarbon media. In accordance with this, the experience with (projecting) LPR-probes in hydrocarbons has been poor. However, most wells produce some formation water which, depending on flow conditions may deposit at the bottom of horizontal pipe sections or may form a water film at the pipe wall. Under such conditions the flush-mounted probe may be fully 'submerged' in corrosive water while the projecting electrodes still have a major portion of the surface area in the non-conducting hydrocarbon.

A test carried out in a wellhead in the Ekofisk Field recently, utilising the flush-mounted three-electrode probe shown in Fig. 3, proved this probe capable of monitoring corrosion rates which were similar to readings obtained with weight-loss coupons and ER-probes. The probe was inserted at the bottom of a horizontal pipe section.

This experience is even more interesting when it is noted that the well produced approx. 5% water only.

Utilising a computer analysis program for current and potential distributions [12], this flush-probe was modelled to calculate the potential distribution over the electrode surfaces in a condition with a thin water film on the electrodes.

Assuming a corrosion rate of 0.1 mm/year, a water film thickness of 1 mm and a water resistivity of 20 ohm cm, the errors introduced in equation (1) above and in the corrosion rate readings was found to be 12%, which is still acceptable.

3. CONCLUSIONS

A variety of monitoring methods, probes and systems are available for process corrosion monitoring and new methods and designs are continuously being developed.

Probes for direct monitoring of metal loss and corrosion rates like weight-loss coupons, ER-probes and LPR-probes are more commonly used than, e.g., galvanic and hydrogen probes which provide qualitative information about the corrosion condition.

During recent years there has been a tendency towards more frequent use of flush-mounted probes replacing projecting types.

The corrosion engineer selecting probes and designing the monitoring system requires thorough knowledge about relevant corrosion problems, and about the process plant itself.

Recent experience with a flush-mounted LPR-probe has proved that this probe can be used in hydrocarbon media even with relatively low water cut.

REFERENCES

[1] D. R. Fincher and A. C. Nestle, *Materials Protection and Performance,* **12,** No. 7 (1973).

[2] D. R. Fincher and A. C. Nestle, Paper SPE 4220, SPE Symposium, Society of Petroleum Engineers AIME, Los Angeles, Calif., Dec. 4–5, 1972.

[3] J. L. Dawson, Corrosion monitoring techniques, in *Corrosion Monitoring in Oil, Petrochemical and Process Industries,* Oyez Scientific and Technical Services Ltd., 1982.

[4] J. M. Bradshaw, Paper No. 44, Corrosion/76, Hyatt-Regency Hotel, Houston, Texas, 1976.

[5] C. J. Houghton and R. V. Westermark, OTC Paper No. 4332, Offshore Technology Conference, Houston, Texas, 1982.

[6] C. J. Houghton, *Interpretation of Corrosion Monitoring Data in Offshore Oil Production,* Oyez Scientific and Technical Services Ltd., 1982.

[7] Corrocean A.S., Trondheim, *The Piginistic System,* Trondheim, Norway, 1981.

[8] M. E. D. Turner, *Probe Design and Application,* Oyez Scientific and Technical Services Ltd., 1982.

[9] M. Stern and A. Geary, *J. Electrochem. Soc.,* **104,** 56 (1957).

[10] R. M. Vennet, *Materials Performance,* No. 8, p. 31, (1977).

[11] Private communications.

[12] R. Strømmen and A. Rødland, Paper No. 241, Corrosion/80, Chicago, Ill., March 1980.

CHAPTER 8

The use of electrochemical noise for corrosion monitoring

J. L. Dawson, D. A. Eden and K. Hladky

The chapter presents an introduction to methods used for electrochemical corrosion monitoring. The emphasis is on the application of electrochemical noise but since this is a novel technique comparisons also made with a.c. impedance and galvanic coupling response analysis.

Electrochemical noise is observed as spontaneous fluctuations of current or potential during a corrosion process. As a practical method of monitoring corrosion we analyse the small amplitude, low frequency change of corrosion potential. The data is used to assess the corrosion rate and to identify the presence of general or localised corrosion processes. The source of the noise is related to the stochastic processes which occur during film breakdown and metal dissolution.

Examples of the use of the electrochemical noise monitoring technique obtained on both laboratory test rigs and on operating plant are presented for a range of corrosion situations. The noise method can be used to measure plant wall corrosion or corrosion on insert probes. A specially designed multi-element probe can provide rate data under extreme conditions, it is particularly useful when used in conjunction with impedance and zero resistance ammetry. This multi-technique approach can be used to identify complex corrosion processes in plant and ensures that the most appropriate monitoring technique(s) are selected for a particular on-line surveillance application.

1. INTRODUCTION

Corrosion monitoring of plant can involve a wide range of techniques and includes the inspection or non-destructive evaluation methods as well as the electrochemical or insert probe techniques. On-line surveillance is used with the objective of providing either (i) engineering data, related to the condition of the plant or (ii) operational data showing change of process condition within the plant. In the former case this forms the basis for safety and maintanance scheduling since radiography and ultrasonic techniques may also be used for trend monitoring in a similar manner as the traditional electrical resistance probes and weight loss coupons. Further details of these conventional techniques can be found in a number of publications and conference proceedings [1–4].

Corrosion is an electrochemical process and, in principle, the electrochemical techniques offer the advantage that they provide instantaneous readings of the process condition; other methods provide integrated rates. Electrochemical measurements do require that the process environment should be a relatively conductive electrolyte, although we have found that data can be obtained in low conductivity situations, such as in condensing conditions as found during acid dewpoint corrosion in exhaust gas systems, provided that suitable insert probes are employed [5]. The most widely used electrochemical technique is the Linear Polarization Resistance Method (LPRM) based on the theoretical analysis of Stern and Geary [6]. The technique is outlined in many standard corrosion textbooks [2]; however, it is important to note that the method assumes that steady state and not transient conditions are present on the insert electrode probes; other sources of error on the measurement method have also been reviewed [7, 8]. Instrumental techniques for LPRM use an external signal, either voltage or current, to polarise or perturb the sensor electrodes some few millivolts from the corrosion potential. The method is essentially a d.c. technique and improvements can be achieved by use of a.c. impedance studies, particularly in lower conductivity environments or where evidence of corrosion mechanisms is required.

Impedance measurements have found favour in research studies and the use of digital techniques has simplified the data acquisition and data interpretation [9, 10]. Analogue techniques can be used in a simplified approach to monitoring corrosion rates by means of the Tangential Impedance Method [3, 9] or by the Two Frequency Method [11]. Impedance methods are used to obtain the charge transfer resistance, R_{CT}, i.e. a measure of the electron transfer process across the interface or double layer, this is then used in a similar manner as the polarisation resistance, R_p, in the Stern–Gearey equation.

$$i_{corr} = \frac{b_a b_c}{2.3(b_a + b_c)} \frac{1}{R_{CT}}$$

However, it should be noted that the Tafel constants b_a and b_c will not be the same as for the d.c. or R_p case. The theory of electrochemical corrosion impedance shows that the value for R_p includes R_{CT} plus other resistances due to the adsorption of reaction intermediates, R_A [10], diffusion and electrocrystallisation relaxation process.

Non-perturbation electrochemical methods include the use of zero resistance ammeters or galvanic coupling response analysis, corrosion potential monitoring and electrochemical noise measurements. Traditionally the corrosion specialist has used galvanic coupling to study dissimilar metal corrosion but recent studies at UMIST [12] have shown that the technique may be used to advantage as a corrosion monitoring method using two electrodes of similar metal or a metal and a reference electrode; further examples of this method will be presented below. Monitoring of the corrosion potential of a plant is a useful technique in establishing the corrosion condition of the equipment [13], the method may also be used to control inhibitor dosage in a similar manner as with LPRM used in water treatment where a feedback control loop is used in the monitoring equipment. Electrochemical noise is of more recent interest in both research studies and as a possible corrosion monitoring technique. Research at UMIST has been directed towards the development of the method for on-line monitoring [14], the initial research being sponsored jointly by MoD and ICI [12]. This monitoring technique could be considered as an extension of the corrosion potential techniques and involves the analysis of the low frequency and small amplitude fluctuations of the potential [15], a brief description of the method is given below.

Since electrochemical noise is a novel approach to corrosion monitoring it is essential that it is evaluated against other techniques and this chapter also describes its use compared to impedance and galvanic coupling analysis in a range of environments.

2. ELECTROCHEMICAL NOISE

Noise is a generic term used to describe the emission or fluctuation of a base signal be it vibration analysis or from electronic devices. In the study of electrochemistry and electrochemical corrosion there are papers going back some twenty years describing the phenomenon, but it is only over the past few years that electrochemical noise has been recognised as a subject-area of considerable interest. The reasons are due to the increasing use of digital equipment which enables the signals emanating from the corroding interface to be analysed in more detail, combined with the recognition of its application in research studies.

The noise signal is ignored in many studies and it is often attributed to fluctuations in the electronic instrumentation, i.e. the potentiostat or chart recorder. Two earlier papers describing studies on aluminium by Haggard and Williams [16] and on iron electrodes by Iverson [17] show that fluctuations in the corrosion potential can provide an insight into the corrosion processes. The mechanism involved can best be described as electrocrystallisation, that is a series of transient measurements of the process associated with a phase change, be it the formation of an oxide film, the deposition of a metallic layer or the dissolution of an alloy. A theoretical treatment of electrochemical noise has been presented by Fleischmann *et al.* [18] and their paper includes an analysis of electrocrystallisation processes based on the statistics of an ensemble of transients and/or stationary states. Their data expressed as functions of current may be presented as a power spectral density plot. Other workers, notably the Paris school [19, 20] and Bertocci of the US National Bureau of Standards [21, 22] have presented spectral density or autocorrelation plots of current noise. At UMIST we have adopted the approach of analysing the corrosion potential noise [15, 23] since this can provide both rate information and evidence of general or localised corrosion and the method may also be used for on-line monitoring.

The corrosion potential is measured using a sensitive voltmeter and the time record so obtained can then be interpreted using spectral analysis [15] either by means of the Fast Fourier Transform (FFT) or the Maximum Entropy Method (MEM) [24]. The noise spectral density is constant at the lowest frequencies, typical of a Poisson distribution, and decreases at higher frequencies. This decrease in amplitude is typical of $1/f^{\alpha}$ noise and the slope of the roll-off varies with the type of corrosion, presumably as the kinetics of the reactions change.

The frequency of the roll-off point appears to be related to the time constant of the process. At the highest frequencies, typically above 1 to 10 Hz, the plot is again horizontal but is now of very small amplitude which in many cases is the noise floor of the instrumentation. Any noise observed at these higher frequencies is probably thermal in origin and has a Gaussian distribution.

Electrochemical corrosion noise appears to be comprised of two components, the first is related to the initiation stage and this is a stochastic process. The equations used are based on a statistical analysis in order to describe the process at the atomic or molecular level. The second component will follow the normal electrochemical kinetic equations, i.e. a deterministic process. Thus in the case of pitting corrosion the overall mechanism involves a crack/heal process; the film rupture often being induced by chloride ion adsorption and this is followed by a re-passivation process [25]. The noise is essentially a series of transients each having a rapid potential drop followed by a slow recovery. For general corrosion the potential-time record is more regular and this gives rise to an increase in the slope of the roll-off. With active corrosion the mechanism is presumed to be one of micro-pit etching [26], the process involving metal dissolution at a moving kink step.

This introduction illustrates that there is a substantial theoretical and practical framework for the application of electrochemical noise as an investigative and monitoring technique, experimental details of the technique and some on-site measurements will now be described.

3. MONITORING EQUIPMENT

3.1 Electrochemical noise

As indicated above, the use of digital measurements is essential if a detailed analysis of the noise from a corrosion process is required. Fig. 1 shows schematically the arrangement used for our research investigations. The signal is measured by a digital voltmeter and the time record obtained is stored via the microprocessor which is also used to analyse and interpret the data. The sensitivity of the voltmeter and capacity of the microprocessor will depend on the accuracy required and the type of corrosion system being monitored. The displays or hard copies are shown as the time record and the spectral density plots, whilst the change of standard deviation of the noise signal, σ, with time shows the use of the method in a corrosion monitoring application. Our previous and present work indicates that the standard deviations or r.m.s. value is proportional to the corrosion or penetration rate.

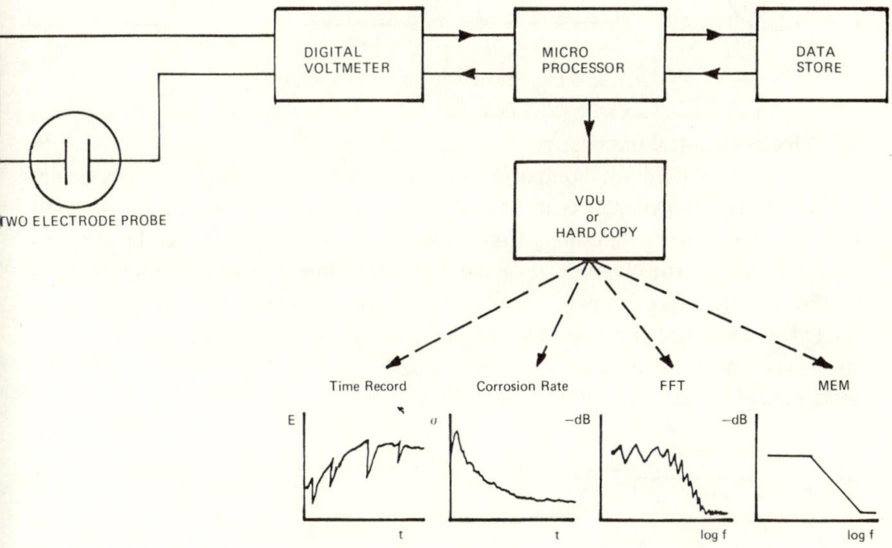

Fig. 1 – Digital method of electrochemical noise analysis.

Analogue instrumentation is particularly suitable for use with a process room chart recorder, as in Fig. 2. The instrument measures the r.m.s. value of the amplitude of the electrochemical noise over a narrow band of frequencies. The centre frequency, the bandwidth and the gain of the circuit are selected to cover a range of corrosion situations. The output is related to the logarithm of the r.m.s. of the noise for a range of amplitude from 1 μV to 10 mV, which approximately corresponds to corrosion rates of 0.1 to 100 mils per year. The main advantage of the technique lies in its inherent capability to detect localised corrosion attack. Such forms of attack give distinguishable 'signatures', thus, although the output is logarithmically scaled to indicate the noise level in decibels, the r.m.s. value also incorporates changes in the d.c. level and use of the instrument makes it possible to monitor phenomenon such as pitting attack, crevice corrosion, etc. Examples of the use of electrochemical noise monitoring in practical situations are given below.

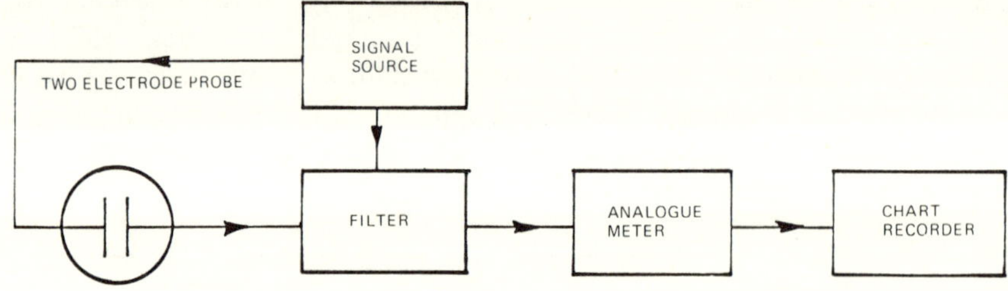

Fig. 2 – Two-frequency impedance monitor.

3.2 Electrochemical impedance

The instrument used for comparing the corrosion rate is shown schematically in Fig. 3. The impedance is measured at two frequencies simultaneously, the high frequency perturbation is first applied to the two electrodes in order to measure the electrolyte resistance. At the same time a second measurement is carried out at a low frequency and the difference between the two readings is related to the corrosion rate. The continuous output makes it possible to detect short-term changes in corrosion rate which accompany pitting or periodic variations in the corrosivity of the environment.

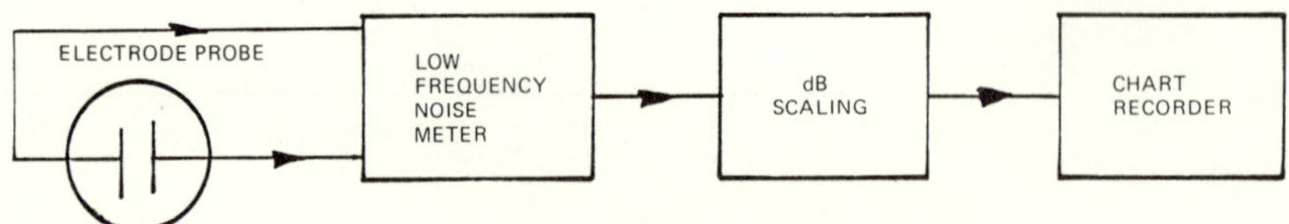

Fig. 3 – Analogue instrumentation for on-line monitoring.

3.3 Zero-resistance ammeter (ZRA)

The meter measures the galvanic current between two electrodes; normally the technique is used to monitor the current between two dissimilar metals but there are advantages in using two 'identical' electrodes. The free corrosion potentials will, however, be different in most cases and will also change with time.

Coupling the two electrodes through the ZRA will force their potentials together, hence the difference in rates, together with any fluctuations, are observed as a current flow. Any short-term fluctuations reflect the localised corrosion events in a similar manner as the electrochemical noise meter, whilst the current magnitude is related to the corrosion current. The output from the instrument is proportional to the logarithm of the absolute value of the ZRA current, Fig. 4, this type of display being convenient for a range of applications since it is often the magnitude of the corrosion rate which is important in on-line monitoring.

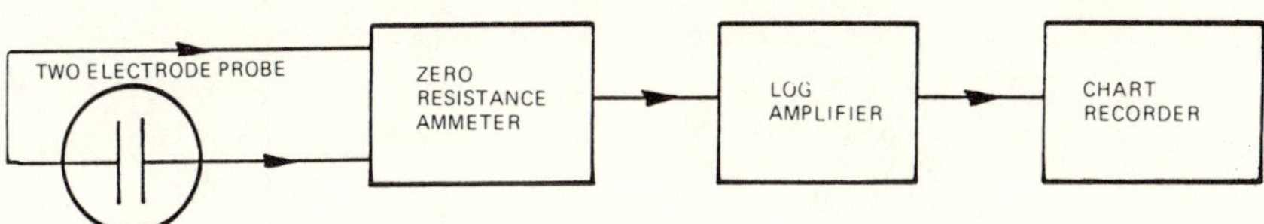

Fig. 4 – Galvanic coupling monitor.

4. PRACTICAL EXAMPLES OF THE TECHNIQUES

The electrochemical noise methods described above have been utilised in the investigation of a variety of systems, including laboratory test rigs and on-line studies of corrosion. These data, as presented below, are not a definitive list of our corrosion monitoring studies but are indicative of some of the present areas of interest at UMIST [12]. Nevertheless they do indicate the broader applications of these novel techniques and in particular they illustrate specific uses where the more conventional methods of linear polarization, electrical resistance probes and weigh loss coupons would either be incapable of providing corrosion information or be lacking in sensitivity.

4.1 Corrosion monitoring in condensed acid conditions in boiler flue gas ducts

These studies form part of a comprehensive investigation into acid dewpoint corrosion in gas exhaust systems [5], the research uses laboratory combustion rigs, pilot plant and on-line boilers. The insert probes developed for the electrochemical studies are constructed in a multi-element laminar form with one face exposed to the gas stream, Fig. 5. This face is flush mounted and the design ensures that the thin film of condensate, either water or acid, provides a conductive path between the metal electrodes. It should be noted that this electrolyte film often has a large resistance which would make conventional d.c. measurements impossible. For some studies the probe can be cooled by an air stream directed on the rear of the electrode assembly; also adjacent pairs of electrodes can be used to measure electrochemical noise, impedance and galvanic coupling current simultaneously on the same probe.

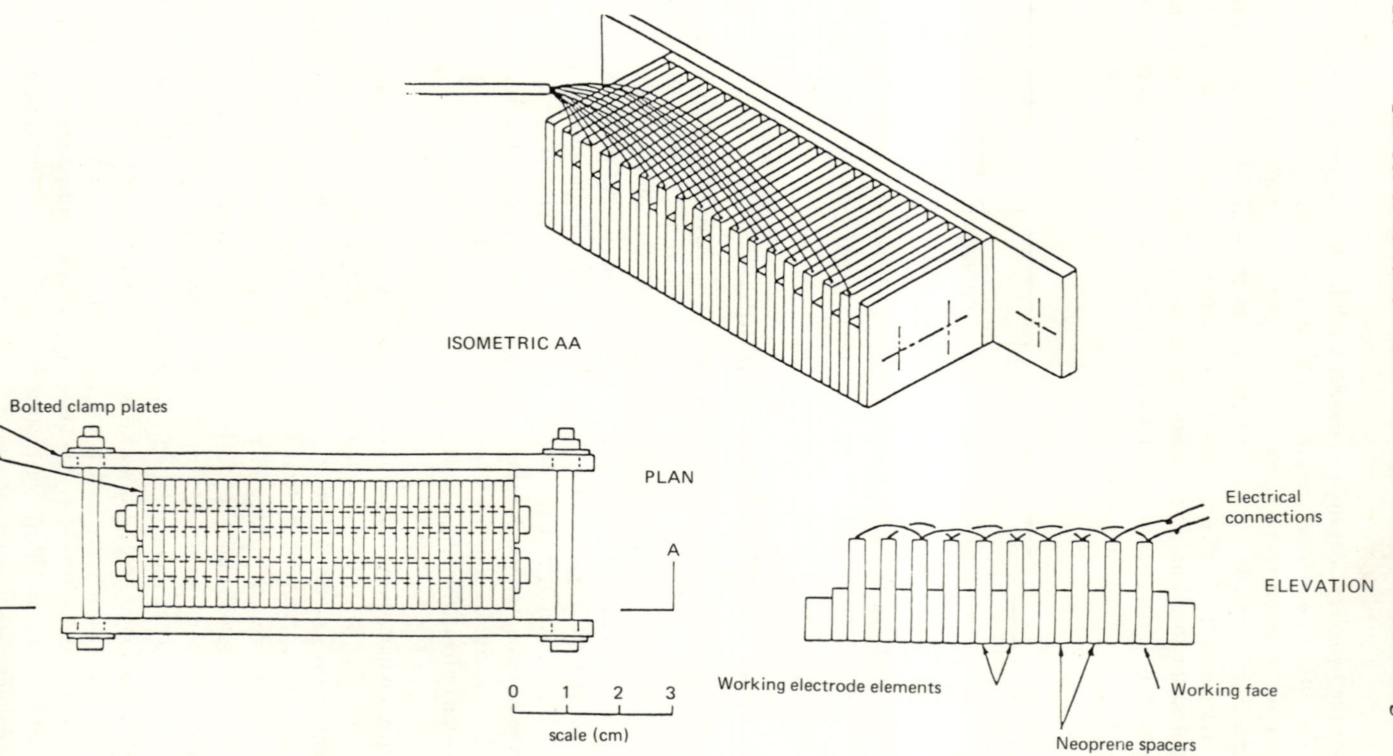

Fig. 5 – Design of multi-element flush mounted probe.

In Fig. 6, the correlation between electrochemical noise, corrosion rate as monitored using conventional a.c. impedance techniques, and weight loss may be seen for the corrosion of mild steel under a thin layer of sulphuric acid at ambient temperature. The data shows that metal passivity is achieved in the concentrated acid over the first few hours, the corrosion decreasing to a low rate during this initial period of exposure. However, as the sulphuric acid is progressively diluted by the adsorption of water vapour from the atmosphere then passivity can no longer be maintained and there is an increase in the corrosion of the mild steel. This figure illustrates the sensitivity of the techniques and shows the correlation between the weight loss, impedance and electrochemical noise. The hygroscopic nature of the corrosion products formed on mild steel is an important aspect of duct work deterioration and most systems (including car exhausts) exhibit off-line corrosion noise which occurs as moisture ingress produces a 'sweating' of the steel; this corrosion is somewhat analogous to a dilute acid or water dewpoint reaction.

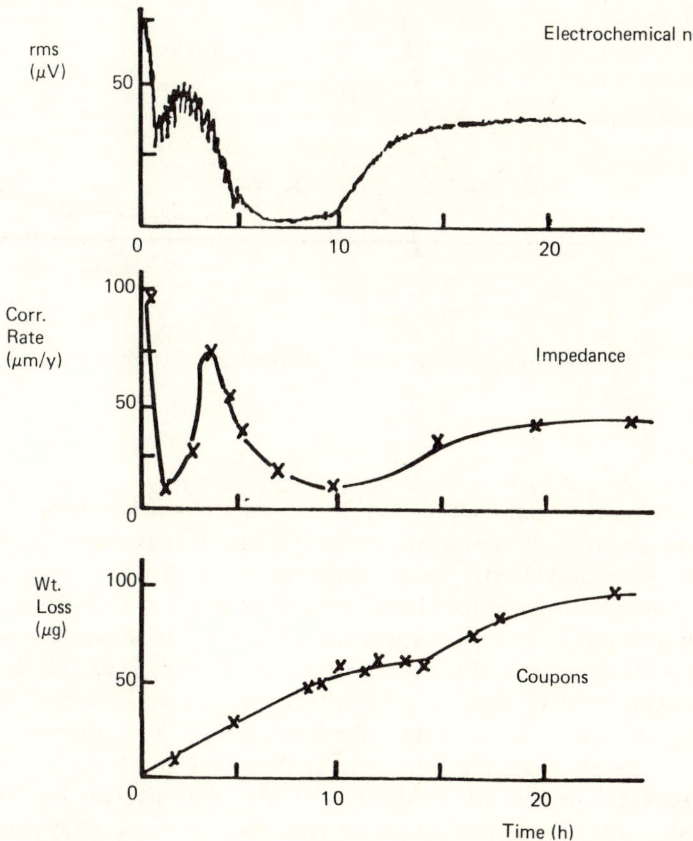

Fig. 6 – Mild steel corrosion under a thin layer of 90°/v sulphuric acid [12].

Typical electrochemical noise from a cooled probe arrangement exposed in a laboratory combustion rig is shown in Fig. 7. The effects of surface temperature and aggressiveness of the condensate are apparent; the corrosion peak due to acid condensation and the lower temperature water dewpoint are both clearly illustrated. These data were obtained using the analogue noise monitoring techniques and the same method was employed to investigate the condensation problems which can occur during plant warm-up and shutdown periods. Fig. 8 shows the monitoring results obtained on probes inserted into the ductwork of a coal-fired boiler system, the graph illustrates changes in electrochemical corrosion rate with temperature and time.

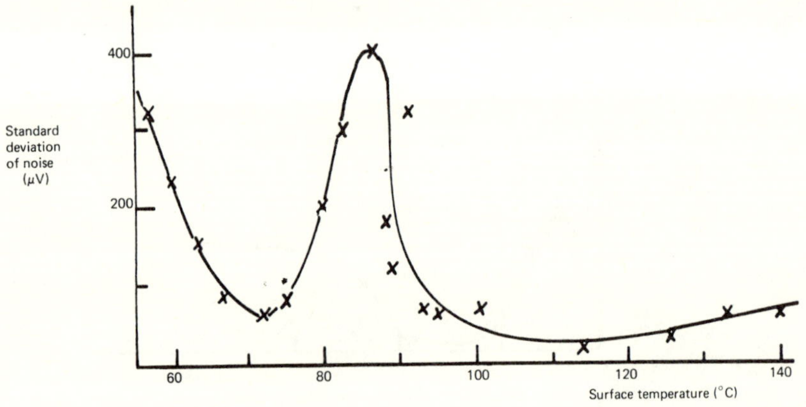

Fig. 7 – Electrochemical noise monitoring in a combustion rig [12].

4.2 Monitoring of active pitting and crevice attack of stainless steels

Studies of pitting and crevice attack of stainless steels in oxidising acidic environments contaminated with halides illustrate the rapid diagnostic uses of the electrochemical noise techniques. Figs. 9 and 10 show typical response when the metal is subject to pitting/crevice attack. The pitting attack causes large fluctuations in the magnitude of the potential in the frequency band of interest, these fluctuations tend to occur at random time intervals, Fig. 9, and are indicative of the stochastic nature of the film breakdown process. Crevice corrosion, although having much the same form as the pitting attack since the initiation processes are similar, often exhibits a regular noise pattern, Fig. 10. These periodic pulses, being apparent on the time record, we have always found to be associated with crevice corrosion.

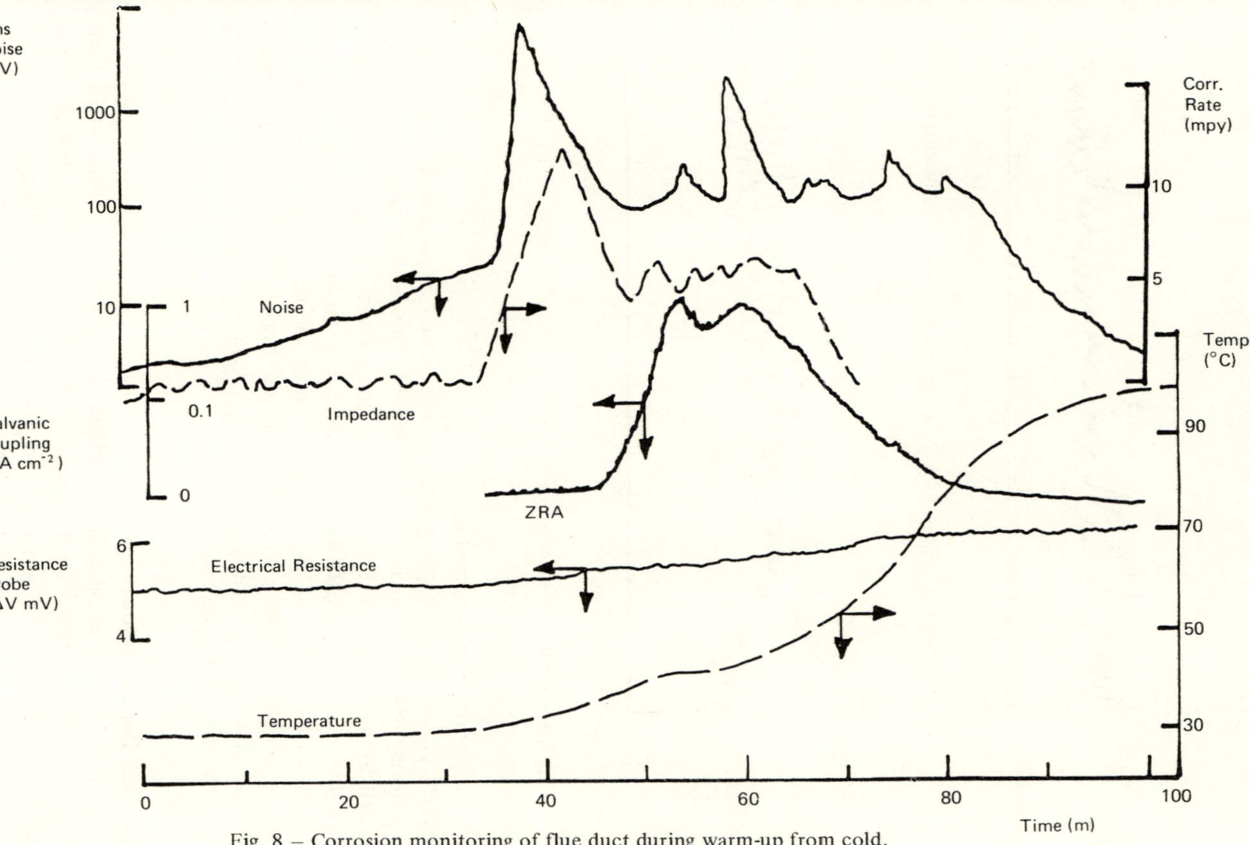

Fig. 8 – Corrosion monitoring of flue duct during warm-up from cold.

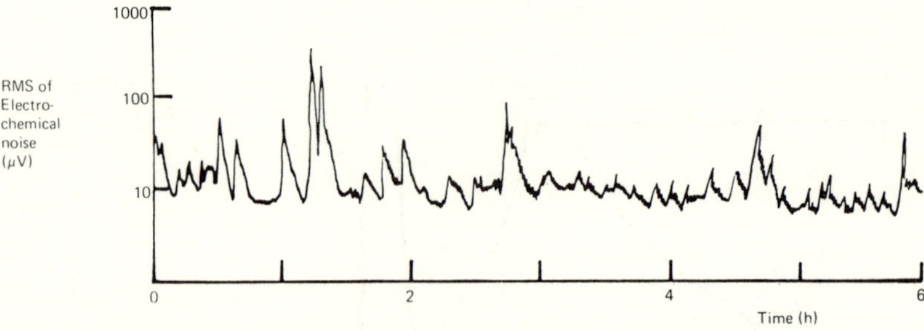

Fig. 9 – Electrochemical noise from 304 stainless steel showing pitting.

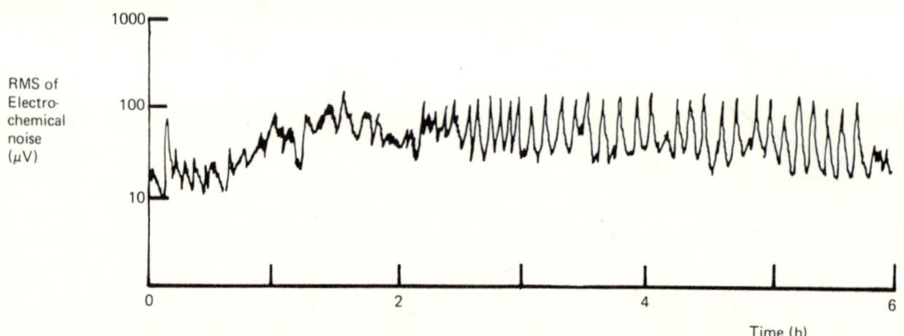

Fig. 10 – Electrochemical noise from 304 stainless steel showing crevice attack.

4.3 Cavitation/erosion corrosion

Studies of aluminium alloy systems in turbulent flow conditions using noise monitoring techniques are illustrated in Fig. 11. The onset of cavitation is accompanied by very large fluctuations in the electrochemical noise output. This is indicative of a film breakdown process which is attributable to the degree and rate of the mechanical damage to the electrode surface. The graph of noise output vs. cavitation number shows a distinctive peak in the region of maximum damage; this graph is analogous to that obtained by weight loss measurements using the same cavitation conditions.

4.4 Inhibitor evaluation

Since the electrochemical noise technique is sensitive to potential fluctuations caused by local depassivation–repassivation phenomena it is ideal for rapid inhibitor evaluation. We have found that susceptibility to pitting attack may be ascertained long before visible pits appear on the test specimen. When a well-inhibited system is monitored using electrochemical noise it will display a very steady corrosion potential, fluctuations being limited to occasional and sporadic 'glitches', probably related to breakdown associated with inclusions, etc. In such

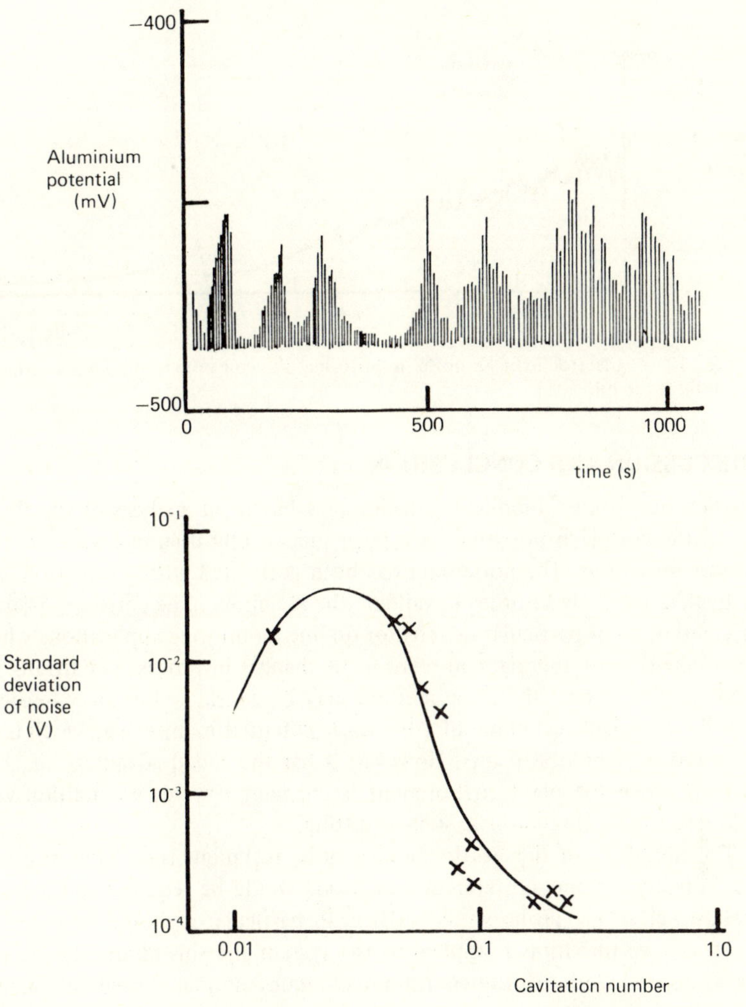

Fig. 11 – Electrochemical noise during cavitation of an aluminium alloy.

a system, with time, the number and frequency of the 'glitches' diminish, Fig. 12. However, in a situation where pitting attack is possible it is usually found that the rest potential fluctuations persist, hence the noise amplitude and other characteristics can indicate the efficiency of the particular formulation. Since this technique does not rely on an imposed electrical perturbation to the test electrodes measurement can be carried out in low conductivity environments, such as cooling waters.

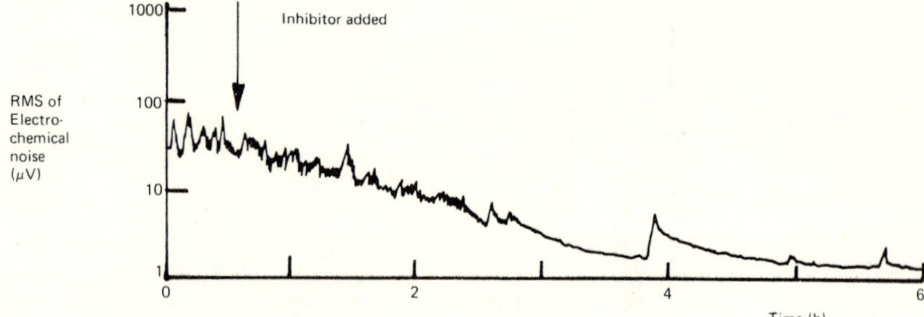

Fig. 12 – Electrochemical noise monitoring showing effect of a passivating inhibitor on mild steel.

5. DISCUSSION AND CONCLUSIONS

Electrochemical noise monitoring techniques based on analysis of the fluctuations of the corrosion potential have been successfully used in a wide variety of corrosion situations. The noise data has been correlated with results from other investigative methods in order to validate the technique. The corrosion potential noise method is of particular benefit for on-line monitoring applications where a rapid evaluation of the plant in relation to changes in process conditions is required. In this respect the noise method may be considered as an extension of the well-established technique of corrossion potential monitoring, which is used on a number of process plants. However, it has the added advantage of clearly indicating when the plant environment is changing towards a condition which would result in localised attack, such as pitting.

The simplicity of the electrochemical noise technique is possibly its greatest asset, in many systems no reference electrode would be required and a conventional two-electrode probe would suffice. In particular it should be noted that no external perturbation is applied to the system and this minimises the problems associated with instrumentation techniques normally used in corrosion monitoring as well as the requirement to assess the condition of the reference electrodes. The noise instrumentation can be either digital or analogue but we have found for most on-line applications the analogue method with a simple strip chart read-out is adequate. The r.m.s. output of the analogue meter provides a good indication of the corrosion rate and the logarithmic scale with the noise amplitude in decibels allows a wide variation of corrosion rates to be measured without having to alter a range switch. The output also displays the sharp discontinuities that are associated with stochastic processes resulting in film breakdown and localised corrosion. Thus susceptibility to localised attack can be ascertained before actual damage occurs, also if the overall rate increases then this provides an indication of the penetration rate.

During the period of research and development of the electrochemical noise method it has been necessary to compare its performance with results from other techniques. Initially this was carried out on laboratory simulation rigs and our general approach or philosophy of corrosion monitoring was to use a wide range of measurement techniques in order to ascertain which of those was most appropriate for a particular situation. This would also be a recommendation for on-line monitoring as in many cases the corrosion processes in the specific circumstance of material/environment combination are not well characterised. Each electrochemical technique provides slightly different corrosion information and it is the combination of the data which often identifies the corrosion mechanisms and allows the most sensitive monitoring method to be employed for the longer-term surveillance.

These monitoring evaluations therefore involve the measurement of electrochemical noise, impedance, galvanic coupling response analysis, linear polarisation resistance and corrosion potential, it is also useful to monitor changes of other parameters such as temperature, pH or specific ion concentration. A multichannel recorder is adequate for following these changes and often some six or more of these different measurements are obtained over a suitable period of time and then a selection of one or more of them is made for the permanent installation. In some cases the conventional corrosion potential measurement may be adequate, whereas in other situations the newer techniques may offer advantages. Measurement of corrosion potential noise via the conventional hard wiring used on plant does not appear to interfere with the signal and there is no evidence of stray 'pick-up' noise from other plant items, presumably because the frequency range of interest is typically less than 1 Hz and this is well below the normal electrical or acoustic range encountered in practice.

The data obrtained with the electrochemical noise and galvanic coupling methods often complement each other, particularly where localised corrosion is imminent or is occurring. Electrochemical impedance and linear polarisation resistance measurements are more appropriate for active or general corrosion situations, linear polarisation finding applications in reasonably conductive electrolytes and where passivity is present. Impedance monitoring using the simplified analogue circuits described previously is especially useful for its rapid response to changes in process environment corrosivity and it is also able to provide data in relatively high resistivity media. Although the two frequency method of measuring the electrochemical impedance can be criticised, in that it does not always provide an accurate value of the charge transfer resistance, we have found that as a practical monitoring technique it provides an adequate indication of the corrosion rate. It can easily distinguish between orders of magnitude in corrosion rate (i.e. 1, 10, 100 mpy); this is because the largest errors in estimating the charge transfer occur when its magnitude is also great, hence since the corrosion rate is proportional to the reciprocal of the charge transfer then variations or errors of between 0.1 and 1 mpy are of no real practical significance for plant surveillance.

The theoretical basis for the use of corrosion potential noise as a monitoring technique is to be found in the analysis of electrochemical noise signals emanating from corroding interfaces. The random fluctuations are the result of local events occurring on the surface. For example, this may be the dissolution of metal at a kink step or grain boundary with the information of a micro etch pit, the electrochemical time constants of such processes are in the range of tens to thousands of seconds, i.e. they are similar in frequency to the electrochemical noise spectra and they may therefore be modelled as inverse electrocrystallisation phenomena. In the case of localised corrosion the noise signal is seen as a series of transients, each the result of a film rupture process which is invariably followed by repassivation, only when the process conditions are sufficiently aggressive is it possible for metal dissolution to continue so as to produce pitting etc. The noise technique is therefore particularly suitable for monitoring alloys based on the transition type metals or aluminium where film breakdown can often result in catastrophic failure. Since process plant is manufactured from such alloys it is in these situations that the combination of techniques can be used to advantage. Electrochemical noise will indicate that changes in process conditions are likely to result in localised corrosion whilst the zero resistance ammetry in combination with impedance will indicate when the increase in corrosion rate is becoming unacceptable.

These recent developments in the use of electrochemical techniques for corrosion monitoring applications, as outlined in this chapter, demonstrate that it is possible to obtain a greater understanding of the complex interactions that occur on plant between the materials of construction and the process environment. This will be of obvious benefit to specialists responsible for material selection or inhibitor formulation development. Perhaps more importantly, they will also enable plant personnel to assess and control the process conditions more closely; this will in turn optimise the overall equipment performance and allow operation much closer to the limits of the materials of construction.

ACKNOWLEDGEMENTS

The authors would like to thank those colleagues in the Corrosion and Protection Centre and in the Corrosion and Protection Centre Industrial Services who are using electrochemical techniques and for their discussion and assistance in providing data and background information on their researches. In particular we would like to thank MoD and ICI for financial assistance during the earlier phases of development of the technique described in this paper.

REFERENCES

[1] W. H. Ailor, (ed.), *Handbook on Corrosion Testing and Evaluation*, Wiley, New York, 1971.

[2] L. C. Shrier, *Corrosion*, Vol. II, 2nd ed., Newnes-Butterworth, London, 1976.

[3] On-line Surveillance and Monitoring of Process Plant Symposium, Sept. 1977, Soc. Chem. Ind., London.

[4] *Industrial Corrosion Monitoring*, Department of Industry Committee on Corrosion, HMSO, 1978.

[5] J. L. Dawson, D. Gearey and W. H. Cox, UK National Corrosion Conference, Nov. 1982, London.

[6] M. Stern, and A. Gearey, *J. Electrochem. Soc.*, **56**, 104 (1957).

[7] L. M. Callow, J. A. Richardson and J. L. Dawson, *Br. Corr. J.*, 123, **11** (1976).

[8] L. M. Callow, J. A. Richardson and J. L. Dawson, *Br. Corr. J.*, 131, **11** (1976).

[9] L. M. Callow, K. Hladky and J. L. Dawson, *Br. Corr. J.*, 20, **15** (1980).

[10] I. Epelboin, C. Gabrielli, M. Keddam and H. Takenouti, ASTM Publication STP 727. p. 150, 1981.

[11] S. Haruyama and T. Tsuru, ASTM Publication STP 727, p. 167, 1981.

[12] Research in progess, Corrosion and Protection Centre, UMIST, Manchester, UK.

[13] P. J. Moreland, Paper No. 12, On Line Surveillance and Monitoring of Process Plant, Sept. 1977, Soc. Chem Ind., London.

[14] UK Patent Application 8200196.

[15] K. Hladky and J. L. Dawson, *Corros. Sci.*, 231, **22** (1982).

[16] T. Haggard and J. R. Williams, *Trans. Farad. Soc., 2288,* **57** (1961).

[17] W. P. Iverson, *J. Electrochem. Soc.*, 617, **115** (1968).

[18] M. Fleischmann, M. Labran, C. Gabrielli and A. Sattar, *Surface Science*, 583, **101** (1980).

[19] U. Bertocci, *J. Electrochem. Soc.*, 931, **127** (1980).

[20] U. Bertocci and J. Kruger, *Surface Science*, 608, **101** (1980).

[21] G. Blanc, I. Epelboin, C. Gabrielli and M. Keddam, *J. Electrochem. Soc.*, 97, **75** (1977).

[22] C. Gabrielli, M. Ksouri and R. Wiart, *J. Electroanal. Chem.*, 233, **86** (1978).

[23] K. Hladky and J. L. Dawson, *Corros. Sci.*, 317, **21** (1981).

[24] J. P. Burg, *Proceedings of 37th Meeting of the Society of Exploration Geophysicists, 1967.*

[25] M. G. S. Ferreira and J. L. Dawson, Paper presented at the 5th Int. Symposium on Passivity, May/June, 1983, Bombannes-Bordeaux, France.

[26] N. Bignold, and M. Fleischmann, *Electrochimica Acta,* 363, **19** (1974).

Part 3
COMPONENTS AND PROCESS MONITORING

CHAPTER 9

Predictive monitoring of process equipment by statistical analysis of ultrasonic survey data

L. E. Wylde and T. Shaw

Costs and time limit the quantity of data that can be generated during an ultrasonic thickness survey. As capital and maintenance costs rise greater emphasis is being placed on the accurate prediction of equipment service life. This chapter considers a statistical technique for predicting minimum overall thickness of process plant suffering from pitting corrosion, from limited ultrasonic thickness surveys. The use of the technique and the constraints it imposes on the ultrasonic survey is demonstrated by its application to surveys on storage tanks, pipelines and ships tanks.

1. INTRODUCTION

For many years ultrasonic techniques have been used to periodically measure remaining wall thicknesses of process plant and vessels. Improvements have continuously been made by instrument manufacturers. Probe design has improved such that pits of diameter less than 1 mm can now be measured relatively accurately.

Whilst accuracy of the ultrasonic measurements has improved it has long been recognised that a fundamental problem exists in predicting equipment life from such measurements. This is a consequence of the fact that in an ultrasonic survey the thickness of only a small proportion of the plant equipment area is actually measured. This does not generally produce a problem in obtaining an indication of average thickness, but such measurements have not been capable of indicating the minimum thickness.

The minimum thickness measured is only the minimum value in the total area seen by the probe. Since this area is only a very small part of the total equipment area there must be a high probability of a value existing in the remainder of the equipment which is less than the value measured by ultrasonics.

It would be possible to obtain this value by an internal examination or by a 100% ultrasonic survey, but in situations where this is impracticable another means must be sought. Provided the probability distribution function for the ultrasonic readings can be defined it should be possible to use this to predict the probability of the existence of values outside the measured range. Pitting corrosion in probabilistic terms can be considered as a random process. It has been shown [1–3] that such processes follow extreme value distributions and Gumbel [4] has developed the theory of extreme value distributions.

The present report considers the development and application of extreme value statistics to the results of ultrasonic surveys. Consideration is given to the relationship between total area and survey area, how the extreme value distribution relates to this, how the data are handled and how minimum thicknesses are predicted. Examples of applications to surveys of process plant are given.

2. APPLICATION OF EXTREME VALUE STATISTICS TO VESSEL SURVEYS

2.1 The statistical theory of extreme values

An excellent review of the extreme value distribution has been given by King [5]. For distributions of the double exponential type

$$\phi_x = e^{-e^{-y}} \tag{1}$$

where ϕ_x is the probability of obtaining a value greater than x and

$$y = \alpha(x - u) \tag{2}$$

y is a reduced extreme value and α and u are scale and location parameters respectively.

From equation (1)

$$y = -\ln(-\ln\phi_x) \tag{3}$$

Thus if ϕ_x can be found y can be calculated. From equation (3)

$$x = u + y^{\alpha - 1} \tag{4}$$

a plot of x against y should yield a straight line if the particular observations considered follow an extreme value distribution. The probability of occurrence (also known as the plotting position) for a value of the variate x_m is

$$\phi_{(xm)} = \frac{m}{n+1} \tag{5}$$

where m is the rank of the mth reading and $n + 1$ is the total number of readings plus one. If the same value of the variate is given by p readings such that the first tank of the variate is m and the last tank is $(m+p-1)$, then the overall ranking of this value is taken to be the geometric mean of m and $(m+p-1)$.

It should be noted that in the derivation of the extreme value distribution the assumption is made that the variate is unlimited in the direction of interest. Pits of infinite depth will not occur in a vessel of finite wall thickness. Whilst the theory sets no limit on the maximum pit depth it does assign a probability of occurrence which diminishes with extreme rapidity as pits in excess of those encountered in practice are considered. This implies that very deep pits can arise very occasionally with a low probability.

2.2 The physical significance of the return period

The return period can be defined as the average number of trials required for an event to occur once. Let us consider the meaning of the return period in relation to process tanks or vessels. In a tank or vessel we are interested in finding the probability of obtaining a value of thickness less than a certain value.

If tanks and vessels are considered, then the return period for obtaining a particular thickness is the number of readings required on average to obtain that thickness once. In a tank, if the individual reading area is a and the tank area of interest is A, then to completely survey the whole tank area of interest and to locate the thinnest point A/a readings are required. Typically in an ultrasonic survey a limited number of readings less than A/a is taken and we are interested in predicting from that limited number of readings what would happen if all A/a readings had been taken. For a tank A/a is the return period and we are interested in finding what value of thickness obtains for that return period. The return period represents an estimate of probable total area required to observe certain specified events.

If ϕ_x is the probability of having a value greater than x in V trials then $1 - \phi_x$ is the probability of having a value less than or equal to x in V trials and thus

$$T_{(x)} = 1/(1 - \phi_{(x)}) \tag{7}$$

where $T_{(x)}$ is the number of observations required to obtain a value less than or equal to x once, i.e. the return period.

We now have a means of relating the return period to a probability. After carrying out ultrasonic measurements we can order those measurements in such a way that the probability ϕ_x of obtaining a thickness value greater than any given value can be calculated. We can extrapolate to the ϕ_x corresponding to a desired T_x and thus predict the thickness value for T_x. For tanks this means we can extrapolate data from the area surveyed to the total area of interest and predict the minimum thickness for that area. We are, therefore, finding the minimum value of thickness predicted for an area A from a number of individual sample areas a.

2.3 Requirements for the methods of measurement

We are attempting to predict from n readings each with an area surveyed a (i.e. an area a seen by the ultrasonic probe on each individual reading) to a tank area A. The theory of extreme values assumes that for each individual sample area the thinnest reading of a range of readings is recorded (i.e. the extreme value for that sample). Historically and in reality, the process of taking spot ultrasonic measurements on a grid pattern produces only a single reading for each individual sample area. The sample area in this case being the area seen by the ultrasound under the probe. This area is usually ill-defined. To record the extreme value for the sample area it has been found necessary to scan the probe over a large sample area, typically 5 × 5 cm, and to record the lowest thickness value for that area.

However, a vast body of data from spot thickness surveys exists. It is possible to use these data by means of an approximation. In a survey consisting of N reading the n thinnest readings can be selected (n is typically greater than 10). Each of the n readings is considered to represent N/n readings.

If the individual sample area is a then the area represented by the n selected readings is

$$a^1 = a \times N/n$$

Extrapolation to the return period is effected by considering the return period as

$$T = A/a^1$$

where A is the area of the tank of interest.

The plotting position for each of the readings is altered from

$$m/(N+1) \quad \text{where} \quad 1 \leqslant m \leqslant N$$

to

$$m^1/(n+1) \quad \text{where} \quad 1 \leqslant m^1 \leqslant n$$

The change in plotting position leads to a correponding and equal change for the return period. Thus despite the fact that the individual readings represent single values and not extremes it is considered valid to use the single readings.

2.4 Limits of extrapolation

The number of readings necessary to fully cover a tank is given by

$$A/a = N$$

where A is the tank area and a the individual sample area.

The limited number of actual readings taken is n. The question may be posed as to how far can we extrapolate from the limited number of readings.

It has been considered by other workers and the extrapolation is possible if

$$n^2 \text{ (or } n^3) \geqslant N$$

This has been examined using data obtained from an actual survey. The survey was designed such that the number of sample areas recorded, 250, and the number of sample areas necessary to cover the entire tank area of interest, 30,600, obeyed the condition

$$n^2 > N$$

The readings were taken on a grid pattern and each individual sample area was 5 X 5 cm. It was possible to analyse the full data set and then further data sets obtained by missing out readings on a regular basis. The thickness predicted for the return period (A/n) was plotted against the number of data points selected.

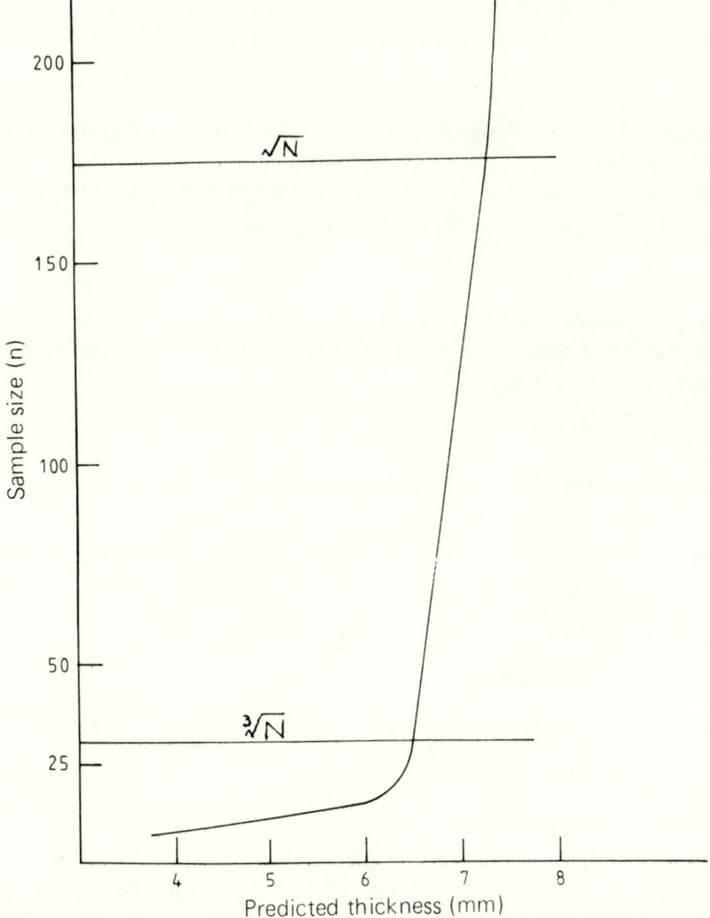

Fig. 1 – Variation of predicted thickness with sample size.

This is shown in Fig. 1. It is apparent from Fig. 1 that for the condition

$$n^2 > N$$

there is only a very slight change in predicted thickness with n. For the condition

$$n^3 > N > n^2$$

there is a slight but perceptible change in predicted thickness of about 1 mm but for the condition

$$n^3 < N$$

there is a rapid change in predicted thickness with n. These results confirm that the number of samples taken should be such that

$$n^2 \text{ (or } n^3) \geqslant N$$

2.5 Application of the extrapolation condition for determining the ultrasonic survey grid size

Let tank area = A, sample area = a, and grid sample = g. The grid area g is the area of tank represented by each sample area a. Thus

$$g = A/n$$

where n is the number of readings each sample area a.

The maximum number of readings possible on a tank of area A with an individual sample area a is

$$N = A/a$$

We require the condition

$$N \leqslant n^2$$

Thus

$$g \leqslant \sqrt{Aa}$$

defines the grid area for a sample area size a. For example, if

$$A = 150{,}000 \text{ cm}^2$$

$$a = 25 \text{ cm}^2$$

$$g \leqslant 44 \times 44 \text{ cm}$$

and 78 readings are required.

In general terms it is found that the additional time spent on the requirement of scanning over a 5 × 5 cm area for each reading is offset by the smaller number of readings required.

3. TREATMENT OF RESULTS

On extreme probability paper the return period scale, the reduced variate scale, and the probability of occurrence are related via equations (1), (3) and (7). The ultrasonic thickness results are ordered and the probability of occurrence is calculated from equation (5) for each thickness value. This is used to calculate y which is plotted against thickness. Reference to equation (4) indicates that if the distribution function for the results is of the double exponential type then a straight line relationship will be observed. If a straight line plot is achieved the results are considered to be obeying extreme value statistics.

If the area of the region of interest covered by the survey is A and the individual sample area is a then the return period

$$T = A/a$$

and the corresponding value of the reduced variate y_T is given by

$$y_T = -\ln\left[-\ln\left(\frac{T-1}{T}\right)\right]$$

Extrapolation of the results to this value of y_T gives the minimum thickness x_T predicted for the vessel.

Manipulation of the results in the above way can be time-consuming. It has been found advantageous to use a computer to analyse the results, test for goodness of fit of the data to extreme value statistics, and to calculate minimum thickness and remaining vessel life.

4. EXAMPLES OF USE

Table 1 – Thickness ranking for vertical tank

Thickness (mm)	Rank	ϕ_x
3.8	64	0.969
4.1	63	0.954
4.3	61.5	0.932
4.6	58.9	0.878
4.8	54.5	0.826
5.1	49.4	0.748
5.3	42.4	0.643
5.6	33	0.500
5.8	22	0.334
6.1	14	0.210
6.4	5.8	0.089
6.6	2	0.030

4.1 Survey on a vertical tank

During 1978 a routine ultrasonic thickness survey of a vertical tank revealed a corroded band on the bottom strake of the tank extending around the tank circumference. A more detailed ultrasonic thickness survey was carried out on the corroded band. The results of this survey were analysed by extreme value statistics and the results are shown in Table 1 and Fig. 2. The thinnest section predicted within the corroded band was 0.28 mm and the life expectancy predicted to be 6 months. The tank was found to be leaking 8 months after the survey was carried out.

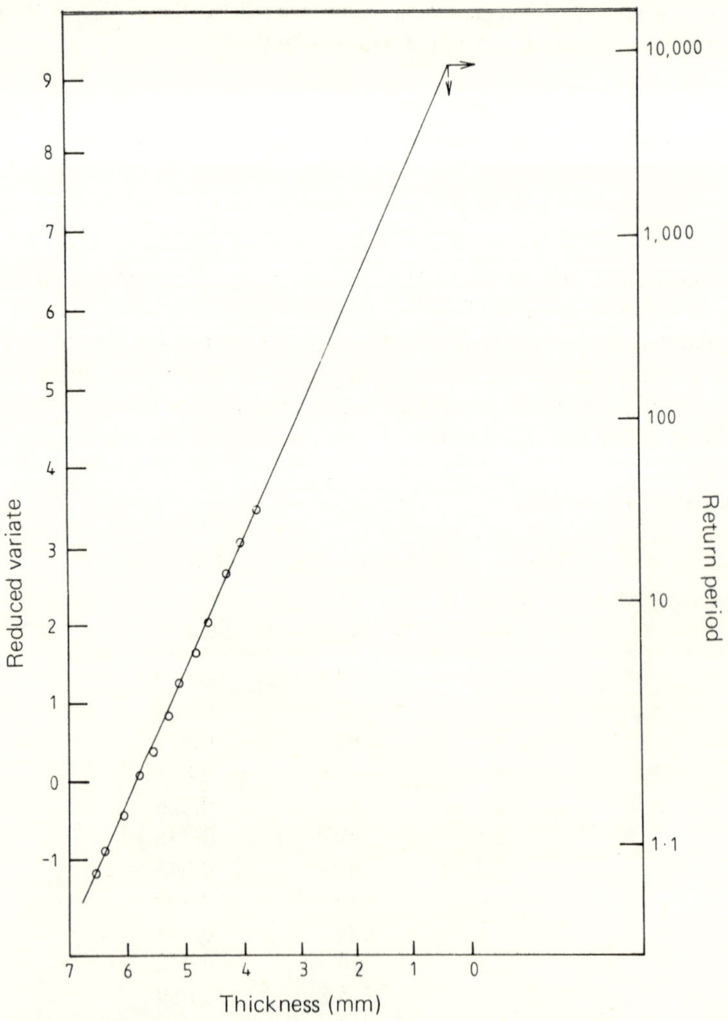

Fig. 2 – Reduced variate as a function of thickness from vertical tank survey.

4.2 Leaking pipeline

Routine examination of part of a 40 m long stainless steel pipeline after removal of lagging showed extensive pitting. An approximately 10 m long sample was removed and examined. The sample was divided into 8 lengths each 1.19 m long and the deepest pit measured for each length. These deepest pits were analysed, using extreme value statistics. The results are shown in Table 2 and graphically in Fig. 3. Examination of Fig. 3 shows the data to be obeying extreme value statistics. For perforation a return period of 12.5 is predicted.

The return period (T) is defined as

$$T = \frac{\text{length for a 4 mm pit}}{\text{sample length}}$$

Thus a perforation would be expected on average every 15 metres. Subsequent examination revealed three perforations in the pipeline.

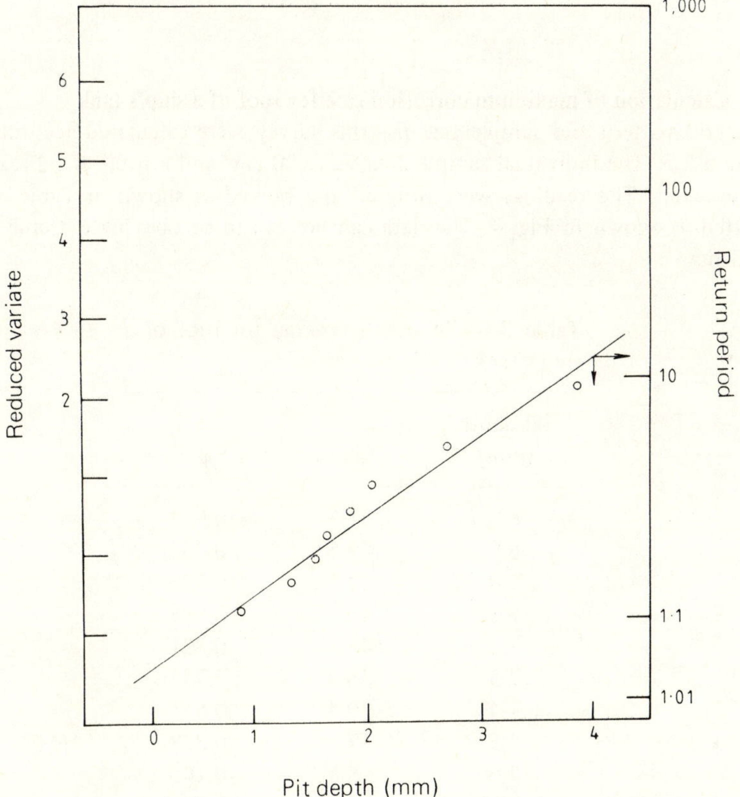

Fig. 3 – Reduced variate as a function of pit depth for pitted pipeline.

Table 2 – Pit depth ranking for stainless steel piping

Pit depth (mm)	Rank	ϕ_x
0.889	1	0.111
1.32	2	0.222
1.57	3	0.333
1.65	4	0.445
1.98	5	0.556
2.03	6	0.667
2.72	7	0.778
3.88	8	0.889

4.3 Calculation of maximum corrosion rate for roof of a ship's tank

The grid pattern and sample size for this survey were calculated according to section 2.5. The individual sample area was 100 cm^2 and a total of 52 readings were taken. The readings were ordered and ranked as shown in Table 3 and plotted as shown in Fig. 4. The data can be seen to be obeying extreme value statistics.

Table 3 – Pit depth ranking for roof of a ship's tank.

Pit depth (mm)	Rank	ϕ_x
6.8	51.5	0.972
6.9	49.5	0.934
7.0	47.5	0.896
7.1	46	0.868
7.2	43.5	0.821
7.3	38.9	0.734
7.4	29.4	0.555
7.5	17.9	0.338
7.6	8.8	0.166
7.8	5	0.094
7.9	2	0.038

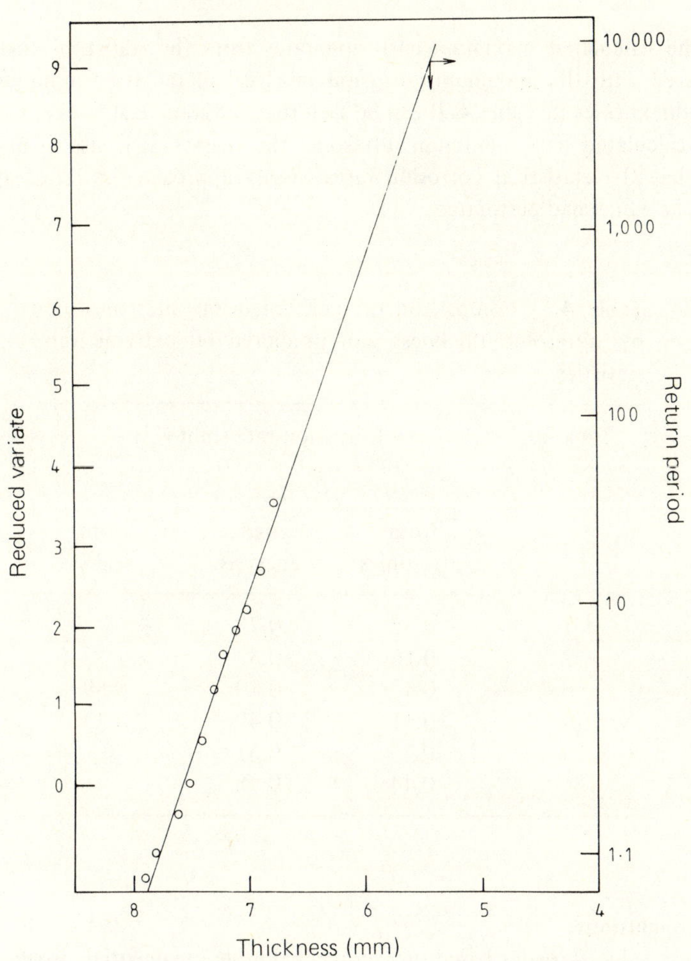

Fig. 4 – Reduced variate as a function of thickness for a ship's tank roof.

The surface area of the roof was 769,824 cm^2 and the return period therefore was 7,698. The thinnest section predicted was 5.4 mm which can be compared with the thinnest value measured of 6.8 mm. The maximum corrosion rate was calculated as 0.2 mm y^{-1}.

4.4 Computer analysis of data from a number of surveys

A number of surveys have been analysed using a computer program based on equations (1) to (5). In all cases the data was found to be obeying extreme value statistics.

The calculated maximum corrosion rates from the statistical analysis are compared with the maximum corrosion rates calculated from minimum ultrasonic thicknesses in Table 4. It can be seen that the statistical rates are ~2 times those calculated from minimum ultrasonic thicknesses. It is also of interest to note that the statistical corrosion rates closely approach the rates calculated for tanks which had perforated.

Table 4 – Comparison of tank corrosion rates measured by ultrasonic thickness and predicted by extreme valve statistics.

Tank no.	Corrosion rate (mm y^{-1})		
	from ultrasonics	from statistical analysis	from leaks
1	0.37	0.73	–
2	0.16	0.38	–
3	0.47	0.89	0.89
4	0.31	0.49	0.48
5	0.2	0.31	
6	0.44	0.78	

4.5 Conclusions

Extreme value statistics based on Gumbel's double exponential distribution can be applied to ultrasonic surveys of corroding tanks. Such surveys should be carried out on a defined sample area. The results obtained are encouraging and indicate that the technique provides a powerful tool of predictive analysis for monitoring process plant condition allowing optimum decisions to be made on timely equipment repair, maintenance and usage whilst meeting the increasingly stringent demands of regulatory legislation.

ACKNOWLEDGEMENTS

The authors wish to thank the Associated Octel Company Limited for permission to publish the paper, and to express their gratitude to John Church who wrote and developed the computer programs and to their colleagues for assistance in preparing the paper.

REFERENCES

[1] H. P. Goddard, *Can. J. Chem. Eng.*, pp. 167–173 (1960).
[2] G. G. Eldridge, *Corrosion*, **13**, pp. 67–76 (1957).
[3] P. M. Aziz, *Corrosion*, **12**, pp. 495–506 (1956).
[4] E. J. Gumbel, *Statistics of Extremes*, Columbia University Press, 1958.
[5] J. R. King, *Probability Charts for Decision Making*, Industrial Press, 1971, Chapter 11.

CHAPTER 10

The vibration diagnostics of a high engine speed, high performance centrifugal compressor

I. Ballo, M. Péteri and P. Tirinda

This chapter deals with a diagnostic monitoring system by which the technical state of a large centrifugal compressor, operating at an ammonia plant, is traced. The compressor consists of three independent coaxial bodies and is driven by a 20 MW performance steam turbine. The normal compressor speed is 11,000 revolutions per minute and the discharge pressure 30 MPa.

1. THE CONCEPTION OF THE MONITORING SYSTEM

Primary data for the monitoring system are provided by the set of piezoelectric accelerometers. At one point on each body of the compressor they scan the acceleration of mechanical vibration in three perpendicular directions: radial, tangential and axial. Their amplified signals are recorded by a tape recorder.

Data thus stored are later processed by a real time analyser that cooperates with a minicomputer. From each registered course a statistically sufficiently reliable estimation of the power spectral density (PSE) is made, in the first stage, within the 0 + 10 kHz frequency band. To essentially lower the data scope, the PSE thus established continues to be processed. Peaks of diagnostic significance are searched automatically in it and compared statistically with those recorded in previous diagnostic measurements.

According to need, the entire system may provide diagnostic information on the examined compressor in one of the following forms:

(a) graphically representing PSE behaviour in a linear or logarithmic display within the entire frequency band or in its part; information of minor significance may be summarily suppressed;
(b) printing out data on diagnostically significant peaks in PSE and comparing them statistically with similar data from preceding measurements;

(c) estimating trends, by regression analysis, in the development of the individual diagnostically significant peaks in the spectrum and representing them in form of a graph or table.

The operation of the entire system is schematically shown in Fig. 1.

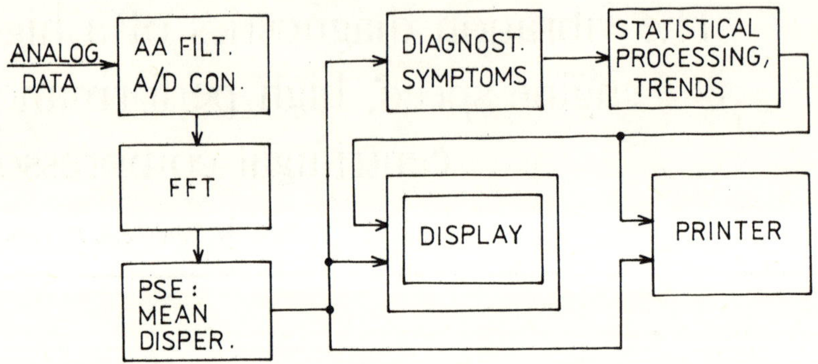

Fig. 1 – Functional scheme of the entire diagnostic system operation.

2. DIAGNOSTIC SYMPTOMS IN PSE

Conspicuous peaks in PSE whose frequency position in course of the PSE is unambiguously and rationally tied up with processes in the examined machine, are considered to be diagnostic features. We shall divide them into synchronous and asynchronous ones, as is done in other papers.

To the synchronous class belong primarily the peak at the rotation frequency $f_{rot} \doteq 183$ Hz. It is usually barely conspicuous, since the rotors applied are, as a rule, very well balanced.

Another group of synchronous symptoms are the peaks at blade frequencies f_{Bi} $(i = 1, 2, \ldots)$

$$f_{Bi} = n_{Bi} \times f_{rot}$$

The number of blades n_{Bi} on the individual (ith) rotor wheel differs because, in the course of the PSE, several peaks of this type are observed. Sometimes they are very conspicuous and are obviously due to the flow feature in the engine. It is noteworthy that, following the exchange of otherwise geometrically congruent rotors, essential deflections in the height of these peaks may be observed in some cases that may be indicative of the sensitivity of this symptom to minor changes in the flow feature in the engine, occasionally to minor changes in the rotor parameters.

In addition to the blade frequencies f_{Bi}, peaks at $2 \times f_{Bi}$ and $3 \times f_{Bi}$ may almost always be registered. In the diagnostic interpretation of PSE they are to be taken into account together with the peaks at blade frequencies.

The group of asynchronous symptoms is numerous in PSE. Significant is the peak at the 1st eigenfrequency of the rotor $f_{eig} \doteq 69$ Hz. Changes in its size may be due to alterations of forces destabilizing the first eigenmode of the rotor bending vibration, hence considerable attention is paid to this symptom. Along with this symptom in PSE, the peaks at $2 \times f_{eig}$, $3 \times f_{eig}$ and $4 \times f_{eig}$, arising through the action of non-linearities in the system, are also traced.

Sometimes there arises a peak in PSE in the neighbourhood of $f_c \doteq 100$ Hz. Experience has shown that its rise is due to the successive impairment of oil seals which, from the dynamic point of view, start to take over the function of virtual bearings. This shortens the effective length of the rotor between the bearings, giving rise to its eigenfrequency in the 100 Hz domain.

A series of significant peaks in PSE at frequencies f_{Dj} ($j = 1, 2, \ldots$) is responsible for the vibration of rotor wheel disks. Their higher number is due to the fact that the individual wheels on one rotor differ in dimensions and that several vibration modes are simultaneously excited on the individual wheels.

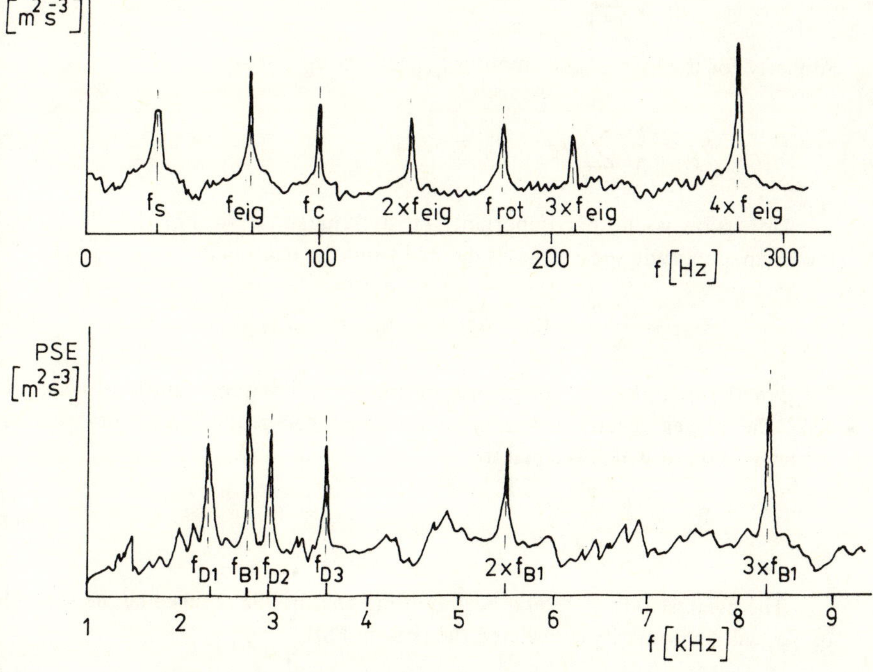

Fig. 2 – Typical PSE behaviour in the examined turbo-compressor.

The dependence has been observed of their frequency position in PSE on the operating speed as well as a rather conspicuous frequency shift when exchanging the engine rotor, although both rotors are geometrically identical. There has so far been no occasion to trace the changes of the frequency position of these peaks when cracks occur on the wheel disks.

Apart from keeping track of the symptoms referred to, the monitoring system also envisages the possible occurrence of additional peaks in PSE that might be due, for example, to various types of the flow destabilisation of the working medium in the engine.

The typical behaviour of the observed PSE is shown in Fig. 2.

3. THE ALGORITHMISATION OF SYMPTOM DETERMINATION IN PSE

Each line of the power spectrum S_{ij}, calculated by applying FFT to the digitalised analog signal, is a (jth) realisation of the random variable S_{ij} whose frequency position in the power spectrum is given by frequency f_i $i=1, 2, \ldots L$. If these realisations amount to a total of N, ($j=1, 2, \ldots N$), then the PSE line at the frequency f_i is calculated as the mean of N individual realisations:

$$\hat{S}_i = \frac{1}{N} \sum_{j=1}^{N} S_{ij} \tag{1}$$

Similarly for the lines of the amplitude spectrum ϑ_{ij}

$$\hat{\vartheta}_i = \frac{1}{N} \sum_{j=1}^{N} \vartheta_{ij} \; ; \quad \vartheta_{ij} = \sqrt{S_{ij}} \tag{2}$$

Dispersion is another important statistic characteristic. For the ith line of the mean amplitude spectrum it is defined from the relation:

$$D_{\vartheta i} = \frac{N}{N-1} (\hat{S}_i - \hat{\vartheta}_i^2) \; ; \quad \sigma_{\vartheta i} = \sqrt{D_{\vartheta i}} \tag{3}$$

For the definition of the algorithm used in defining the significant peaks in PSE, the power spectrum density of the noise background is of importance, the estimation of which is equal to

$$\hat{S}_N = \frac{1}{L} \sum_{i=1}^{L} \hat{S}_i \tag{4}$$

The greatest out of several neighbouring lines in PSE, exceeding the tenfold of S_N, will be primarily considered the peak in PSE.

$$\hat{S}_P \geqslant 10 \hat{S}_N \tag{5}$$

Further, those lines in PSE that are tied up with some of the outstanding frequencies are referred to in the second part of this paper, even if they do not satisfy relation (5).

Should the lines adjacent to the one defined as the peak, differ from the peak by less than 2 dB, a correction of the peak height is carried out that follows from the known picket-fence effect [1, 2].

The height of peak ϑ_P in the amplitude spectrum and the dispersion $D_{\vartheta P}$ pertaining to it are defined by an entirely analogous procedure.

4. STATISTICAL COMPARISON OF PEAKS FROM TWO MEASUREMENTS

In the preceding sections of the paper a description has been given of the way of processing and evaluating a signal that is recorded in course of the diagnostic examination of the engine. Denote the time datum, that is the hour and date of this examination, by symbol τ_1. Denote all quantities measured and calculated for this time, by index 1.

After a definite space of time, say one week, the next diagnostic examination of the engine is carried out. Denote the date of this examination by τ_2 and furnish all quantities connected with this examination, with index 2.

For both the first and second diagnostic examination the symptoms in PSE may be defined together with their statistic characteristics, as referred to in the preceding sections of this paper. From the diagnostic viewpoint it is of importance to decide whether the difference in data from both measurements are of statistic significance or not.

The first two characteristics to be assessed from this aspect are the dispersion of the traced symptom in the amplitude spectrum. As far as relation (3)

$$1 \leqslant \frac{D_{\vartheta P1}}{D_{\vartheta P2}} < F_{\alpha/2}\,(N_1-1\,;\; N_2-1) \tag{6}$$

is satisfied, where $F_{\alpha/2}\ (k_1;\ k_2)$ is the critical value of F-distribution on the significance level α, the difference between both dispersions is statistically but little significant. As far as $N_1 = N_2 = 20$, and $\alpha = 0.05$, we have

$$F_{0.025}\ (19;19)\ \doteq\ 2.53$$

In this case both dispersions differ with statistic significance, if the larger of them is greater than the 2.53 multiple of the smaller.

If both dispersions are found not to differ with statistic significance, we step up to testing the statistic significance of height difference of peaks ϑ_{P1}, ϑ_{P2} in the amplitude spectrum, established in course of both measurements. To be able to carry out this test, we set up the testing quantity [3]

$$t = \frac{\vartheta_{P1} - \vartheta_{P2}}{\sqrt{D_d}}$$

$$D_d = \frac{N_1 + N_2}{N_1 N_2} \cdot \frac{D_{\vartheta P1}(N_1 - 1) + D_{\vartheta P2}(N_2 - 1)}{N_1 + N_2 - 2} \tag{7}$$

Both defined peak heights will differ with little statistic significance, if

$$|t| < t_{\alpha/2}(N_1 + N_2 - 2) \tag{8}$$

where $t_{\alpha/2}\ (N_1 + N_2 - 2)$ is the critical value of the Student distribution on the significance level α. For the parameters previously used, we shall have

$$t_{0.025}(38) \doteq 2.025$$

5. CONCLUSION

A diagnostic system has been described in the paper for monitoring the operation of a centrifugal compressor running under relatively high parameters. It will probably be possible to maintain the conception of such a system also in monitoring the operation of other, similar engines. In interpreting both measured and recorded data, their statistic evaluation, by which the developmental trends of the individual symptoms in PSE may also be traced, has appeared to be a significant factor.

REFERENCES

[1] N. Thrane, The discrete Fourier transform and FFT analysers, *B & K Techn. Rev.*, No. 1 (1979).

[2] A. Papoulis, *The Fourier Integral and its Applications*, McGraw-Hill, New York, 1962.

[3] I. W. Burr, *Applied Statistical Methods*, Academic Press, New York, 1974.

CHAPTER 11

Valve condition monitoring

D. Harrison and A. Heath

The chapter describes an investigation of techniques to monitor the condition of ball valves. The work shows that the variation of valve operating torque with position is a useful indicator of the operational state of the valve and that the torque position plot can be regarded as a valve signature. Changes in the signature with operating conditions and during life tests of valves are discussed. Experiments to monitor valve leakage have confirmed that acoustic emission is a useful technique.

1. INTRODUCTION

Valves, in particular actuated shutdown valves, are key components throughout the petroleum and chemical industries. The failure of a valve to operate can lead to potentially hazardous situations, lost production and plant downtime. At present high reliability is ensured by planned programmes of valve maintenance and by duplication of components using two or three valves in series.

Particularly severe problems are encountered offshore where the combination of high maintenance costs, weight restrictions on production platforms, problems of maintaining and replacing subsea valves and the cost of lost production leads operators to seek ways of knowing that a valve will operate satisfactorily when required.

Monitoring of valve performance if successful would lead to cost savings through reduced duplication of valve components, a reduction in maintenance costs, and improved reliability and operator confidence.

2. OBJECTIVES

The objectives of the work programme are to determine and prove methods of monitoring the condition of valves, in particular actuated emergency shutdown valves, in order to predict continued satisfactory operation and predict valve failure before it occurs. The two principal modes of valve failure are:

(1) Failure of the valve to open or close when required.
(2) Valve leakage, either when closed with a differential pressure across the valve, or leakage to the surroundings through the valve stem.

Priority has been given to developing methods of detecting the first failure mode, that of the valve failing to operate, and a limited amount of effort has been devoted to non-intrusive methods of detecting valve leakage. The first part of the study has concentrated on monitoring the performance of pneumatically actuated floating ball valves. The field of study is being widened and the performance of actuated trunnion mounted ball valves and gate valves is to be investigated.

A three-stage experimental programme has been adopted.

(1) To investigate valve operational parameters, such as operating torque, operating time and position, to determine which parameters give a useful indication of valve performance, and to determine the effect of changes in valve operating conditions such as flow rate and pressure on the measured parameters.
(2) To correlate changes in valve performance with valve condition and likelihood of failure and to prove the techniques by long-term life testing of valves under simulated operating conditions.
(3) To prove the techniques developed in operational situations.

This chapter describes work to date on the first two objectives.

3. TEST RIG, VALVES AND INSTRUMENTATION

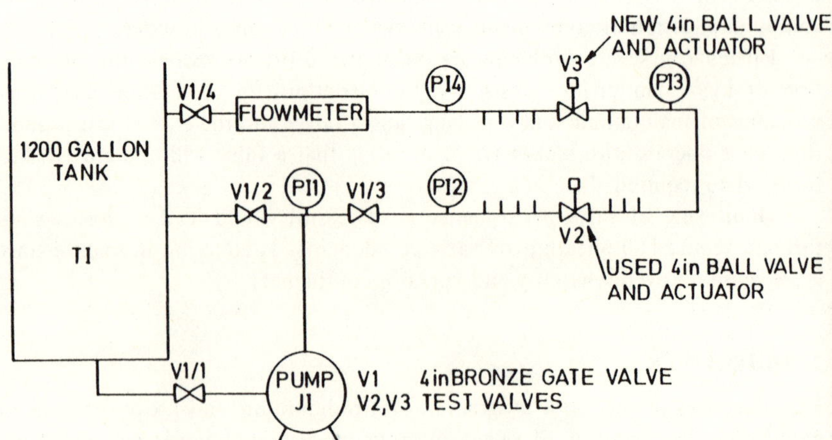

Fig. 1 – Valve condition monitoring rig.

3.1 Test rig

The rig has been designed to enable 4″ diameter actuated valves to be operated, on water, at typical flowrates of 0.025 m^3/s and at operating pressure of up to 4 bar. A simplified flow diagram of the rig is shown in Fig. 1. A centrifugal pump J1 is used to circulate water from the tank T1 through the pipe loop at pressures of up to 4 bar and flowrates up to 0.04 m^3/s. Gate valves V1/2, V1/3, V2/4 are used to vary fluid flow rates and test valve operating pressures. A number of pneumatically actuated ball valves V2, V3 are mounted in the loop for evaluation. The test valves can be cycled automatically or operated manually.

3.2 Test valves

Test work has been carried out on 4″ Class 150 floating ball valves with PTFE seats operated by a rack and pinion type single acting pneumatic actuator with spring action to close.

3.3 Instrumentation

Fig. 2 shows a schematic view of an instrumented valve. Upstream and downstream fluid pressures are monitored by pressure transducers mounted adjacent

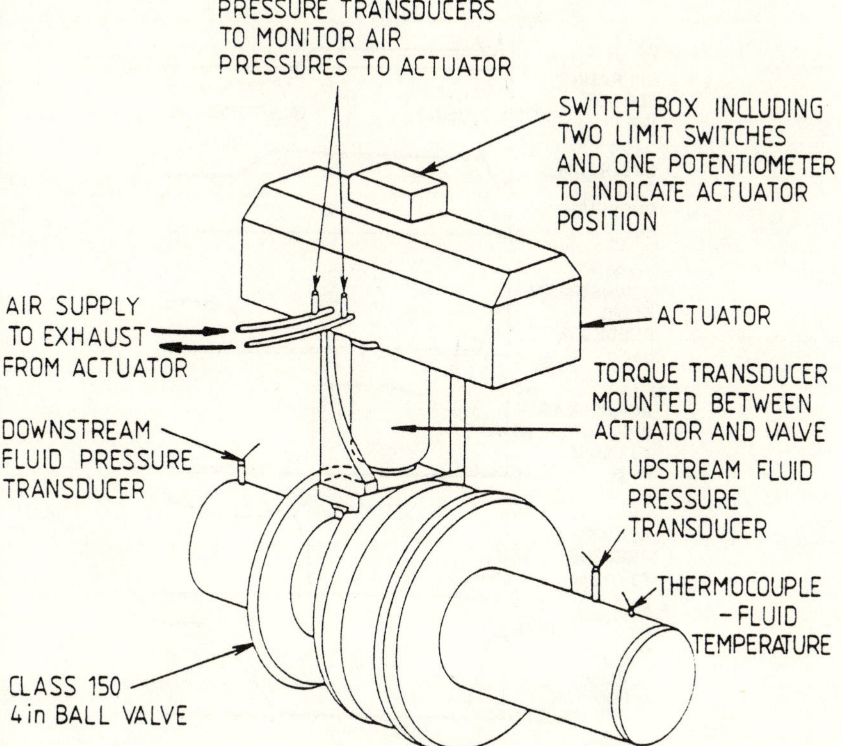

Fig. 2 – Schematic view of instrumented valve.

to the valve. Valve operating torque is measured using a torque transducer mounted between the actuator and valve. Valve (actuator) position is monitored using a potentiometer and valve operating time measured using two cam-operated limit switches.

Actuator air pressure is measured using pressure transducers mounted adjacent to the actuator. Fluid temperature is also measured. Valve response is recorded using a quick response multichannel chart recorder and X, Y plotter.

4. EXPERIMENTAL RESULTS

4.1 Comparison of performance of old and new valves

An initial investigation was carried out to compare the performance of a valve which had a history of failing to close on a separator water service at 70°C with a similar new valve. Operating parameters – valve operating torque, upstream and downstream fluid pressure, actuator air pressure and valve position during operation were displayed using a multichannel quick response recorder.

A typical trace during a valve operating cycle is shown in Fig. 3.

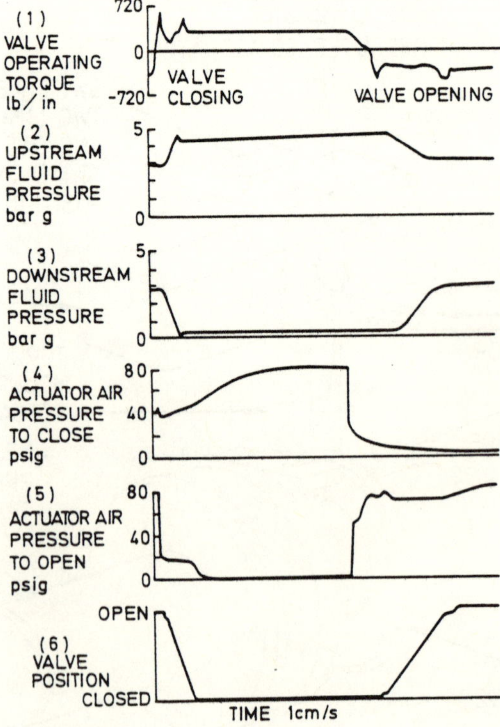

Fig. 3 – Chart recorder output during valve operation.

It was apparent from the traces that valve torque was yielding useful information and giving significantly different results for the two valves.

A useful and convenient method of presenting and comparing data was found to be in the form of an X, Y plot of valve torque against operating position.

The plot enables comparisons of valve performance under differing operating condition to be readily made, the area enclosed by the plot giving a direct measure of the energy required to operate the valve. The plot can be considered to be a 'valve signature'. Comparative plots for the two valves are shown in Fig. 4 under identical operating conditions.

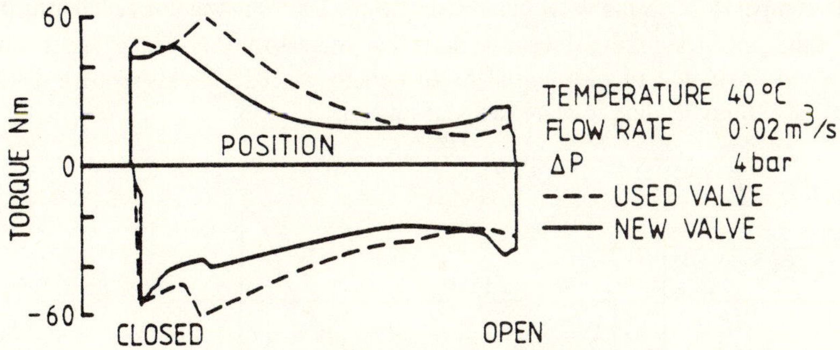

Fig. 4 – Signatures of old and new valves.

4.2 Investigation of valve performance under varying operating conditions

Valve performance was studied under varying operating conditions. The following operating conditions were varied: flow rate, operating pressure with the valve open, differential pressure when the valve was closed and valve operating temperature.

Increasing differential pressure across the valve when closed was found to increase valve operating torque near the closed position, see Fig. 5. Valve performance near the open position was not effected.

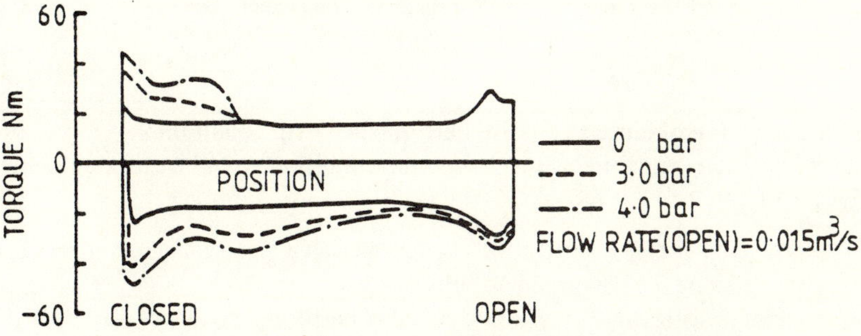

Fig. 5 – Valve signatures at varying closed differential pressures.

Increasing flow rate through the valve in the open position was found to lead to an increase in the operating torque of the valve when partially closed. This increased operating torque is caused by higher differential pressures at high flow rates.

The operating pressure of the valve when open was found to have little effect on valve operating torque.

Valve operating temperature was found to have a very significant effect on valve operating torque, particularly, under flow conditions, when the valve was near the open position. Fig. 6 shows comparative valve signatures under flow conditions. The effect of temperature on torque is immediately apparent. At high temperature available actuator torque of 100 Nm was exceeded causing the valve to stick in the open position. The increase in operating torque with temperature is due to differential expansion of the PTFE seats relative to the steel valve body leading to increased seat loadings.

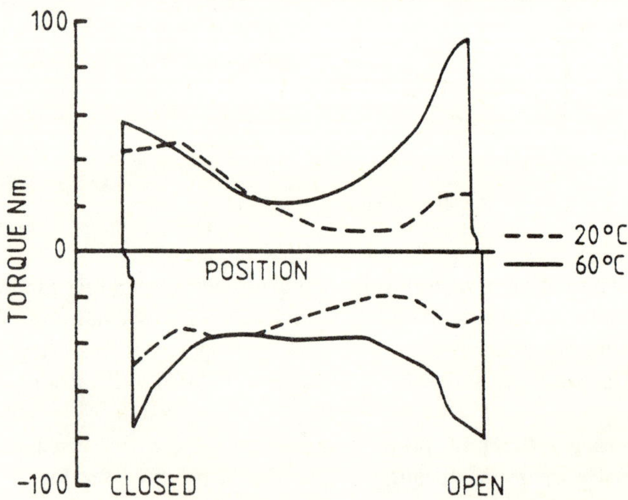

Fig. 6 – Valve signatures at varying temperatures.

4.3 Investigation of changes in valve performance with condition

Changes in valve signature with changes in condition were investigated in two experiments.

(i) A cycling test investigating changes in valve performance, due to wear, caused by a large number of operations.

(ii) An investigation of changes in valve performance due to the time the valve is left standing since the last operation.

During the life cycling test significant changes in valve performance were detected. Fig. 7 compares valve signatures under no-flow conditions at increasing numbers of valve operations from new. The decrease in operating torque due to seat wear and loss of seat pre-load is apparent. After 600 cycles valve operating torque had halved and the valve was beginning to leak. Fig. 8 compares valve signatures under flow conditions. Valve operating torque near the open position is similar to that at no flow conditions. Torque to close the valve under flow conditions can be seen to rise marginally as the valve deteriorates. Under flow conditions differential pressure across the valve contributed to the seat loading and the increasing operating torque can be ascribed to deterioration of the valve ball leading to an increase in the coefficient of friction between the valve ball and seats. The changing shape of the signature under flow conditions can be seen to be an indicator of valve condition.

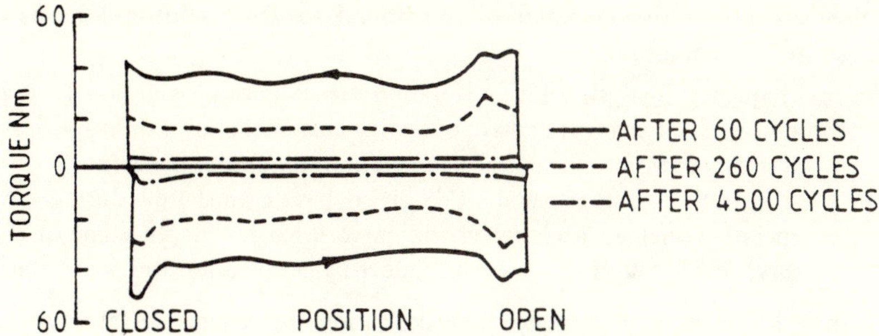

Fig. 7 – Valve signatures at no-flow conditions.

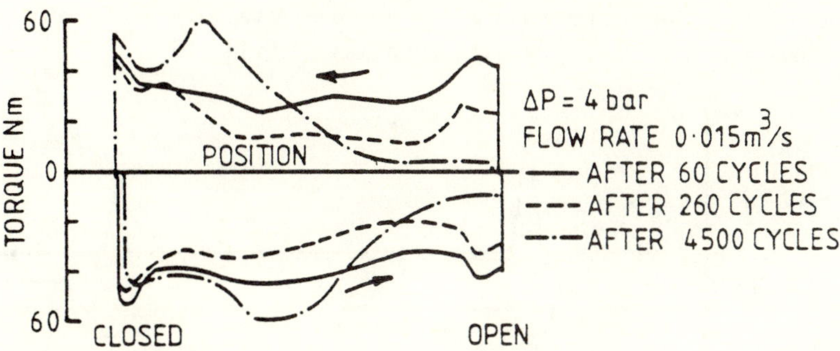

Fig. 8 – Valve signatures under flow conditions.

Initial investigations of changes in valve operating torque with the time that the valve has been inoperative show that valve torque increases with the time since the last operation. Fig. 9 shows valve signatures under no flow conditions at varying delay times.

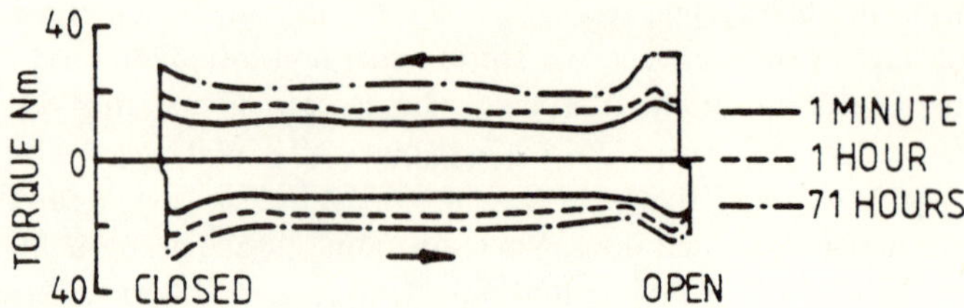

Fig. 9 – Valve signatures at no-flow conditions with varying time since last operation.

5. VALVE VIBRATION IN OPERATION AS AN INDICATOR OF CONDITION

Valve vibration in operation was investigated as an indicator of condition. Under no-flow conditions two techniques were found to give useful indications of changes in valve condition.

(1) Frequency analysis of the vibration signal during operation gave an indication of changes in valve operating characteristics. For instance the development of a squeak during operation was clearly apparent.

(2) Presentation of the overall RMS signal level against time during valve operation yielded a characteristic curve. Changes in the shape of this curve due for instance to a valve juddering in operation were observed.

Though the above techniques gave indications of changes in valve characteristics, relating the results to valve condition is difficult.

Most valves are likely to be monitored under flow conditions. Under these conditions throttled flow through the valve generated broad band white noise which swamped mechanical vibration of the valve. A typical plot of RMS signal level against position during valve operation is shown in Fig. 10.

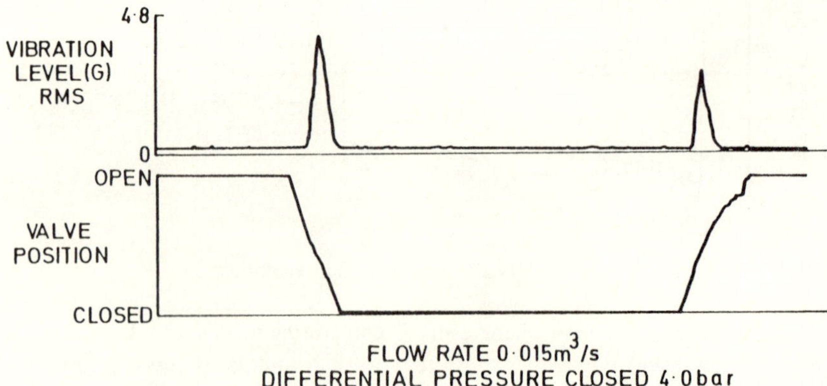

Fig. 10 – Vibration response of valve under flow conditions.

The following indications of valve performance can be inferred from the signal.

(1) The time of valve operation, being the time that significant throttling noise can be detected from the valve.
(2) That the valve was operated, a pulse of throttling noise having been observed.
(3) That the valve has closed and is not exhibiting gross leakage, the vibration signal having fallen to a low level.
(4) That the valve has opened to a position approaching the fully open position, the vibration signals having fallen to a low level.

In the closed position vibration in the frequency range 6–20 kHz was found to be an indicator of leakage. Following the approach of Thompson [1] the vibration signal from an accelerometer mounted on the valve body was analysed. Leakage was found to give a characteristic spectrum, increasing leakage leading to increased signal level but the frequency of characteristic peaks remaining unchanged. The technique was sensitive to background noise levels and in the presence of pump noise was limited to the detection of leaks greater than 1 l/min.

Acoustic emission, detecting signals due to leakage typically in the 100–500 kHz frequency range is being evaluated. Owing to lower levels of background noise this technique in tests to date appear promising detecting leak rates down to 0.5 l/min and is being evaluated further.

6. FURTHER WORK

Work to date has been limited to a study of pneumatically actuated floating ball valves. The investigation is being extended and will include evaluation of further floating ball valves, trunnion mounted ball valves and gate valves using pneumatic, electric and hydraulic actuators. Methods of detecting valve leakage will be investigated in detail. To facilitate this work a larger test rig has been constructed capable of handling a range of test fluids at higher pressures and incorporating a data logging facility to aid analysis of the results. The investigation is to be extended to include field trials on large operational valves.

7. CONCLUSIONS

(1) The torque displacement signature of a valve has been identified as a useful indicator of valve condition.
(2) The torque required to operate floating ball valves incorporating PTFE seats increases with increasing temperature.
(3) Changes in torque near the open position were an indicator of valve wear.

(4) Acoustic emission is a useful technique to detect valve leakage.

ACKNOWLEDGEMENTS

Permission to publish this Paper has been given by The British Petroleum Company plc.

REFERENCES

[1] G. Thompson, The Maintenance of Process Control Valves, *Chartered Mechanical Engineer,* June (1982).

CHAPTER 12

Application of adaptive noise cancellation to the condition monitoring of rolling element bearings

C. C. Tan and B. Dawson

1. INTRODUCTION

The vibration condition monitoring of rolling element bearings for the prediction of bearing condition is now an important requirement of most rotating machinery condition monitoring programmes. However, a disturbing number of faults still remain undetected until a significant amount of damage has occurred, especially in those situations where the bearing signature is corrupted by a high background noise (vibration) level.

Adaptive noise cancellation (ANC) is a technique that has been used very successfully in other fields [1] to significantly reduce corrupting noise. Its application to the enhancement of vibration signatures from rotating machinery is therefore an attractive proposition.

For application to the condition monitoring of bearings the technique requires two signal inputs, namely (i) a primary input containing both the bearing signature and corrupting background vibration, and (ii) a reference input consisting of a vibration signature related to the background vibration but not containing any bearing signature.

Given the availability of suitable primary and reference signals the basis of the ANC technique is that the reference signal is adaptively filtered and subtracted from the primary input resulting in significant cancellation of the noise from the signal plus noise primary input.

The feasibility of using the ANC technique to remove background vibration from a bearing signature has been successfully demonstrated by the authors [2] using a computer simulation study and tests on a simple open bearing test rig in a closely controlled laboratory situation. Whilst the results from the tests were very encouraging the primary and reference inputs were obtained from a rig specifically designed so that the measured signals satisfied the basic signal requirements of the ANC technique.

Chaturvedi and Thomas [3] have demonstrated with artificial data that statistical and spectral analysis techniques which fail to detect and diagnose faults because of a poor signal to noise ratio can be made more effective by using the ANC technique. They also applied the technique to a bearing test rig [4] and obtained some interesting results which indirectly indicated that the success of the direct ANC technique is dependent upon judicial positioning of the primary and reference transducers.

In this paper the authors describe the application of the ANC technique to the detection of a seeded bearing defect in a rolling element bearing located in a housing in a laboratory test rig. Background noise was introduced via a meshing gear arrangement and the primary and reference signals were both measured simultaneously at a number of different locations.

It is demonstrated that successful application of the technique is dependent upon the positioning of the primary and reference transducers and is therefore machine-dependent. A method of obtaining the primary and reference signals is presented which overcomes this limitation.

2. CONCEPT OF ADAPTIVE NOISE CANCELLATION

The general concept of adaptive noise cancellation has been described in detail by Widrow *et al.* [1]. Its application to the monitoring of vibration signatures from machinery is described schematically in Fig. 1(a), where it is assumed (i) a primary input is available which consists of a signal s correlated to the original bearing defect and corrupted by noise (vibration) n_0, and (ii) a reference input is also available which consists of noise n_1 correlated to the corrupting noise n_0 but not related to the signal.

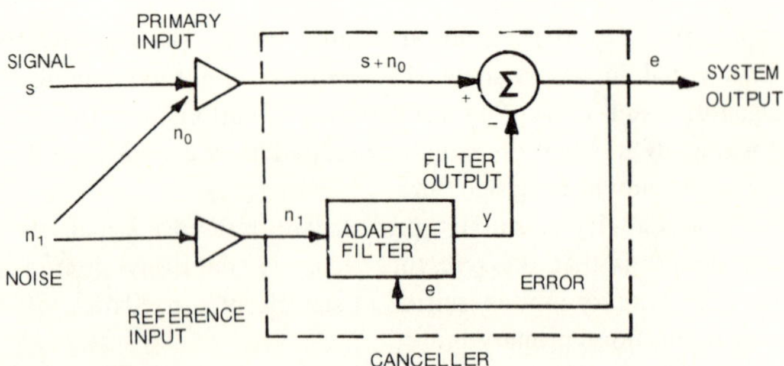

Fig. 1(a) – ANC concept.

The objective of the ANC technique is to produce a system output that is a best fit in the least square sense to the signal s. In practice, this is achieved by feeding back the adaptive filter with the system output and adjusting the filter through an LMS adaptive algorithm to minimise the total system output power.

Thus, the system output effectively serves as the error signal for the adaptive process and this leads to effective cancellation of the noise n_0 with the ANC output converging to the original signal s.

2.1 LMS adapative filter

The concept of the LMS adaptive filter is illustrated schematically in Fig. 1(b). It comprises basically (i) an adaptive linear combiner, and (ii) the LMS adaptive algorithm. The function of the combiner is to weight and sum a set of input signals to produce an output signal.

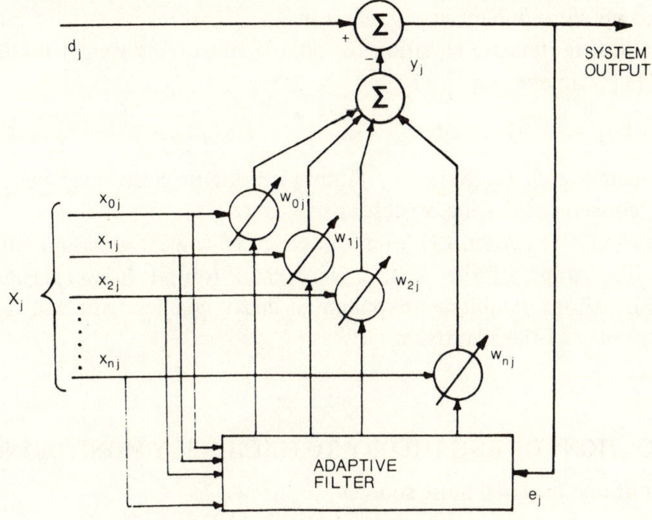

Fig. 1(b) – LMS adaptive filter.

Defining the input signal vector X_j and the weight vector W_j in the form:

$$X_j = \begin{bmatrix} x_{0j} \\ x_{1j} \\ \cdot \\ \cdot \\ \cdot \\ \cdot \\ \cdot \\ \cdot \\ \cdot \\ \cdot \\ x_{nj} \end{bmatrix} \quad \text{and} \quad W_j = \begin{bmatrix} w_{0j} \\ w_{1j} \\ \cdot \\ \cdot \\ \cdot \\ \cdot \\ \cdot \\ \cdot \\ \cdot \\ \cdot \\ w_{nj} \end{bmatrix}$$

where the input signal components are assumed to appear simultaneously on all the input lines at the discrete time denoted by the subscript j, then

$$y_j = X_j^T W_j = W_j^T X_j \tag{1}$$

Defining the system error e_j as the difference between the desired response d_j and the actual response y_j

$$e_j = d_j - W_j^T X_j \tag{2}$$

where d_j is the primary input to the system, X_j is the reference input and the purpose of the adaptive algorithm is to adjust the weights of the adapative linear combiner to minimise the mean square error.

The following iterative equation for adjustment of the weight vectors may be derived [1], namely:

$$W_{j+1} = W_j + 2\mu e_j X_j \tag{3}$$

where the paramter μ (convergency factor) is a factor controlling the stability and rate of convergence of the process.

Application of equation (3) leads to convergence after a certain number of iterations. The graph of the mean square error (output power) against the number of iterations exhibits an exponential decay characteristic and is termed the learning curve of the adaptive process.

3. IMPLICATIONS OF ANC THEORY TO MACHINERY MONITORING

3.1 Effect of uncorrelated noise sources

For machinery monitoring applications of the ANC technique it is most probable that uncorrelated noises will also be present in the primary and reference signals. In this case the noise canceller may be represented as shown in Fig. 2. The primary input consists of a signal s_j plus a sum of the two noises m_{0j} and n_j. The reference input consists of a sum of two other noises m_{1j} and $n_j{*}h(j)$ where the noise $n_j{*}h(j)$ is the noise n_j propagated through a channel with a transfer function $H(z)$ and impulse response $h(j)$. The noises n_j and $n_j{*}h(j)$ have a common origin and are correlated with each other and are assumed to be uncorrelated with s_j. The noises m_{0j} and m_{1j} are uncorrelated with each other, and with n_j and $n_j{*}h(j)$. Assuming all the noise propagation paths to be equivalent to linear time invariant filters, Widrow *et al.* have analysed the characteristics of the noise canceller shown in Fig. 2.

Defining $\rho_{\text{out}}(Z)$ as the primary S/N density ratio, $\rho_{\text{pri}}(Z)$ as the S/N density at the output, $A(Z)$ and $B(Z)$ as the ratio of the spectra of uncorrelated to correlated noise at the primary and reference input respectively, they derived the following expression for the ratio of the S/N density ratio at the output to the S/N density ratio at the primary input, namely:

$$\frac{\rho_{out}(Z)}{\rho_{pri}(Z)} = \frac{[A(Z)+1]\,[B(Z)+1]}{A(Z)+A(Z)\,B(Z)+B(Z)} \tag{4}$$

from which it is apparent that the ability of a noise cancelling system to reduce noise is limited by the uncorrelated-to-correlated noise density ratios at the primary and reference inputs. The smaller are $A(Z)$ and $B(Z)$ the greater $\rho_{out}(Z)/\rho_{pri}(Z)$ and the more effective the action of the canceller.

Thus, as shown by Widrow *et al.*,

$$\text{for small } A(Z),\ \frac{\rho_{out}(Z)}{\rho_{pri}(Z)} = \frac{1+B(Z)}{B(Z)} \tag{5}$$

$$\text{for small } B(Z)\ \frac{\rho_{out}(Z)}{\rho_{pri}(Z)} = \frac{1+A(Z)}{A(Z)} \tag{6}$$

$$\text{and for both small } A(Z) \text{ and } B(Z)\ \frac{\rho_{out}(Z)}{\rho_{pri}(Z)} = \frac{1}{A(Z)+B(Z)} \tag{7}$$

For the ideal canceller for $A(Z)$ and $B(Z)$ equal to zero there is an infinite improvement and complete removal of the correlated noise. In practice for small $A(Z)$ and $B(Z)$ the finite length of the filter and misadjustment caused by gradient estimation noise in the adaptive process also limits the system performance.

In addition, in machinery applications it may well be impossible to obtain a reference signal devoid of any signal component. Thus, it is important to establish the effect upon the noise cancelletion of low-level signal components in the reference input and this is discussed in the following section.

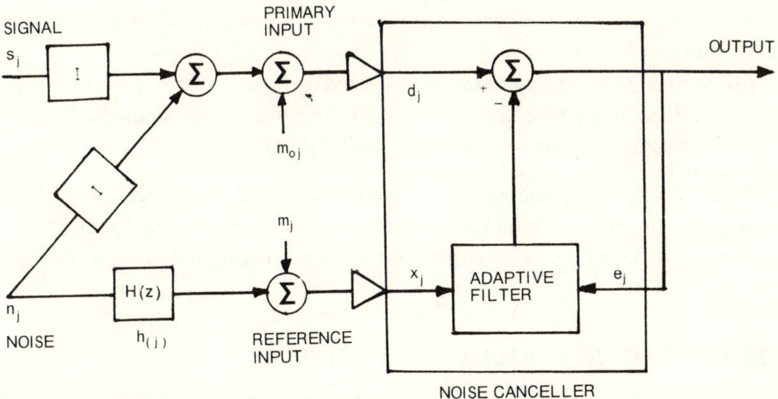

Fig. 2 – Adaptive noise canceller with correlated and uncorrelated noise in the primary and reference inputs.

3.2 Effect of low level signal components in reference input

The effect of low level signal components in the reference source has been analysed by Widrow *et al.* who showed that assuming the adaptive solution to be unconstrained, and the noise in the primary and reference input to be mutually correlated, the signal to noise density ratio at the noise canceller output is simply the reciprocal at all frequencies of the signal to noise density ratio at the reference input, i.e.

$$\rho_{\text{out}}(Z) = \frac{1}{\rho_{\text{ref}}(Z)} \tag{8}$$

Also defining signal distortion $D(Z)$ as the dimensionless ratio of the spectrum of the output signal component propagated through the adaptive filter to the spectrum of the signal component at the primary input then it may be shown that

$$D(Z) = \frac{\rho_{\text{ref}}(Z)}{\rho_{\text{pri}}(Z)} \tag{9}$$

This shows that low signal distortion results from a high signal to noise density ratio at the primary input and a low signal to noise density ratio at the reference input.

Widrow *et al.* also determined the following relationship for the canceller output noise spectrum, namely:

$$\delta_{\text{output noise}}(Z) = \delta_{\text{nn}}(Z)\,[\rho_{\text{ref}}(Z)]\,[\rho_{\text{pri}}(Z)] \tag{10}$$

where $\delta_{\text{nn}}(Z)$ represents the power density spectrum of the correlated reference noise

Equation 10 implies

(i) output noise spectrum depends on input noise spectrum.
(ii) that if the S/N density ratio at the reference input is low, the noise will be low, and
(iii) if the S/N density ratio in the primary input is low, the filter will be trained most effectively to cancel the noise rather than the signal and consequently the output noise will be low.

Thus, whilst the presence of a signal component in the reference signal is undesirable, the ANC technique is not rendered inoperable by their presence except for the special case where the reference signal contains no noise component.

4. BEARING DEFECT SIMULATION TECHNIQUE

Bearings operating under designed working conditions normally have a long life, often measured in years. Consequently, research into the development of condition-monitoring techniques for the detection of impending bearing failure, rely upon some form of accelerated failure test rig or the technique of seeding a bearing with a defect assumed to simulate the required failure condition.

In this work the second approach was adopted and a seeded line defect, assumed to simulate a fatigue spall, was made on the bearing outer race using a spark erosion process. The technique gave good control over both the shape and size of defect.

Details of the test bearing together with the calculated defect impact frequencies are given in Table 1.

Table 1 Test bearing specification and outer race defect frequencies.

Bearing Specification

Type: RHP 6202 TNH (detachable cage type)
Mean diameter: 31.25 mm; Ball diameter: 6.0 mm;
No. of balls 8; Rotation speed 1400 rev/min

Assumed contact angle (deg)	0	5	10	15	20	25
Impact frequency of a defect on the outer race (Hz)	75.4	75.7	75.7	76.0	76.5	77.1

5. TEST RIG

Details of the test rig are shown in Fig. 3. A radial load was applied to the test bearing by means of out-of-balance masses attached to two pulley wheels located either side of the test bearing. The bearing shaft was connected by a flexible coupling to a constant speed motor and driven at a speed of 1400 revolutions per minute. Background vibration was introduced by means of a gear mesh system. The characteristics of the meshing gears and their meshing frequencies in relation to the shaft rotation are shown in Table 2.

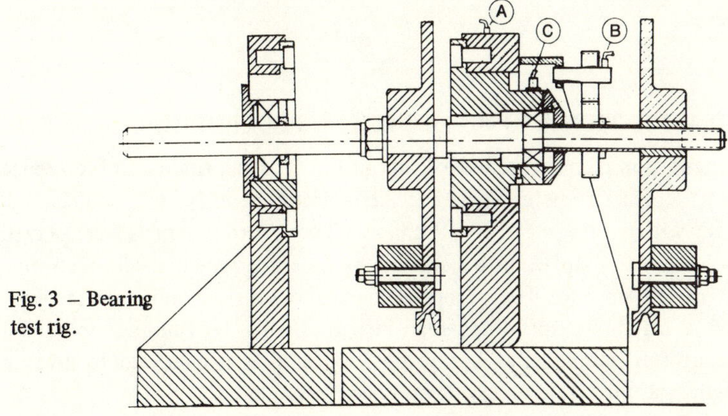

Fig. 3 – Bearing test rig.

Table 2 – Gear mesh frequencies.

Components	Orders w.r.t drive shaft	Frequency (Hz)
F_i shaft fundamental speed	1	23.33
F_{gi} driving gear meshing frequency	51	1190
$F_{gi} \pm F_i$ side bands, due to eccentricity in gear	51 ± 1	(1213.3) (1166.8)
.	.	.
.	.	.
.	.	.
.	.	.
.	.	.
.	.	.
$F_{gi} \pm 10\,F_i$	51 ± 10	(1423.3) (956.7)
F_{go} driven gear meshing frequency	36	840
$F_{go} \pm F_i$ side bands due to eccentricity in gear	36 ± 1	(863.3) (816.7)
.	.	.
.	.	.
.	.	.
.	.	.
.	.	.
.	.	.
$F_{go} \pm 10\,F_i$	36 ± 10	(1073.3) (606.7)

6. INSTRUMENTATION AND MEASUREMENT PROCEDURE

The instrumentation is shown schematically in Fig. 4. The primary and reference signals from the accelerometers were recorded simultaneously on a two-channel tape recorder with a tape speed of 19 cm/sec. The recorded signals were played back at a speed of 2.4 cm/sec into a second two-channel tape recorder recording at a speed of 19 cm/sec. The recorded signals on the second tape recorder were played at a reduced speed and read into the PDP 11/24 minicomputer via an A–D converter sampling at a rate of 250 samples per second to give an effective signal sampling rate of 12.8K samples per second.

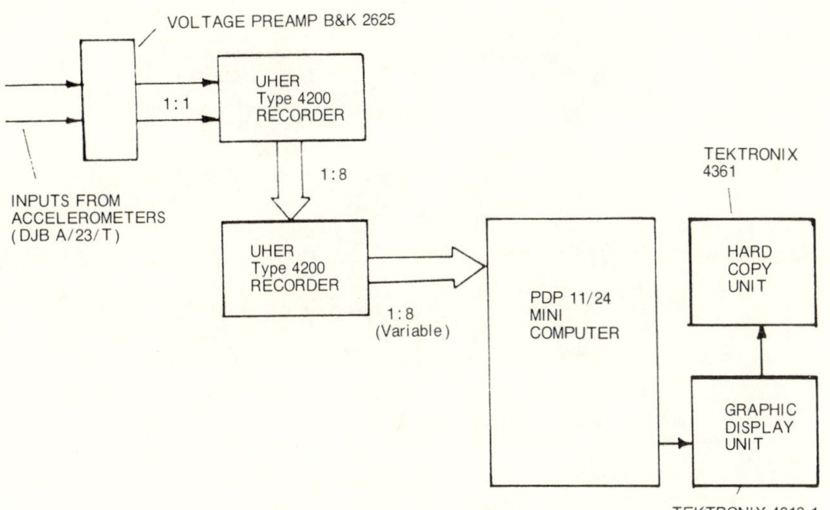

Fig. 4 – Test instrumentation.

The primary, reference, original bearing signatures and the ANC output could be displayed on a Tektronix graphics display unit and a copy obtained via a hard copy unit.

7. ANC TESTS

In order to test the effect upon the ANC output of different primary and reference signals a series of five tests were performed for different combinations of primary and reference transducer locations.

Referring to Fig. 3 for Test No. 1 the primary signal was measured at position C on the outer surface of the housing assembly and the reference signal was measured at position B on the gear pin of the gear mesh system.

For Test No. 2 the primary signal was again measured at position C on the outer surface of the housing assembly and the reference signal was measured at position A on the bearing assembly support block.

For Test No. 3 the primary signal was measured via a series of offset-balls located in a hole drilled through the housing as indicated for accelerometer No. 1 in Fig. 5 and the reference signal was measured at position B on the gear pin.

For Test No. 4 the primary signal was measured at the same position as described in Test No. 3 and the reference signal was measured at position A on the bearing assembly support block.

For Test No. 5 both accelerometers were located on the housing surface as shown in Fig. 5. The primary signal was measured via a series of offset-balls located in a hole drilled through the housing and the reference signal was measured on the housing surface.

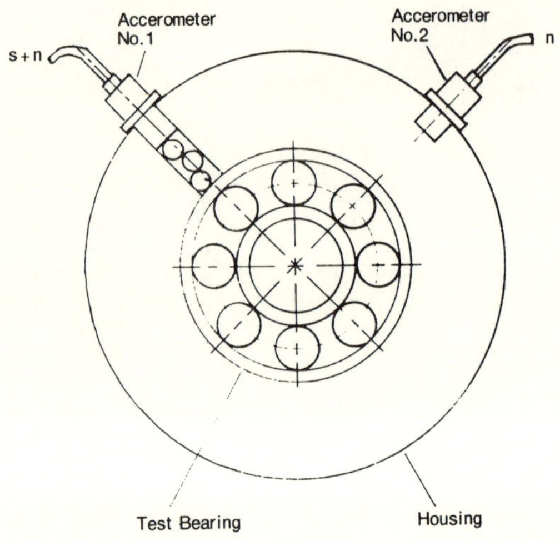

Fig. 5 – Accelerometers location for Test no. 5.

8. DISCUSSION OF RESULTS

It was considered that locating the primary transducer directly on the bearing housing and the reference transducer as close as possible to the noise source at position B in Fig. 3 would result in successful enhancement of the bearing defect signature.

Figs. 6(i) and (ii) shows the reference and primary signals respectively and Figs. 6(iii) and (iv) illustrate successive 1000 samples of the ANC output for this measurement configuration.

The ANC output is disappointing, in that whilst impact responses are now clearly visible, a regular cyclic defect impact frequency at approximately 76 Hz is not detectable. On examination of the reference signal an impact response of 76 Hz can be seen which indicates a signal correlation between the primary and reference inputs and hence some signal distortion would be expected.

Fig. 7 shows a similar set of results but for the reference transducer located on the bearing assembly support block at position A shown in Fig. 3. The ANC output again indicates impact responses but no cyclic impact response due to a defect on the outer race.

In this case it was considered that little of the gear mesh noise would be propagated to the reference transducer and hence effective cancellation would not occur. In addition, for both of the above cases, (i) the primary and reference tranducers are well separated and hence uncorrelated noises may be present, and (ii) the S/N density ratio at the primary input may be small which is contrary to the requirements specified in section 3 for low signal distortion.

REFERENCE AND PRIMARY SIGNALS

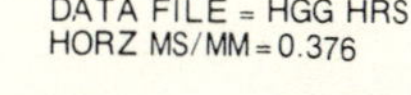

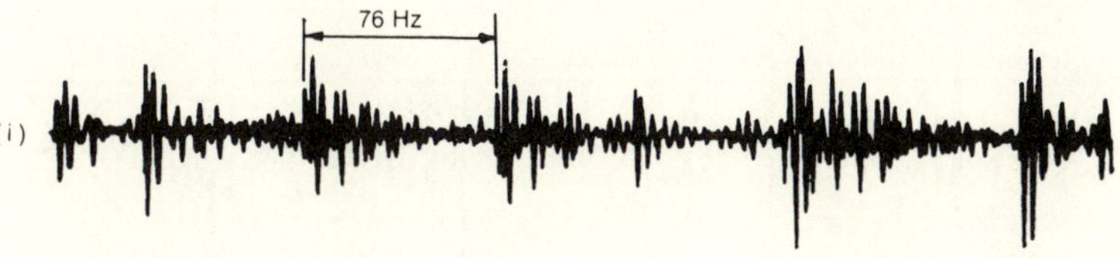

REFERENCE INPUT - Position B

PRIMARY INPUT - Position C

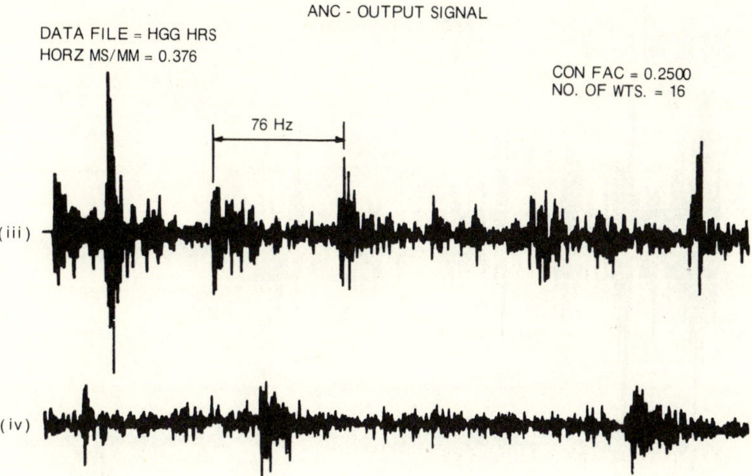

Fig. 6 – Results of Test no. 1.

REFERENCE AND PRIMARY SIGNALS

DATA FILE = H2G. HRS
HORZ MS/MM = 0.376

(i)

REFERENCE INPUT - Position A

(ii)

PRIMARY INPUT - Position C

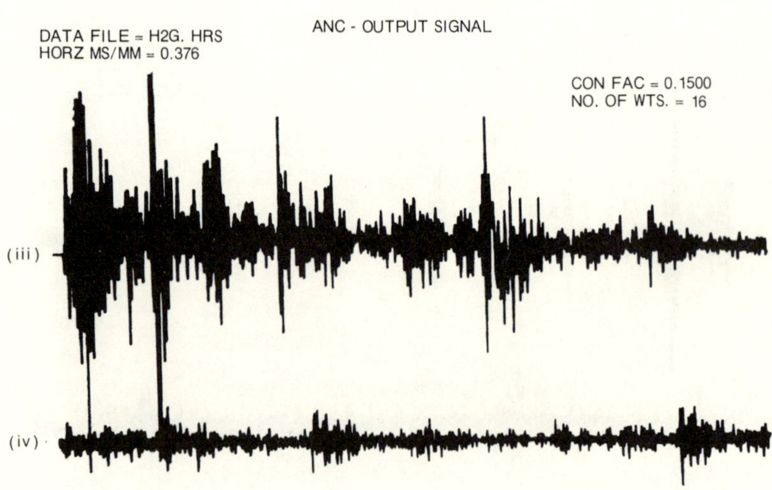

Fig. 7 – Results of Test no. 2.

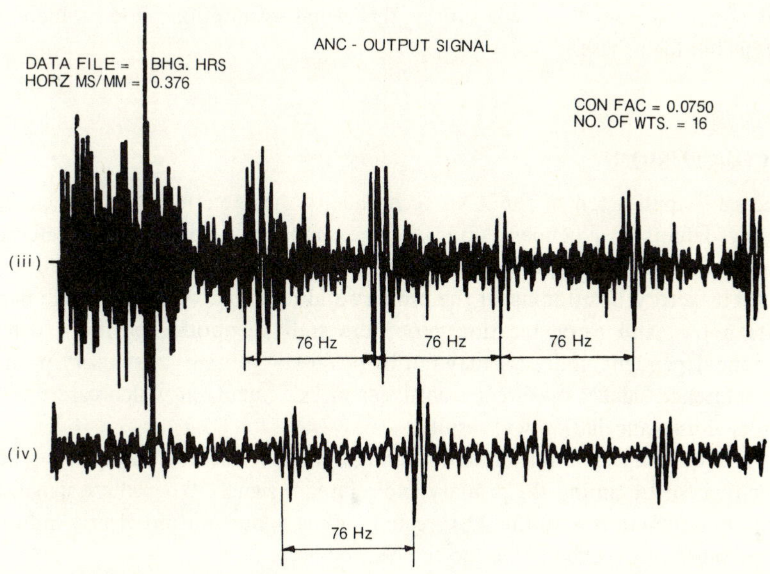

Fig. 8 – Results of Test no. 5.

In order to increase the primary S/N density ratio the offset ball configuration shown in Fig. 5 was adopted for measurement of the primary signal.

Tests 3 and 4 were performed with the offset ball primary signal measurement arrangement but with the reference signal obtained as for Tests 1 and 2 respectively. For both these cases no significant improvement in the ANC output was observed in comparison to the outputs obtained from Tests 1 and 2.

On further consideration it was thought that uncorrelated noises in the primary and reference signals may be the cause of the ineffectiveness of the noise canceller.

Fig. 8 shows the results obtained with the primary and reference transducers both located in close proximity on the bearing housing as shown in Fig 5. In this case a 76 Hz cyclic impact frequency is visible indicating a defect on the outer race of the bearing.

The success of this arrangement is attributed to the primary and reference signals satisfying the basic ANC signal requirements in that (i) the S/N density ratio at the primary input is high, (ii) for a small defect and high noise density the S/N density ratio at the reference input is small, and (iii) the noise input to the primary and reference inputs are highly correlated.

The measurement technique is machine independent but suffers from the disadvantage that it requires a hole to be drilled through the bearing housing and this may prove to be a limitation for certain applications.

Further work is required to develop a non-intrusive technique that will yield signal outputs that will satisfy the signal assumptions upon which the ANC technique is based.

9. CONCLUSION

Successful application of the ANC technique to the detection of rolling element bearing defects is dependent upon the measurement of primary and reference signals which satisfy the basic signal assumptions inherent in the ANC theory.

Satisfactory positioning of the primary and reference transducers will necessitate a trial and error location procedure and the optimal positions will be machine-dependent. Indeed it may not be possible to obtain satisfactory primary and reference signals by direct transducer measurements in which case unsatisfactory noise cancellation will result.

In this chapter a machine-independent measurement technique has been presented for obtaining the primary and reference signals. When these signals are used in conjunction with the ANC technique the system output clearly indicates the presence of a seeded defect on the bearing outer race.

Whilst these results are most encouraging the measurement technique requires a hole to be drilled through the bearing housing and this may not always be possible in certain applications.

Further work is required to develop a non-intrusive technique for obtaining satisfactory primary and reference signals.

REFERENCES

[1] B. Widrow, J. R. Glover, J. M. McCool, J. Kaunitz, C. S. Williams, R. H. Hearn, J. R. Zeidler, E. Pond and R. C. Goodlin, Adaptive noise cancelling: principles and applications, *Proc. IEEE,* **63,** No. 12 (1975).

[2] C. C. Tan and B. Dawson, An adaptive noise cancellation technique for the enhancement of vibration signatures from rolling element bearings: a feasibility study, *Proceedings of the 2nd IMEKO Symposium on Technical Diagnostics, London, 1981*.

[3] G. K. Chaturvedi and D. W. Thomas, Adaptive noise cancelling and condition monitoring, *Jnl. of Sound and Vibration,* **76**(3), 391–405 (1981).

[4] G. K. Chaturvedi and D. W. Thomas, Bearing fault detection using adaptive noise cancelling, *Trans ASME 280,* **104,** (1982).

Part 4
SYSTEM MONITORING – POWER SYSTEMS

CHAPTER 13

Condition monitoring at Bulls Bridge power station

K. N. Addrison and M. L. G. Hill

1. INTRODUCTION

1.1 The plant

The Station chosen for the trial was Bulls Bridge Gas Turbine Station, sited near London Airport (see Fig. 1.1). Bulls Bridge contains four 70 MW sets, each 70 MW unit being powered by four Industrial Olympus gas generators, two at either end of a central alternator (see Fig. 1.2). At each end of the alternator, power is supplied via clutch, to a shaft on which is mounted two power turbines, each driven by a single Olympus gas generator. Thus gas paths are separate between intake and final exhaust, and therefore each gas generator/power turbine assembly can be analysed without being unduly affected by associated plant.

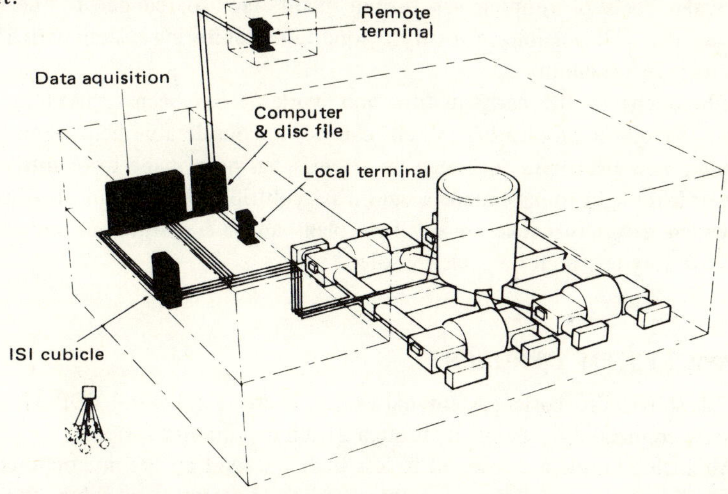

Fig. 1.1 – The installation of the data processor and diagnostic system in the CEGB Bulls Bridge gas turbine power station.

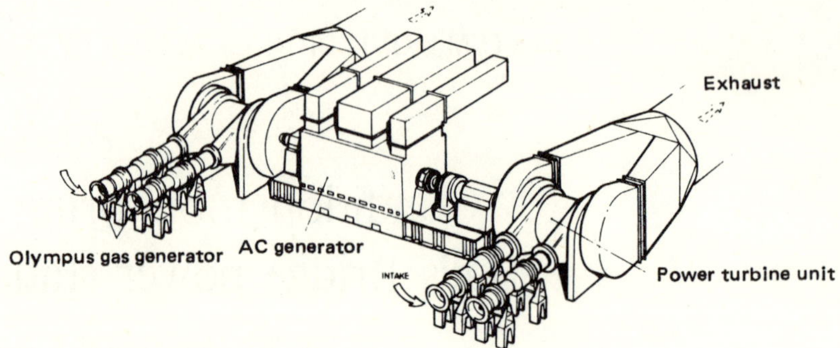

Fig. 1.2 – The Bulls Bridge station contains four Rolls-Royce FS80 generating sets, each comprising four Olympus gas generators, two power turbines and a Brush 70 MW alternator, as indicated in this layout drawing.

1.2 Why condition monitoring?

The station was built as a peak lopping and rapid response plant to augment the nuclear and fossil fuelled larger and less responsive plant. In this role, there is a considerable benefit to be gained from high availability and good starting reliability. The potential advantages to be gained from a successful plant condition monitoring could be, and were quantified. Using the CEGB and equipment suppliers' costings current at the time of the decision, it transpired that with high starting reliability, an increase of 1% in the availability of a four-unit station would justify the provision of condition monitoring equipment. Even now, when costs of running gas turbine plant have soared due to fuel cost increases, it is still possible to justify a condition monitoring system with a 2 to 3% increase in availability.

Subsequent to the early justification work, it has been realised that an efficient condition monitoring system can be of considerable help when commissioning new plant and, although no attempt has been made to quantify this aspect, it is thought to have made a significant contribution to diagnosis of problems which naturally occur on any new plant under construction, and which may prove very costly, if not quickly solved.

2. BASIC SYSTEM APPROACH

Fig. 2.1 shows the basic system hardware. There is a central PDP 11 minicomputer connected to the plant through 21 microcomputer systems.

All analog inputs are scanned in less than a second by the microcomputers and significant changes only are transmitted to the central machine, thus the microcomputers relieve the central machine of many routine and time-consuming functions. The microcomputers also check that the gas generators are in 'steady state' before taking 'condition monitoring' snapshot readings after each 5 minutes of running on load.

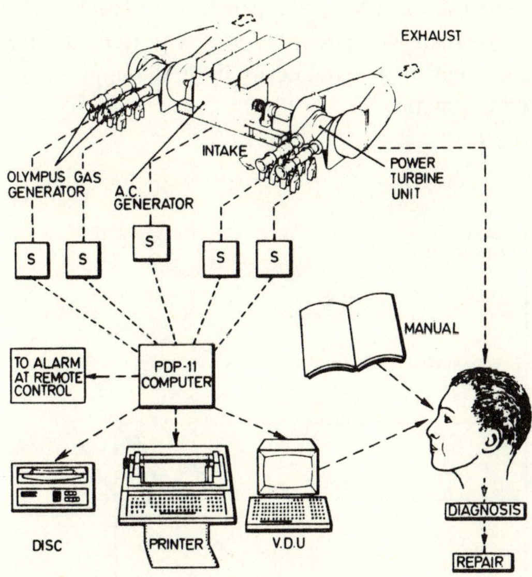

Fig. 2.1 – Condition monitoring – data flow and decision process.

Alarms and plant state contacts, together with sequence information, are scanned by the microcomputer and these are time tagged and stored with a resolution of 10 ms.

We therefore have a comprehensive and up-to-date description of the state of the plant available to the central computer from the peripheral microcomputers.

Thus the central machine is left free to concentrate on assembling the plant history into a form easily assimilable by an operative, who may not necessarily be a gas turbine expert. In order to make a large mass of information easily assimilable, it is first, where necessary, brought to ISO standard conditions and then corrected to a standard power. This operation is best explained by reference to Fig. 2.2. When the installation is newly commissioned or recommissioned a baseline is established for each relevant parameter as a function of the low pressure compressor speed (N1). This is represented by line AB which is held in the computer as the 5 constants of a quartic equation. The value of this parameter at a particular N1 (n) near the full power point (G) is taken as representative of the plant in a satisfactory mechanical condition. When the plant changes, the characteristic baseline represented by AB may change to a new, roughly parallel characteristic such as DE. A spot reading now taken at any running N1 would be represented by C. This value C is normalised to the standard (n) along a quartic equation DE held in the computer and this new value F is held in the computer as indicative of the state of the plant, now somewhat deteriorated. Repeating this process will give a plant history for each parameter at ISO conditions and at a constant N1. The method is seen to be less than 100% rigorous, but it is simple enough to be handled by limited computing facilities, and it has now been proven in practice.

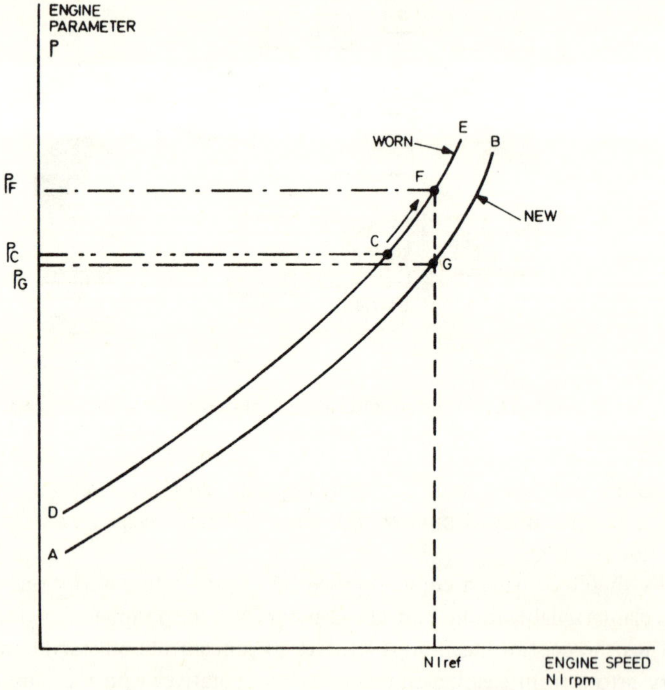

Fig. 2.2 – Correction factors to engine data.

A series of corrected points such as F can now be plotted against time and can indicate progressive or step changes which can be used in fault diagnosis work. Fig. 2.3 shows how such a graph might appear, showing the point G, at the origin and the point F after say, 4 hours running.

This then is the basic automatic information gathering system which is designed to reduce a mass of information to a useful form.

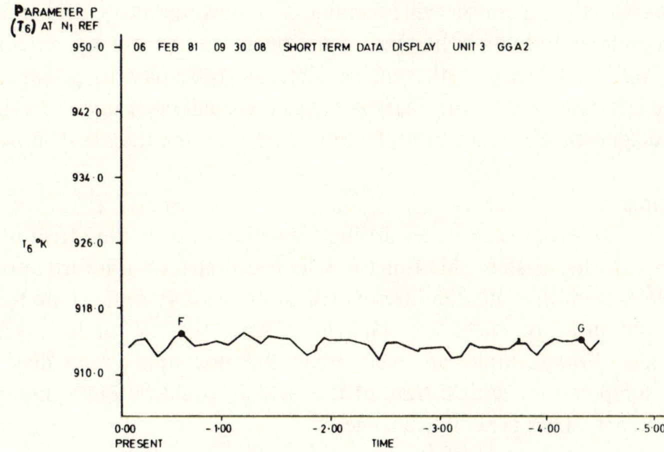

Fig. 2.3 – Trend line of plant parameter.

On receipt of indication that significant changes are occurring, the operator will use a diagnostic chart to ascertain the prime cause of the deterioration. To illustrate the method, a much simplified example is shown in Fig. 2.4 where the plus sign indicates an increasing trend, the minus sign indicates a decreasing trend and (N.C.) means no change. On finding that T6 exhaust temperature

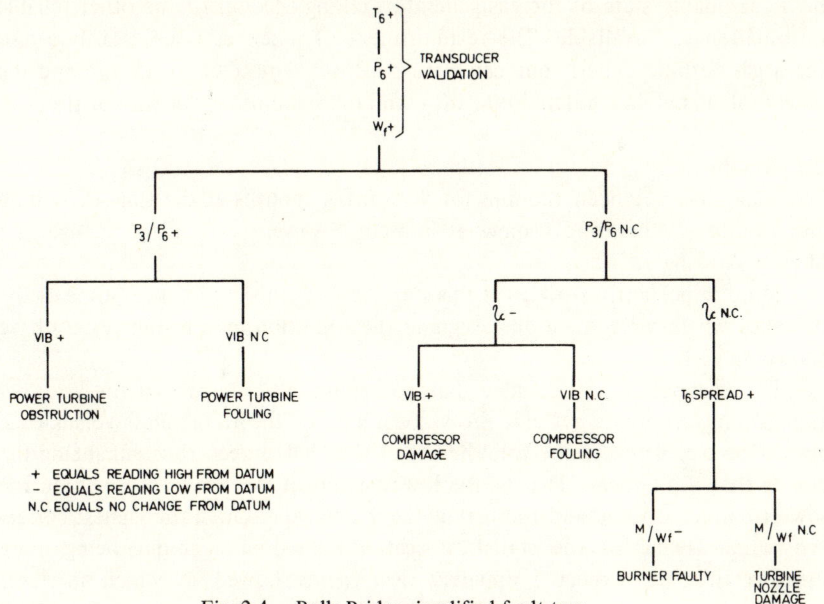

Fig. 2 4 – Bulls Bridge simplified fault tree.

has increased, the operator will examine the pressure ratio plot and decide which branch to follow. If he finds that the pressure ratio has increased the operator will then examine the engine vibration/time plot to progress further down the left-hand side of the chart and thus eventually end up with a diagnosis. The real diagnostic charts are considerably more complex than that shown here.

2.1 Failures

One of the interesting side issues arising from this work is the effect of failures of the monitoring system. Malfunctions here can give false information. However, it does seem that most failures which could possibly mislead are transducer failures, and most of these are self-evident since they result in unacceptable step changes. For example, an open circuit thermocouple circuit may indicate a rise in temperature, with a 'rate of rise' which would be quite impossible to attribute to actual gas generator changes.

Drifts in transducer performance are of course possible and do happen. In this case indication should arise from the diagnostic charts, since the chances are very high that the drift occurring would not agree with patterns that should appear on gas generator deterioration.

However, having established by the most appropriate method that a transducer failure has occurred, it means that the information and original baseline have been lost, since the baseline carries the characteristics of the transducer as well as those of the engine. In order to overcome this problem, software facilities are available to modify baselines to agree with what would appear to be a reasonable state of the gas generator, taking account of the other reliable information still available. The re-adjustment of baselines is necessarily a task for a gas turbine expert, but can be carried out remote from the site and the newly calculated data entered into the computer by another operator at site.

2.2 Results

The plant has not been running for very many months at the time of writing and results are therefore somewhat limited; however, some plant changes are already showing up.

More importantly perhaps at this stage, and for this conference, some of the mistakes which were made in designing the condition monitoring system have shown up early.

For instance, it was decided that, to maximise the impact of the information coming from the system, the vertical axis of the trend plots versus time should be scaled to use the total height of the VDU screen thus enhancing the trends that did appear. That is, the lowest point in the Y axis represented the lowest corrected value and the top of the Y axis represented the highest corrected value. Owing to the statistical scatter reduction technique being more efficient than was expected, apparent clear trends showed up, which on closer analysis, represented parameter changes of perhaps 0.1%, commensurate with the repeatability of the readings. To avoid false impressions being given by this 'microscopic' view of changes, the scaling has now been changed to relate to sensible limits for any parameter rather than the changes actually occurring. It is not intended to imply that no other errors have been made, but this one serves to illustrate how the over-zealous use of highly sophisticated and discriminative monitoring techniques can cause problems in diagnosis unless intelligent use is made of the information and unless the equipment is flexible enough to accept changes.

3 IMPLEMENTATION OF THE ENGINE HEALTH MONITORING REQUIREMENTS – SOME DETAILS

The monitoring requirements fall into two distinct classes requiring differing approaches in the way the plant data is handled and interpreted.

First is that of starting up the equipment and bringing it to a state where electrical power can be delivered to the grid system. Here the monitoring emphasis lies in detecting plant conditions which may mean an inability to complete a successful start, and since this is a peak lopping power generation station, ability to start quickly and with certainty, is of paramount importance.

The second approach is that of running the equipment efficiently during normal electrical power generation. Here the emphasis is in detecting plant deterioration so that ample warning can be given to the operator, of conditions which ultimately will affect the immediate availability of plant for its peak demand role.

So while the first operation concentrates on the use of limited events, the second relies heavily on the statistical interpretation of plant transducer data.

The transducer fit required to cover each of these requirements is in some ways quite different. Looking first at the plant run-up, failure here often involves the protection interlocks not operating at the required time, this is usually brought about by failure of some item of plant to achieve a desired working state within a specified time from start initiation. Failure to achieve working state can come about for a variety of reasons, but can be due to steady deterioration or sudden failure. The basic monitoring during run-up is the comparison of plant state at various specific times from start initiation. We have a signature for each gas turbine for these plant state events, both at the initial commissioning and for the last start. The system is so arranged to look for changes in the timing of those plant states which would hazard any future start requirement. The start signature is one of gas turbine speed on the Y axis, versus time from start initiation on the X axis, with plant state deduced from various logic conditions Fig. 3.1.

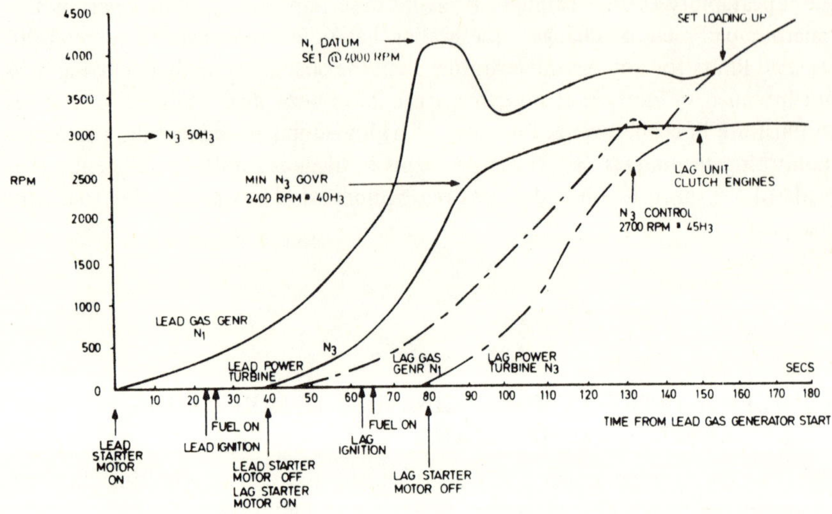

Fig. 3.1 – Typical plant run-up signature.

Deterioration which results in slower start cycles are thus highlighted enabling corrective action to be taken before failure occurs. Often unpredictable and intermittent failure (say dirty relay contacts) occur and in this case the equipment proves its worth by ensuring quick and accurate diagnosis and repair.

We also have a signature for run-down of the plant which is much simpler, where we aim to detect any incipient main bearing problem or light mechanical fouling Fig. 3.2. The analogue plant transducers are all operational during

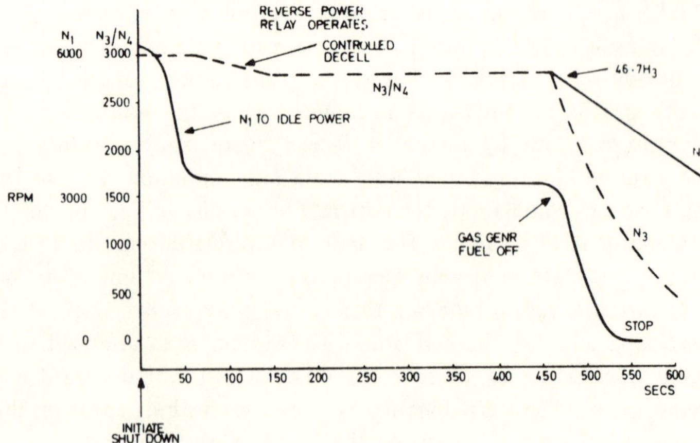

Fig. 3.2 – Typical cooling and run-down signature.

run-up and run-down and whilst not normally used in the mode of operation, the equipment can produce on command for instance, a run-down signature of gas turbine speed and gas turbine vibration amplitude versus time, if vibration problems are suspected. It can be seen that, used in this manner, a powerful diagnostic tool is available (Fig. 3.3).

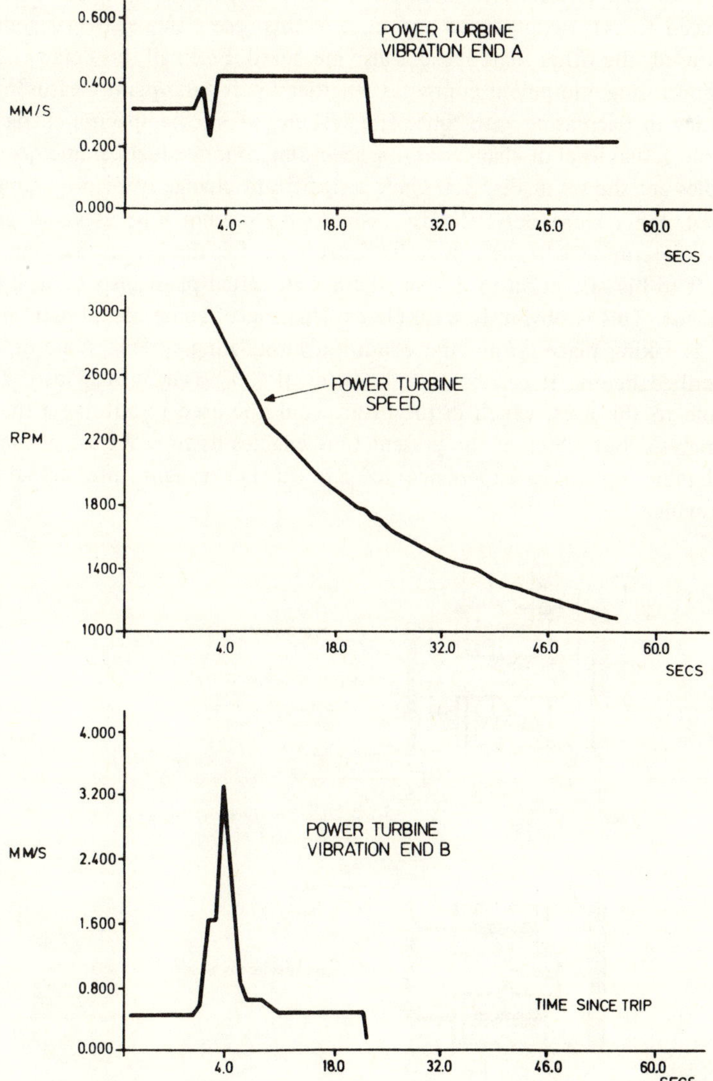

Fig. 3.3 – Actual plant run-down with vibration read out.

Turning to the second part of the condition monitoring, a wide variety of analogue plant transducers must now be used to detect plant state and hence deterioration, but once again this deterioration is detected by comparing the plant state now with its historical condition, including that at the time of commissioning. Thus, pressure, temperature, speeds, fuel flow etc., are all measured throughout the gas turbine as shown in Fig. 3.4 as well as vibration levels, and various electrical parameters. All these parameter measurements are referenced to N1 engine speed and since in this case a two shaft gas generator unit is used, the other shaft speed is also measured. From all this data the condition monitoring equipment computes whether a particular plant measurement is increasing or decreasing with time. The system looks for a specific change with time, the actual level of change varying according to individual parameters. Some examples are shown in Fig. 3.5. Once a significant change of plant parameter is detected, the system alerts the user, who at the present time uses a diagnostic tree to determine what the probable causes are for the plant changing state. It is left to his discretion to decide whether an actual plant inspection is called for or not. This is obviously a simple explanation of some complex processing which is taking place within the condition monitoring system, some of which is described below. It can be seen, however, that a powerful diagnostic tool is available to the user, which in the future could be used to automate the fault tree analysis, but which at the present time enables us to make use of the users considerable operating experience to gain further insight into actual plant performance.

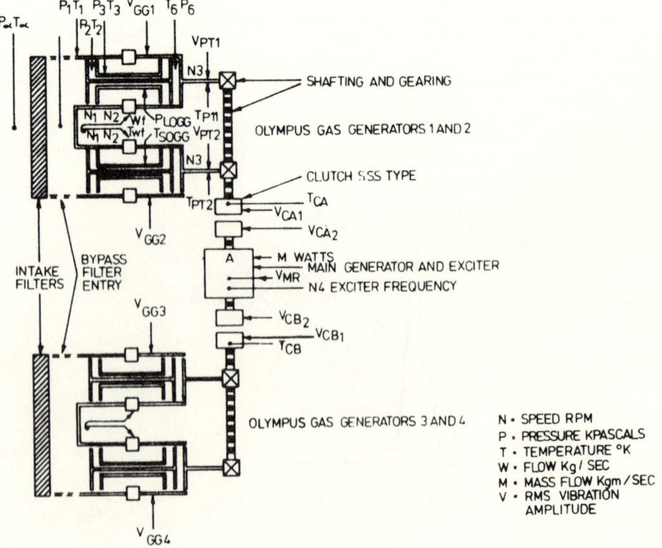

Fig. 3.4 – Transducer points one generator set.

T6 SPREAD
200
120
%
-40
-200
-5·00
-4·00
-3·00
-2·00
-1·00
0·00
HOURS
26 APR 81 14·31 52 SHORT TERM DATA DISPLAY
FUEL FLOW
5·00
3·20
%
1·40
-0·40
-2·20
-4·00
- 5·00
- 4·00
- 3·00
- 2·00
- 1·00
0·00
HOURS
26 APR 81 14·32·30 SHORT TERM DATA DISPLAY
N2 / N1 CHARACTERISTIC
5·00
3·40
%
1·00
0·20
- 1·40
- 5·00
- 4·00
- 3·00
- 2·00
- 1·00
0·00
HOURS

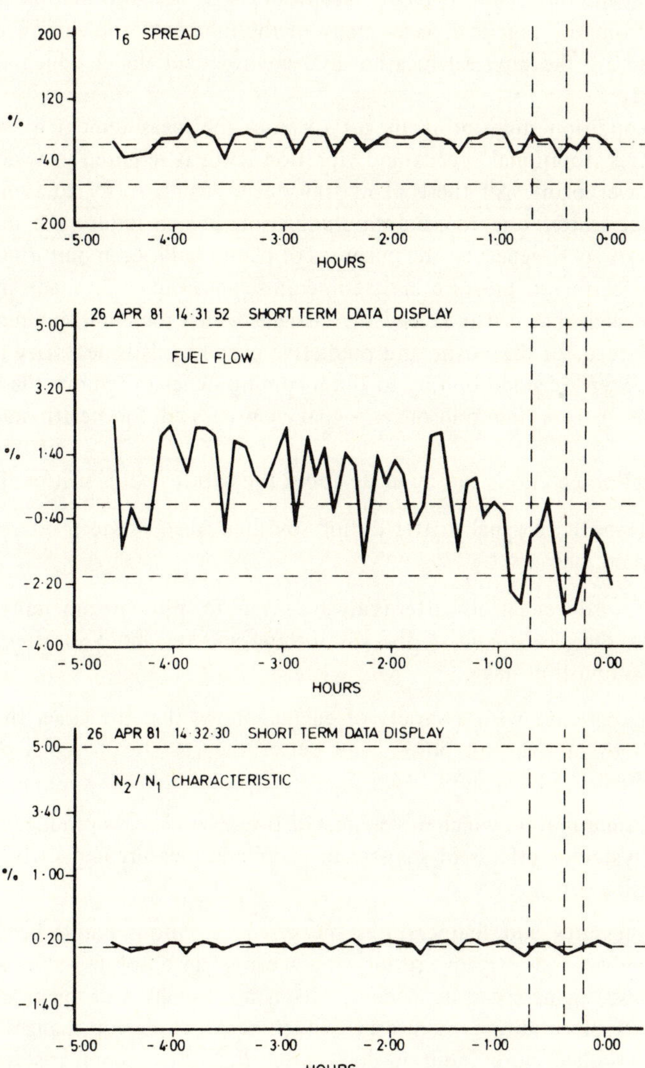

Fig. 3.5.

4 PROBLEMS WITH PLANT DATA SCATTER, TRANSDUCER VARIATION FAILURE AND INACCURACIES ETC.

The correct interpretation of plant state from transducer information cannot be divorced from the practical, as so many of the inherent problems encountered are created by the physical location and environment under which the plant must operate.

For condition monitoring in gas turbines the measurement of pressures, temperatures, rotational speeds and vibration levels is required in various parts of the power plant. All these plant transducer signals suffer to some extent from signal scatter due to random fluctuations which reduce the immediate usefulness of such signals to the purposes of plant condition monitoring. Indeed the signal scatter on pressure and temperature measurements made in the gas stream are such that it can exceed the changes which the condition monitoring seeks to detect, for diagnostic and predictive purposes. It is necessary therefore to apply extensive conditioning to the incoming signal to remove the worst of this scatter before the transducer signal is processed for health monitoring purposes.

Several observations can be made about the nature of this scatter. They are:

(1) Transducer signal scatter during so-called steady state plant conditions is normal.

(2) Transducer signal scatter is always severe for those parameters measured in the gas stream of the gas turbine and the more so after the fuel combustion stage.

(3) Experience with a variety of engines shows that the larger the engine, both in size and power, the better the gas stream signal parameter quality.

(4) Prime movers which are subjected to power changes produce reversible hysteresis effects of gas stream parameter measurements which have a time dependency.

(5) Pressures and temperatures measured at various points in the prime mover's gas stream are location sensitive in absolute value terms and also engine speed dependent. This is not usually a problem as such, in actual condition monitoring which seeks to detect changes from an 'installed norm', but it does mean that direct comparison between prime movers is difficult, and transducer replacement is more complicated.

(6) Condition monitoring involving prediction (i.e. a time series) can only be performed on data gathered whilst the prime power can be regarded as being in a steady state condition, that is developing constant power for a predefined period.

An example for a typical engine is shown in Fig. 4.1 of turbine outlet temperature degC versus gas generator speed (rpm). In this example one or two of the sample data points could be rejected on the grounds of exceeding a 'reasonable maximum rate of change' credibility check, but the majority of the sample data points are valid by any normal applicable criteria. It will be seen that the scatter is about 6 degC in 600 degC which is too high for condition monitoring purposes on this particular parameter and engine, where a change of 5 degC is considered to be of monitoring importance.

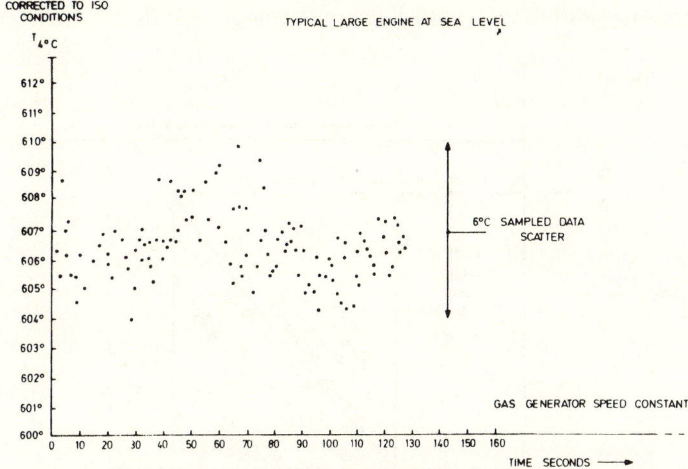

Fig. 4 1 – Scatter on sampled data.

It therefore has to be accepted that the scatter of raw sampled data is too high, and has no meaning in the monitoring context. To enable condition monitoring to proceed, methods of reducing scatter must be adopted.

Even in engines known to be in good condition this data scatter has been observed to be consistent in overall amplitude. No significance can be attached to this scatter which appears to be random. Equivalent data from widely differing engines displays the same pattern.

No detailed analysis has yet been conducted as to the cause of the scatter and whether any useful information is contained therein. Scatter can obscure trends and therefore for the purpose of this sytem it is removed.

Summarising therefore:

(1) If the scatter had meaning in the dynamic sense the plant would be running erratically but we know from practical observation of prime parameters that it is in steady state and that this is *not* the case.

(2) If the scatter had a meaning in the condition monitoring sense, then rapid deterioration of the plant would be indicated, but such data is obtained from plant in 'good' condition, therefore, *no* rapid deterioration is present.

The problems of validating the plant transducer information can be grouped under the following headings.

4.1 Valid window of operation – see Fig. 4.2

Data is only accepted for condition monitoring provided the gas generators are operating within their normal power regime.

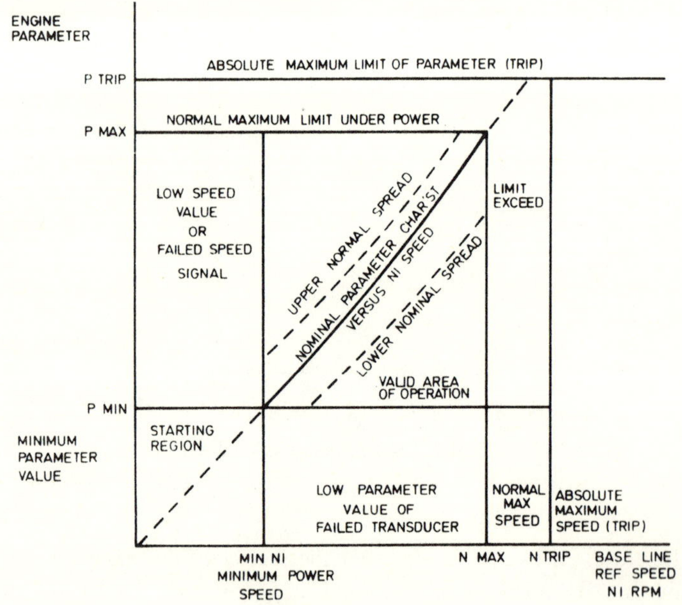

Fig. 4.2 – Valid window of operation.

The data is now ready for two corrections to be applied.

4.2 Correction to ISO conditions

All internal engine transducer data are corrected to a standard ambient intake pressure and temperature. It is assumed that provided actual data is correct, the correction factors as applied do not introduce any inaccuracies other than those due to scaling and resolution, although the correction factors are non-linear. Experience has shown this to be justified.

4.3 Transposition of plant data to a fixed base line reference point

Data from the plant are gathered over a variety of operating conditions, but always at a nominal steady state condition. This is not ideal for the purposes of condition monitoring, where in the ideal case, plant would operate under fixed conditions for long periods of time giving the best data base for future predictive work. In the normal peak lopping mode, these long steady state conditions will not normally occur. Data must therefore be gathered from the plant whenever possible to improve the data base population related to the fixed base line reference point. That is, incoming data must be transferred via the appropriate base line reference, using the techniques described.

Treatment of the data in this manner does however introduce changes which can appear in the short-term as false condition monitoring (EHM) trends; fortunately by careful examination of initial results it has been possible to obtain a balance between such false changes, such that they can be made to cancel one another out. Fig. 4.3 shows the various effects and the balance obtained.

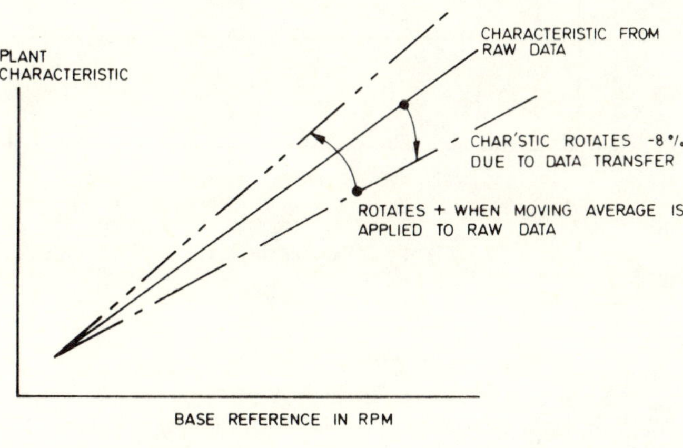

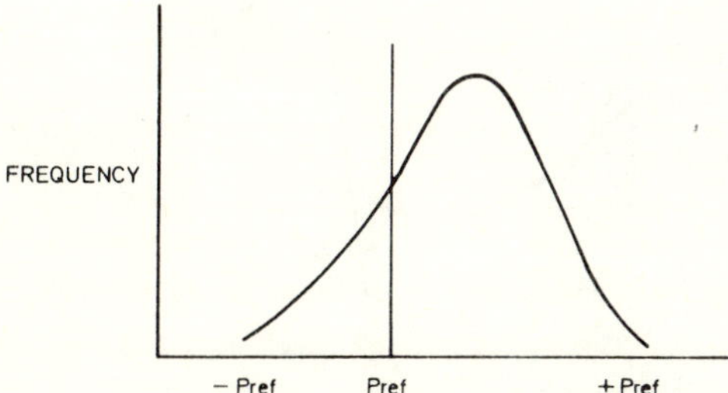

Fig. 4.3 – Zoning of data and use of moving average.

4.4 Establishment of base characteristic for the gas generator

During the commissioning of the power plant time has been taken to establish base characteristics for the gas generator and other appropriate plant parameters. To do this the prime mover is run at a series of fixed power settings giving sufficient time for the EHM system to gather enough data for a realistic characteristic to be assembled. The typical result is shown in Fig. 4.4. It is from this characteristic that the future prediction of EHM changes in prime mover will be made. The effect of a failed or failing transducer is shown in Fig. 4.5. It will be seen that it is easy for the system to detect this automatically.

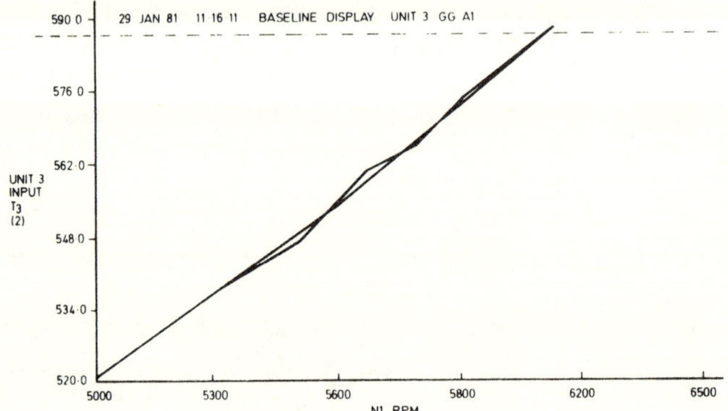

Fig. 4.4 – Baseline display.

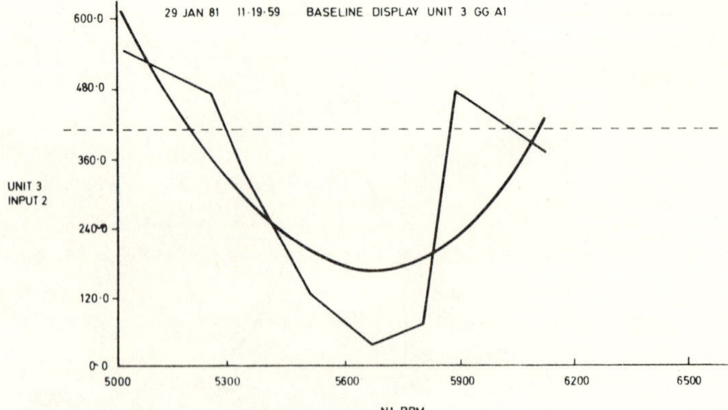

Fig. 4.5 – Baseline display failed transducer.

4.5 Accuracy and repeatability

EHM relies on comparing historical plant data with current plant data rather than to any absolute value, absolute accuracy of result does not enter into the comparison but repeatability is of prime importance. In designing and implementing the system therefore great pains have been taken with hardware, transducers, etc. to ensure repeatable results. This repeatability requirement also extends to any mathamatical treatment in the software of the system, but here it is easier to achieve by increasing basic accuracy.

4.6 Treatment of data to establish condition monitoring changes

General – For the purpose of plant condition monitoring it is required that the plant be in a steady state condition during the measurement of all appropriate plant parameters. The frequency with which the plant achieves steady state condition, depends upon the criteria chosen to define steady state running. These criteria must therefore be flexible and be adjusted to suit the running regime.

Short-term trend – At Bulls Bridge the average powered run is about 20 minutes and the short-term trend is geared to this pattern of usage. Sufficient data must be gathered during this time on all relevant parameters to be able to suppress the effects of noise and signal scatter, and thus produce good plant data for the short-term trend analysis programme.

Typical plant usage and short-term trend – The typical plant usage will determine at which points in the plant operating characteristics the majority of the data will be gathered. This is of interest because it dictates the type of plant data most likely gathered, and therefore available for condition monitoring purposes. Generally at Bulls Bridge the plant will be running near maximum power for intervals of 10 minutes to one hour. It is difficult, given this plant usage, to obtain a suitable sample population. The frequent starts and stops produce apparent changes in prime movers which are false and therefore have to be removed from the EHM predictions. The false change in engine parameters we have called trend sweeping and it is effectively removed by adjusting the sample population size, and overlapping the population of plant data samples, for two or more powered runs. Fig. 4.6 shows the technique. To compute the so-called short-term trend from this data it has been found that a standard linear regression technique is sufficient, with the population samples sizes used.

Long term trend – Using short-term trend data, long-term trends are computed and geared to at least 1/100 of the expected mean life of the plant. In an engine with a time to major overhaul of 2000 hours, a meaningful long-term trend time base would be 10–20 hours of operation. Assuming individual short-term trend minimum population of at least 30, will give a population greater than 100

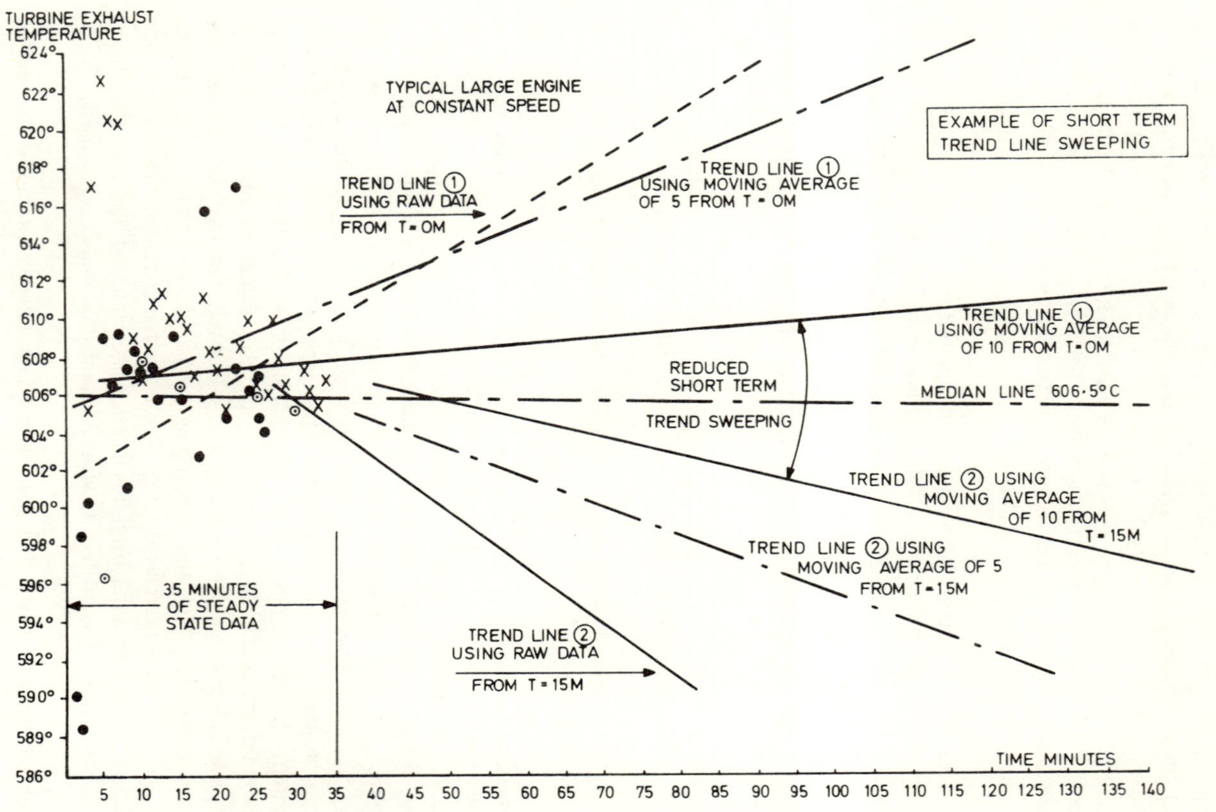

Fig. 4.6 – Overlay of short-term trends to prevent trend sweeping.

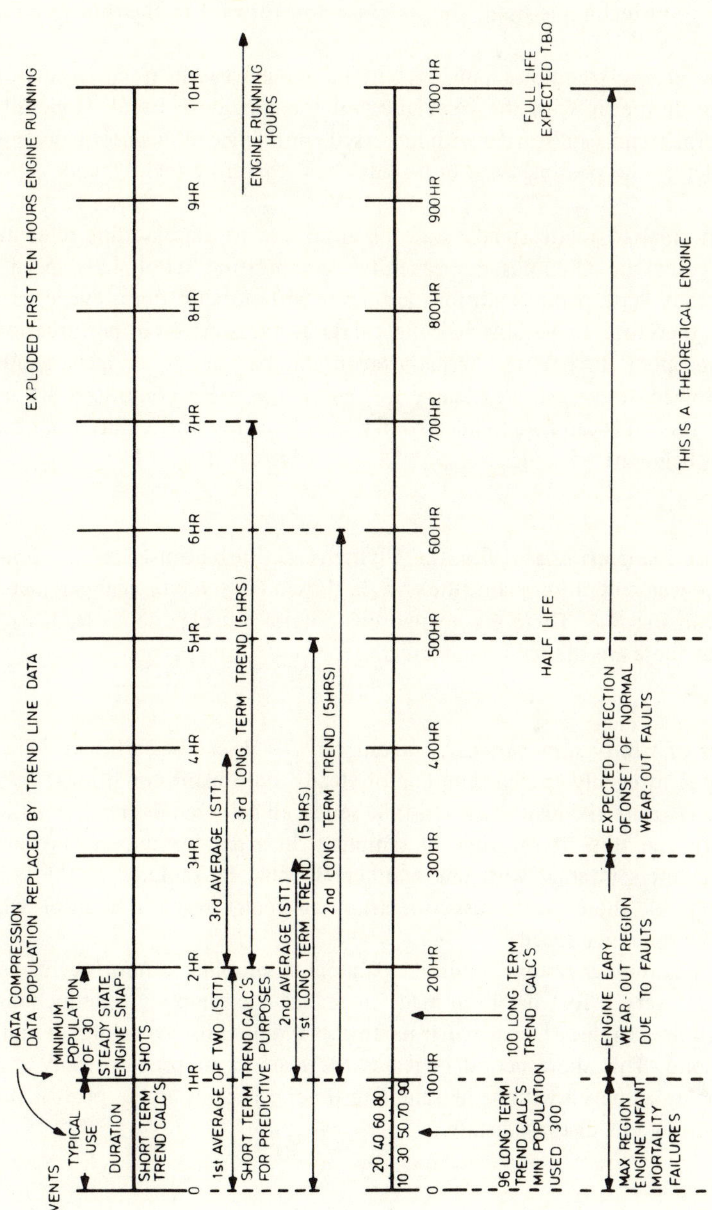

Fig. 4.7 – Timebases for short- and long-term trends.

for the long-term trend, increasing confidence in the accuracy of the standard linear regression technique used, and its extension forward in the time (see Fig. 4.7).

Obviously any trend sweeping present in the short-term trend calculation will largely disappear with the calculation of the long-term trend. Hopefully, the long-term trend can be used with increased confidence in diagnostic decision tables to determine possible wear-out modes, and therefore maintenance action decisions.

It is normal to relate steady state plant criteria to the base line reference parameter. For the Olympus gas generator low pressure spool shaft speed is chosen. Steady state plant conditions are never without short-term cyclic variations, and therefore to be able to collect data at all, some small perturbations must be accepted. In practical terms a favourable base reference speed should not be allowed to vary more than 0.1%, but if operating conditions are unfavourable then this can be extended to 0.4%, without too much scatter in other parameters being introduced.

Steady state plant criteria – For the Olympus gas turbine used as the prime mover in power generation plant, the criteria chosen for some typical parameters are shown in Fig. 4.8. There are no rigorous criteria but experience leads us to believe that these are the right parameters.

The timing of steady state parameter measurement – It can be seen that there is no guarantee of evenly spaced sampling of steady state plant conditions. To be able to reconstitute the plant data as a time series, all gathered data must be time tagged with real time at the time of sampling. In a gas turbine engine where many parameters interact with one another all must be sampled at the same time, to provide the correct cross-correlation, all being tagged in addition with the base line reference speed.

With a scanning system, multiple parameter sampling cannot be instantaneous, but every effort has been made to reduce the sampling time including those multiple samples of a particular parameter required for averaging, to under half a second. This short period increases the chances of pseudo-steady state running being achieved within the sampling time, especially if the plant is normally operating in a disturbed state.

Plant parameters for condition monitoring purposes – For gas turbines, plant condition monitoring requires the measurement of all the main pressure and temperature tapping points, together with the relevant shaft speeds. In addition to these parameters it is important to measure intake pressure temperature, and air mass flow, together with exhaust temperature and pressure. If a free turbine is in use, then free turbine speed and where possible torque should also be measured. A typical list of engine plant parameters for a condition monitoring snapshot are shown in Fig. 4.9 and is the one used at Bulls Bridge Power Station.

Item	Engine parameter	Active power range and valid range	Max. rate of change of valid signal	Small signal variation regarded as steady state	Parameter shift and EHM significance	Notes
1	Low pressure spool speed N_1 rpm	6200 rpm to 5500 rpm 1500–2750 hertz	1500 Hz/sec	± 10 Hz	Baseline	The power range is taken as 60% to 100% normally giving an N_1 range of 6200 to 5500 rpm.
2	High pressure spool speed N_2 rpm	7700 rpm to 7100 rpm 1500–2750 hertz	1500 Hz/sec	± 10 Hz	± 3% to ± 5% ± 200 rpm to + 450 rpm ± 30 rpm S/S	Used as temporary baseline if N_1 fails.
3	Low pressure compressor exit temperature T_2	385 °K to 265 °K	250 °K/sec	± 2 °K	± 5 °K	
4	High pressure compressor exit temperature T_3	590 °K to 540 °K	300 °K/sec	± 7 °K	± 10 °K	
5	HP compressor exit pressure P_3	950K pascals 750K pascals	300 Kps/sec	± 6 kPa	−3% to −5%	
6	LP compressor exit pressure P_2	220K pascals 195K pascals	150 Kps/sec	± 2 kPa	−3% to −5%	
7	Power turbine entry pressure P_6	270K pascals 210K pascals	100 Kps/sec	± 5 kPa	± 3% to ± 5%	
8	Power turbine entry temp. T_6 mean	875 °K to 750 °K	400 °K/sec	± 10 °K	+ 10 °K @ N_1 ref.	Difficult signal to process due to wide scatter but is good indicator of combustion chamber problems.
9	Power turbine entry temp. T_6 individual	875 °K to 750 °K	400 °K/sec	± 10 °K	Deviation ± 35 °K max. rate of change 50 °K/3 hours	Deviation is from arithmetic mean of 8 individual readings. Rate of charge is for running hours on load.
10	Fuel flow W_f	65,000K watts to 45,000K watts	1,000K watts/sec	± 200K watts	± 3% @ N_1 ref.	
11	Intake depression for calculation M_1	0 to 3.45 kPa	Not checked	± 2 kPa	± 3% @ N_1 ref.	
12	All vibration transducers	N/A	N/A	N/A	+ 3 mm/sec above reference	

Fig. 4.8 – Steady state plant criteria.

Item	Transducer description	Per Olympus gas generation	Per set i.e. of 4 engines	Power station total	Use of transducer input in system not all patameters are trended – see individual notes
1	Intake plenum chamber pressure (P_1) 10–14.7 psia	–	2	8	This input is used to normalise engine base line data. ISO condition is P_{1S} = 14.7 psia = 101.3K pascals.
2	Plenum chamber temperature (T_1) –30 to +40°C	–	2	8	This input is used to normalise engine base line data. ISO condition is T_{1S} = +15°C = 288 °K.
3	LP compressor delivery pressure P_2	1	4	16	A trends parameter. Used in compressor condition monitoring.
4	LP compressor delivery temperature T_2	1	4	16	A trends parameter. Used in compressor condition monitoring.
5	HP compressor delivery pressure P_3	1	4	16	As for 3 above.
6	HP compressor delivery tempersture T_3	2	8	32	As for 4 above. See Fig. 4.5.
7	GG exhaust pressure – manifolded P_6	1	4	32	A trends parameter (individual unit power indicator).
8	GG exhaust temperature T_6	8	32	128	Both individual T_6 and T_6 mean will be trended.
9	LP compressor speed N_1	1	4	16	Engine base line parameter, to which all data is related in 'trends' and condition monitoring
10	HP cpmpressor speed n_2	1	4	16	Related to N_1 speed. Used as a back-up base line. This parameter also trended.
11	Fuel volume flow W_f	1	4	16	A trended parameter.
12	Fuel temperature T_{wf}	1	4	16	A trended parameter.
13	Fuel filter differential pressure P_{FFD}	2	4	16	Limited condition only.
14	Lubricating oil pressure P_{LOGG}	1	4	16	Limit condition, min. value – also used in run-up profile. (Trend if machine-time permits.)

15	Scavenge oil temperature (tank) gas generator T_{SOGG}	1	4	16	Limit condition. (Trend if machine-time permits.)
16	Vibration gas generator V_{GGI}	1	4	16	Limit condition. (Trends parameter vibration level vs N_1 rpm uncorrected. Average output. Calibrated on site.)
17	Anti-icing air on/off logic signal	–	2	8	Used to indicate state to condition monitoring calculations.
18	Breather pressure engine air system P_{AB}	1	4	16	Limit condition, also trends compared to installed value.
19	Power turbine speed N_3	–	2	8	Used in run-up, run-down profile, used in vibration trending.
20	Power turbine 1 bearing vibration T_{PT1A}	–	2	8	Limit condition. (Trend if machine-time permits).
21	Power turbine 1 bearing temperature T_{PT1B}	–	2	8	Limit condition. (Trend if machine-time permits).
22	Power turbine 2 bearing vibration T_{PT2A}	–	2	8	Limit condition. (Trend if machine-time permits.)
23	Power turbine 2 bearing temperature T_{PT2B}	–	2	8	Limit condition.
24	Power turbine 1 thrust bearing temperature T_{PT1C}	–	2	8	Limit condition.
25a	Power turbine 2 thrust bearing temperature T_{PT2C}	–	2	8	Limit condition.
25b	Thrust bearing wear	–	2	8	Limit condition.
26	Power turbine 1 cooling air temperature T_{PT1D}	–	2	8	Limit condition.

Item	Transducer description	Per Olympus gas generation	Per set i.e. of 4 engines	Power station total	Use of transducer input in system not all parameters are trended – see individual notes
27	Power turbine 2 cooling air temperature T_{PT2D}	–	2	8	Limit condition.
28	Genarator power MW MW_A	–	2	4	Used in condition monitoring and trends calculation.
29	Generator reactive power $MVAR_A$	–	1	4	Limit condition also used in condition monitoring and trends calculation.
30	Generator volts V_{GA}	–	1	4	Limit conditions only.
31	Main rotor bearing temperature	–	1	4	Limit conditions only.
32	Main rotor bearing vibration	–	1	4	Limit conditions on average output. (Trend if machine-time permits.)
33	Rotor bearing (exciter end) vibration	–	1	4	Limit conditions only.
34	Clutch bearing temperature end A T_{CA}	–	1	4	This parameter trended if machine time permits.
35	Clutch bearing temperature end B T_{CB}	–	1	4	This parameter trended if machine time permits.
36	Clutch bearing vibration end A input V_{CAI}	–	1	4	Trended if machine time permits.
37	Clutch bearing vibration end A output V_{cao}	–	1	4	Trended if machine time permits.
38	Clutch bearing vibration end B input V_{cbi}	–	1	4	Trended if machine time permits.
39	Clutch bearing vibration end B output V_{cbo}	–	1	4	Trended if machine time permits.

40	Lubricating oil pressure (alternator and power turbine) P_{2alt}	–	1	4	This parameter limit conditions only.
41	Generator cooling air outlet temperature T_{GA}	–	1	4	Limit only.
42	Generator transform winding temperature HV	–	1	4	Limit only.
43	Generator transformer winding temperature LV	–	1	4	Limit only.
44	Fuel tank level for station	–	–	4	Limit only.
45	Busbar volts (A, B, C, D) B_{VA}	–	–	4	
46	Busbar frequency	–	–	4	
47	Station barometric pressure $P\alpha$ (P Station 12–15.5 psia)	–	–	1	For use in EHM and trend calculations.
48	Station external air temperature $T\alpha$	–	–	1	For use in EHM and trend calculations.
49	Air flow W_1 (intake diff. pressure) P_{MI}	1	4	16	Trends parameter.
50	Exciter Frequency (N4) 400 cps	–	1	4	Used in run-down profile.

Fig. 4.9 – List of transducers used.

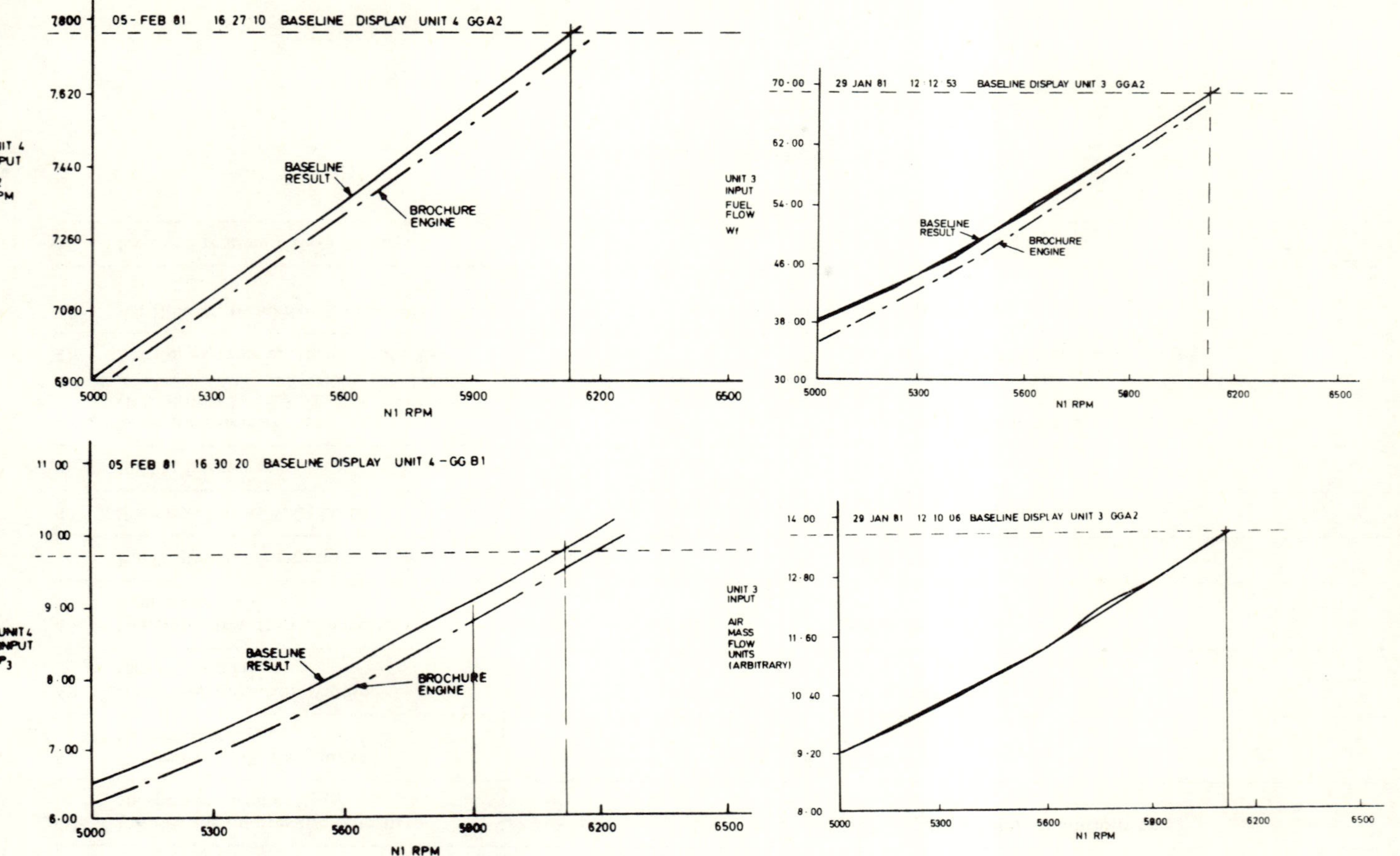
05 - FEB 81 16 27 10 BASELINE DISPLAY UNIT 4 GGA2
UNIT 4 INPUT N_2 RPM
7800
7620
7440
7260
7080
6900
BASELINE RESULT
BROCHURE ENGINE
5000 5300 5600 5900 6200 6500
N1 RPM
29 JAN 81 12 12 53 BASELINE DISPLAY UNIT 3 GGA2
UNIT 3 INPUT FUEL FLOW Wf
70·00
62·00
54·00
46·00
38·00
30·00
BASELINE RESULT
BROCHURE ENGINE
5000 5300 5600 5900 6200 6500
N1 RPM
05 FEB 81 16 30 20 BASELINE DISPLAY UNIT 4 - GG B1
UNIT 4 INPUT P_3
11·00
10·00
9·00
8·00
7·00
6·00
BASELINE RESULT
BROCHURE ENGINE
5000 5300 5600 5900 6200 6500
N1 RPM
29 JAN 81 12 10 06 BASELINE DISPLAY UNIT 3 GGA2
UNIT 3 INPUT AIR MASS FLOW UNITS (ARBITRARY)
14·00
12·80
11·60
10·40
9·20
8·00
5000 5300 5600 5900 6200 6500
N1 RPM

5 THE RESULTS OBTAINED TO DATE

5.1 Basic datum commissioning characteristics

The first results obtained from the system have been very encouraging in two aspects. Firstly the mathematical treatment of data appears to be working well despite the relatively noisy plant environment and, secondly, it is proving straightforward for the operating staff to understand the plant data as presented to them.

The system was first used to establish the base line characteristic for each generator set as it was commissioned, and several typical characteristics are shown in Fig. 5.1. These characteristics are shown for interest in two parts, firstly as a set of straight lines joining the datum points at which the engine was run on a constant power setting, and secondly, the stored datum characteristic generated from these points using a quartic regression technique. The resultant curve is shown also with the brochure engine characteristic, and although we make no claim to absolute accuracy, it is pleasing in fact that the base lines lie within a small margin of brochure engine characteristics.

These characteristics from the reference 'engine' within the condition monitoring system. There being 16 such reference 'engines'. Fig. 5.2 shows the base line obtained when a transducer has failed. In this case an intermittently open circuit input lead has resulted in a large scatter and as can be seen, it is perfectly obvious that we have a faulty measurement and the operator can reject this information. On demand from the operating staff, the system can display any predetermined characteristic. It has proved to be an invaluable commissioning

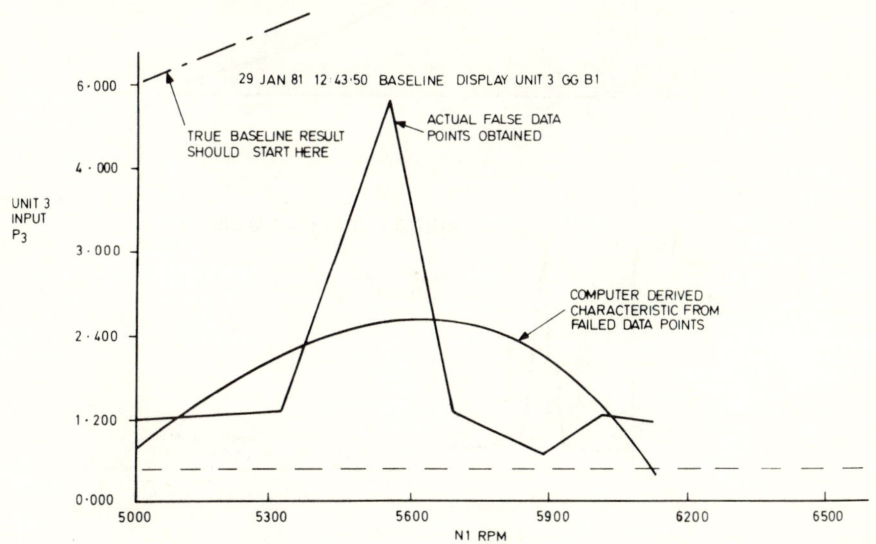

Fig. 5.2 – Baseline result failed transducer.

and trouble-shooting aid, and to illustrate this point a vibration signature on plant run-down is shown for one generating set in Fig. 5.3. The vibration level shown here is the RMS power turbine case level. (It is fairly obvious that one end has a problem, and indeed this proved to be correct during subsequent examination.) During commissioning of the various gas turbines enough engine running hours were achieved to be able to print out some short-term trends in the measured parameters.

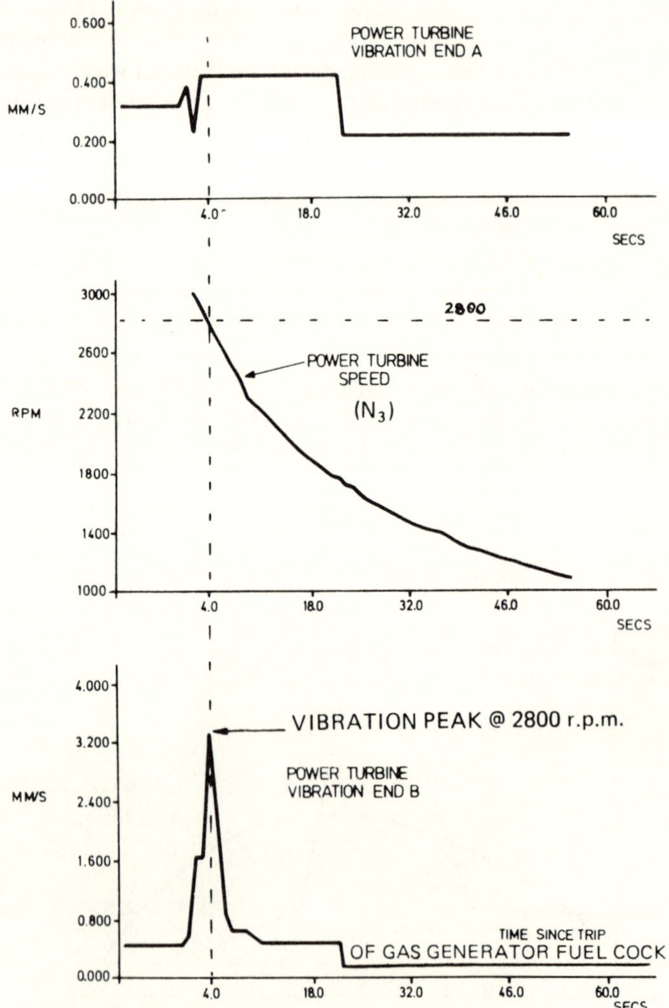

Fig. 5.3 – Actual plant run-down with vibration read-out showing vibration problem on end B.

5.2 Before and after an initial compressor wash

Two of the generator sets had at this point run sufficient hours for several short-term trends to be collated together. These results indicated that a compressor wash might be required on two of the gas generators. Figs. 5.4 and 5.5 show the before and after situation. A clear indication is obtained here of the resolution of the condition monitoring system, as can be readily seen for the step changes in the various plant parameters. In the actual system the operator types into the system an 'M' for maintenance when any is carried out. This is to help with any future diagnostic work when more dramatic change in plant health may be taking place.

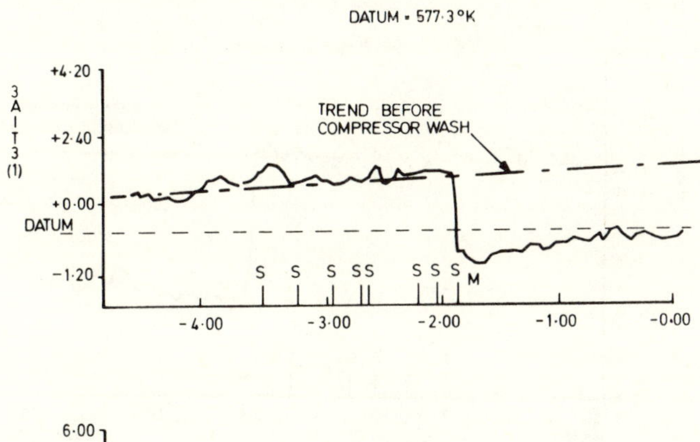

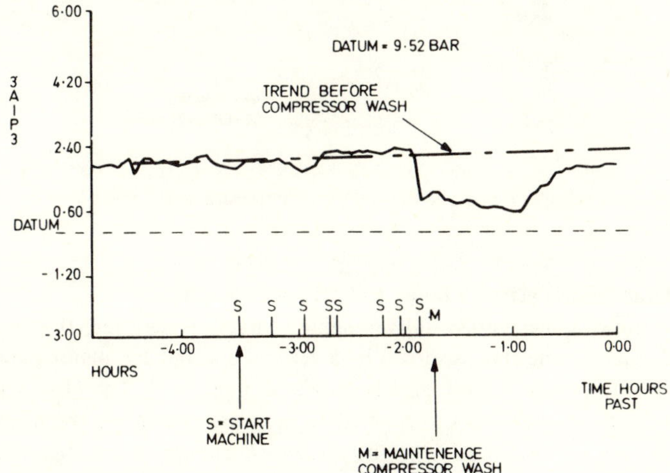

Fig. 5.4 – Before/after compressor wash.

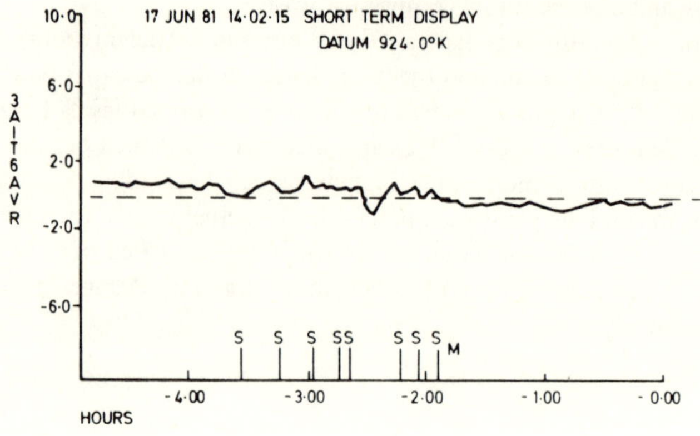

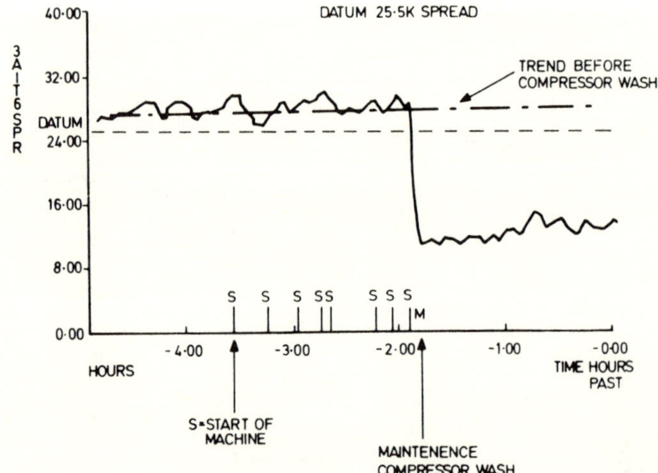

Fig. 5.5 – Before/after compressor wash.

5.3 Characteristics after 10 hours and 50 hours running

Fig. 5.6 shows some representative engine characteristics after 10 hours running on load. Each engine start is shown by a vertical line and any maintenance action taken by the power station staff is shown as a suffix M. The plant parameters are indicated on the left, the time, date etc., are shown on the heading. The print-outs are in 'time past' from the present. There are no significant parameter trends at this point. This is to be expected from a plant where even with the number of fast start-ups experienced, it is only 0.5% into its expected life. At this point in the life of a plant, one is still looking for infant mortality and any such effects are expected to be quite dramatic in terms of parameter change. The next major check on plant change will be after 50 hours running, when it is expected that some noticeable changes will be taking place.

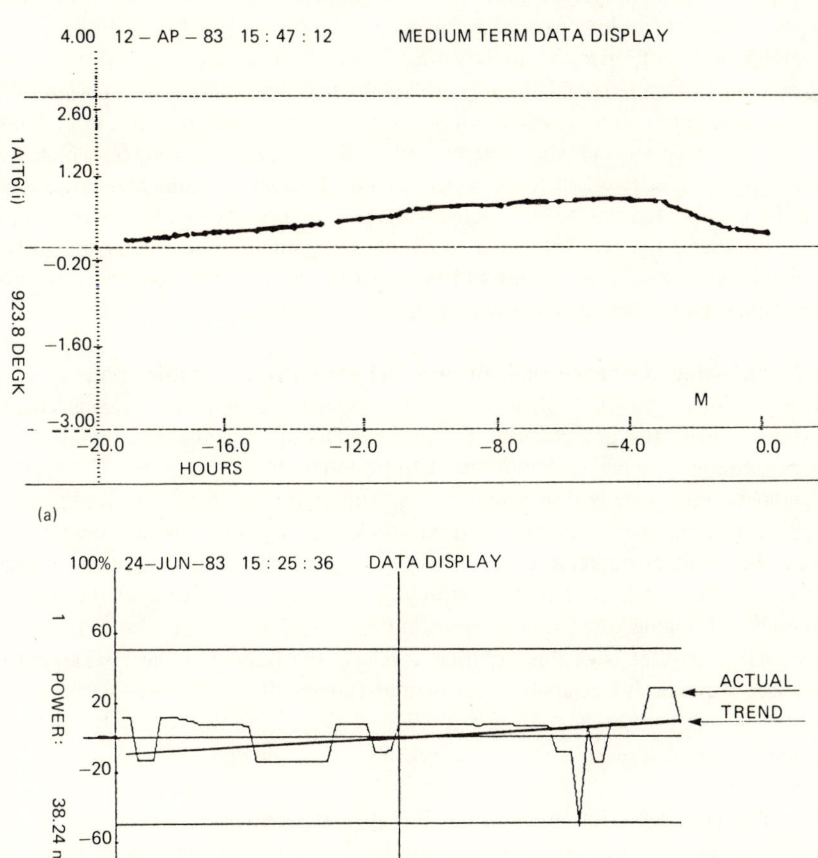

Fig. 5.6 – (a) Exhaust temperature vs. time. (b) power output vs. time.

6. THE FUTURE

6.1 Improving the understanding of the plan

At Bulls Bridge we have extensive monitoring of 15 high power gas turbines and associated electrical generating equipment whose performance is monitored by hundreds of transducers. We have a unique opportunity therefore to gather data for a much better understanding of plant performance and change with usage. Instead of the usual data logger output we have a more condensed form of plant information so that the amount of data available does not overwhelm the potential investigator. Like reliability engineering, condition monitoring relies on understanding past performance and applying this to the future. In a complex situation like the Bulls Bridge Power Station this necessarily means close cooperation between the user, the plant manufacturer, and, for want of a better description, the condition monitoring engineer. Our first lesson at Bulls Bridge has been an immediate reappraisal of those engine parameters which are speed/power sensitive and those which are not. Changed running regimes caused by the unexpected pattern in electrical demand at present being experienced in the UK, mean that the expected usage of the plant has changed significantly, and relatively long power runs will be replaced by very short power runs with consequential increase in starts and stops.

6.2 Improving the diagnostic fault trees and possible future automation

The present diagnostic fault trees have been generated by making changes to an extensive Rolls-Royce computer model of the Olympus engine. Not unnaturally, experience is proving that these need to be modified. An example of a typical diagnostic fault tree is shown in Fig. 2.4. This it will be seen, involves the user calling up a number of parameter trends on his display VDU and deciding which way the plant characteristics are 'moving'. During the initial use of the equipment this is useful in that it is improving the user's knowledge of the system as well as bringing to light any errors, omissions and inconsistencies. It is expected, however, that with this experience gained and built into the system, automation of the fault tree analysis can be undertaken with confidence, without the system causing dissatisfaction due to false alarms etc. In turn of course this will mean a reduced level of participation on the part of the user.

6.3 Improved detection methods possible with experience

We consider ourselves to be still on the learning curve in the overall detection and analysis of plant condition. At the present time the monitoring undertaken for instance of the vibration aspects of the plant is simple. With the knowledge gained here from collected data we consider that advances can and will be made, both in the siting of the relevant transducers and the application of more advanced mathematical techniques to the interpretation of their data. It seems most probable that considerable computation will take place at the transducer itself, or the nearest convenient outstation, using one of the available dedicated microcomputers. In fact, with costs falling so rapidly for such devices, the point cannot be far away where it will be economic to place or dedicate a microcomputer with every vibration transducer. This will overcome many of the current problems of trying to work with complex formulae in real time.

With experience gained, it is hoped to be able to reduce the cost of the transducer fit to the plant to be monitored, unless carefully planned these costs can quickly become excessive. As previously stated, at Bulls Bridge we have used a transducer fit selected for repeatability and stability rather than absolute accuracy, because condition monitoring is essentially a historical comparison process rather than an absolute measurement. The temptation is to add more transducers to measure yet more parameters – for instance power turbine shaft torque would have been very useful – but we hope with the experience gained here to make the presently fitted transducers work harder for us. We do not claim that we foresee a situation where the transducer fit will be simple but rather one of balance maintained between desirable knowledge and affordable working.

CHAPTER 14

On-line monitoring of the true corrosion rate in power plants

E. Diacci, R. Rizzi and C. Ronchetti

1. INTRODUCTION

On-line monitoring of chemical and physical-chemical parameters of power plants has involved a continuous improvement of the analytical systems. The aim of such measurement is the detection of the anomalous behaviour of the operating conditions in order to point out its sources, suggest the most convenient intervention and avoid or minimise the material damage, which is the worst effect of anomalous operating conditions.

This kind of measurement, however, gives only indirect information on the material corrosion that can be deduced only by previous laboratory tests or in-service results, whereas the measurement of the corrosion rate gives straight information on the phenomenon. For this reason, systems capable of supplying this information have been developed in recent years. Such systems make use of different methodologies and consist of a probe inserted in the plant and of an instrument for the conversion of the probe signals into corrosion rate.

Current systems exploit the polarisation resistance of a given electrode or the electric resistance of a wire that becomes thinner and thinner. In the latter case the measurement gives the value of the metal loss between two successive measurements from which the average corrosion rate can be calculated. The polarisation resistance is, instead, an electrochemical parameter obtained from the polarisation of a metal electrode in the vicinity of the corrosion potential [1].

This paper will describe the performance of such systems that make use of electrochemical methodologies for corrosion rate measurements.

2. MEASUREMENT CONSISTENCY

As it turns out of the above mentioned characteristics, the corrosion rate measurement depends on the probe and instrument performances. In both cases, good understanding of the phenomenon (as far as materials and chemistry are concerned) and good knowledge of the operating conditions of the plant are essential for the probe and instrument design. For instance, the coefficients necessary for the instrument to calculate the corrosion rate from the polarisation resistance value are closely related to the corrosion phenomenon. The same corrosion phenomenon both effects the electrode polarisation method (e.g. application of a current ramp, of a current or potential step) and the law of current signal processing. In fact, instruments that are based on microprocessors capable of sophisticated and realistic processing are becoming more and more attractive.

The probe and instrument designs depend on the operating conditions of the plant. For instance, insulator and gasket materials are chosen depending on pressure and temperature; hydraulic conditions are binding on the electrode geometry; proper systems for the rejection of electric noise are required in the instrument; high purity of water (low conductivity) is binding on the performances of both volumetric and amperometric circuits of the corrosimeter and it is therefore necessary to give due account to the formation of both a dishomogeneous potential field on the surfaces of the measurement electrode and streaming potentials by the liquid flow.

In a few words it can be said that the systems available at present on the market are unable to carry out measurements either in high purity water or high temperature and pressure conditions.

In more detail, the corrosimeters usually supply information on the environment aggressivenesss and on its trend; only in simple cases do they supply the true value of corrosion rate.

Various approaches were made at CISE to meet the above-mentioned requirements. Understanding of the corrosion phenomenon and of related parameters has been gained during long-term research activities on both corrosion of different materials and chemistry of power plants. Such researches are performed on laboratory plants capable of simulating true in-service conditions and by means of measurement and failure-analysis activities on power plants. Compliance with the hydraulic conditions on the electrode surfaces, which is impossible using the current three-rod probe, is obtained using tubular electrodes, where feasible and necessary.

The problem of high water resistivity is overcome by means of convenient developments of commercial corrosimeters and by computer-aided design of the probes, thanks to the availability of finite element computer codes that permit in-depth understanding of potential field and current vectors inside the probes [2]. Fig. 1 shows a typical example. The same computer code permits the assessment of the potential behaviour of an electrode surface as a function of its length (see Fig. 2), water conductivity (see Fig. 3) and corrosion rate (see Fig. 4), while unvarying the other parameters. After defining the probe design parameters by means of the computer code it is then possible to assess the behaviour of the two values, which are basic for the consistency of the results, i.e. the ratio between measured and true polarisation resistance and the ratio between the maximum and measured potential. It is also possible to identify the best operating conditions of the probe as for water conductivity and corrosion rate (see Fig. 5). Finally, a set of calibration curves, capable of supplying the true corrosion rate from the measured polarisation resistance value are plotted as functions of water conductivity in operating conditions (see Fig. 6).

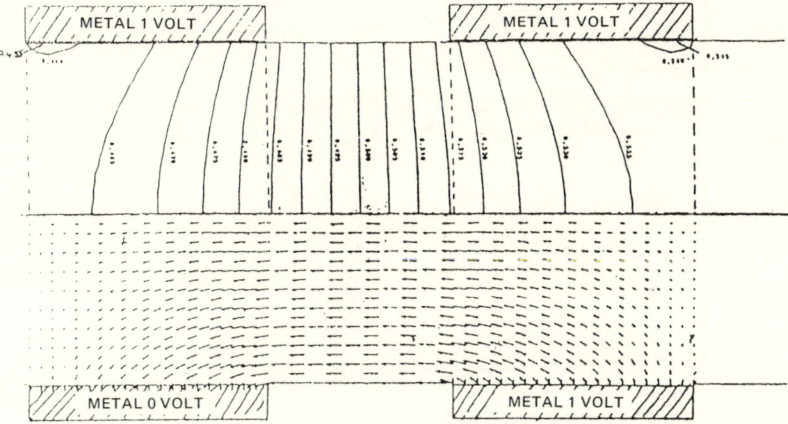

Fig. 1 – Equipotential lines and current vectors for a metal with high corrosion rate.

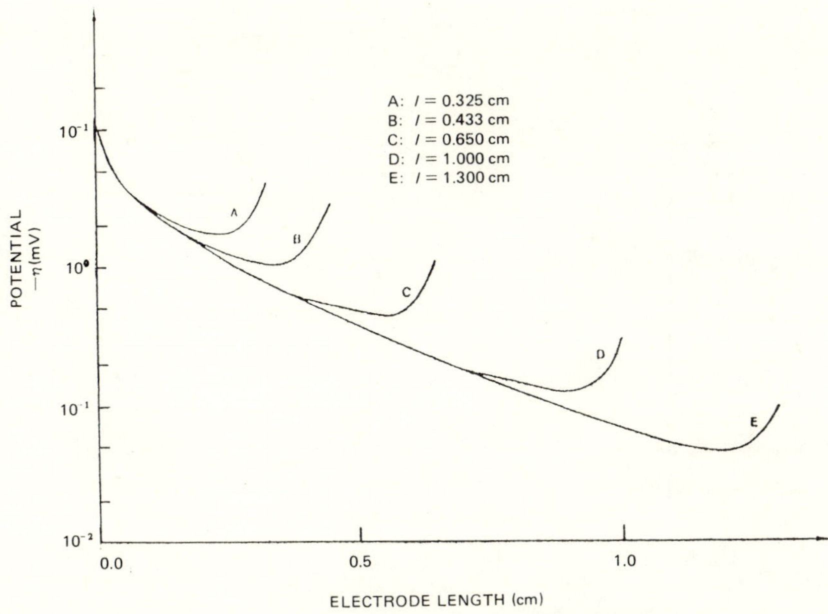

Fig. 2 – Working electrode potential as a function of its length.

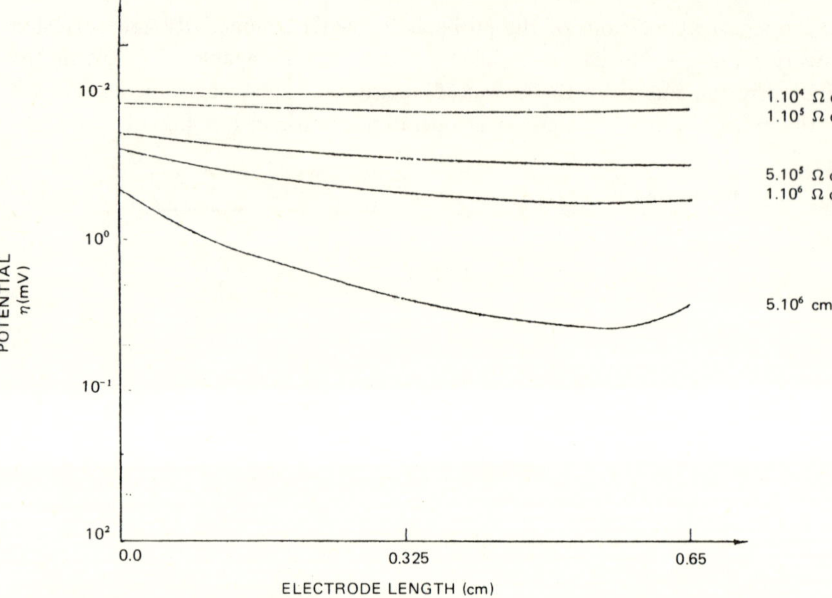

Fig. 3 – Working electrode potential for several water resistivities with the same corrosion rate ($\bar{r} = 0.1$ mdd).

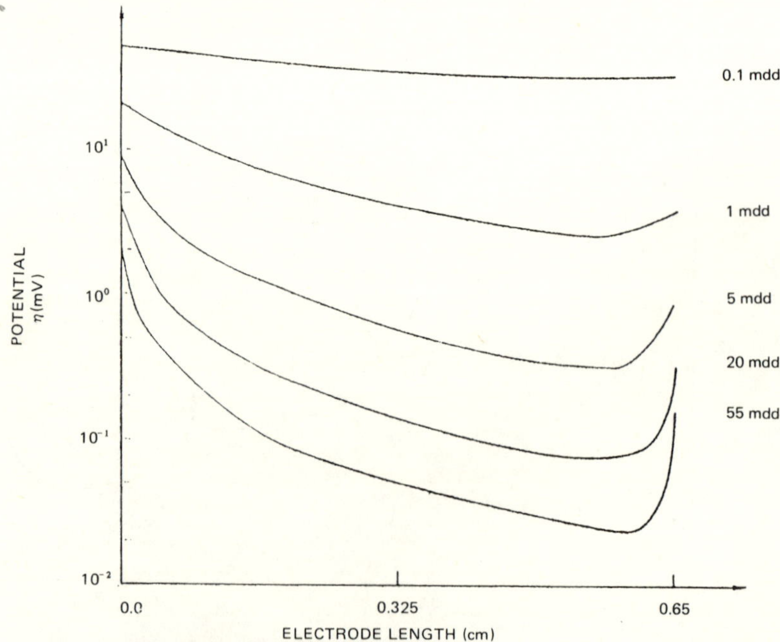

Fig. 4 – Working electrode potential for several corrosion rates with the same water resistivities ($\theta_{H_2O} = 5\ 10^5\ \Omega$cm).

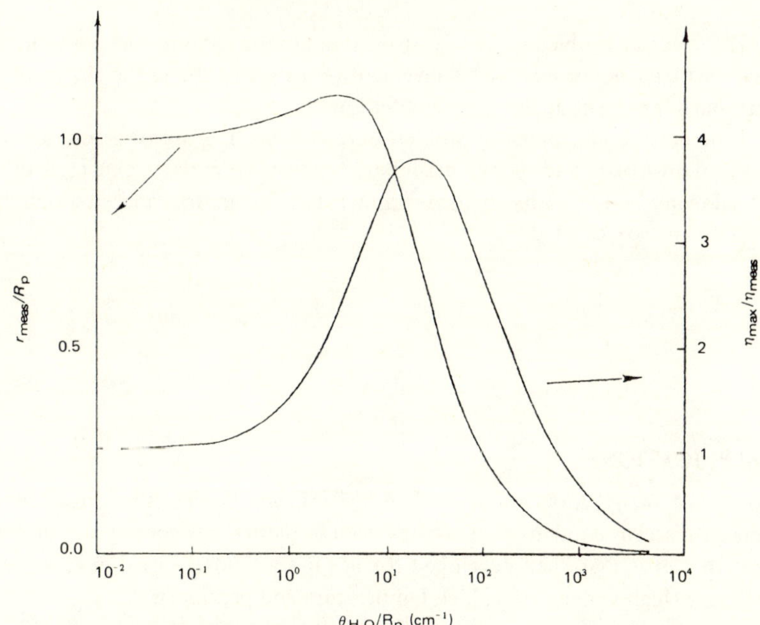

Fig. 5 – Parameters r_{meas}/R_p and η_{max}/η_{meas} as function of θ_{H_2O}/R_p ratio.

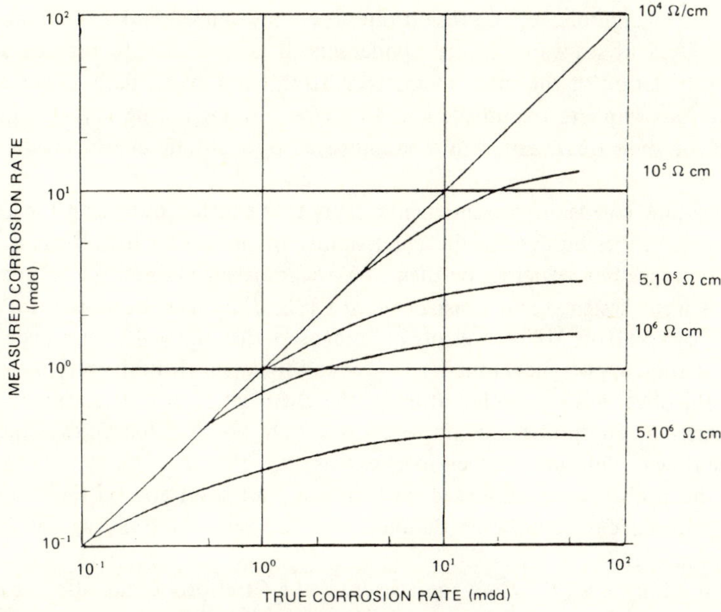

Fig. 6 – Relation between the measured corrosion rate and the true corrosion rate for different water conductivity (electrode length = 0.65 cm).

The electric problems, that is, streaming potentials and noises due to high water resistivity, are solved by a convenient geometry of the probe and by proper electronic adjustment at the corrosimeter inlet.

Temperature and pressure problems are met making use of a suitable technology of insulator and gasket materials. Further progress is expected in this field allowing good results in harder and harder geometric and environmental conditions to be obtained.

3. APPLICATIONS

Two of the many applications studied at CISE will be described in detail: the former, on steam condenser tubes, has been exploited for some years in power plant; the latter has been developed for use in particularly hard environmental conditions (high water purity, high temperature and pressure).

3.1 Steam condensers [3]

In order to minimise the corrosion rate in copper–nickel and aluminium–brass tubes, used in sea-water cooled condensers, it is necessary to provide a good initial oxidation of the tubes themselves. To this aim, particularly important are the ferrous sulphate treatments and Taprogge cleaning during start-ups. Optimisation of such treatment would be improved by real-time checks on their efficiency.

On-line corrosion measurements cannot be carried out using the current three-rod probes because of the impossibility of simulating the effects of water and Taprogge ball velocities on them. To overcome such bonds, special corrosion probes were designed and constructed at CISE. They have been used for about four years in 160 MW power plant condensers that showed such a high corrosion of the copper–nickel tubes as to require their substitution every 3–4 years. The objective was the optimisation of the chemical sea-water treatment during the start-up. The probes, of simple construction, are installed on the tube sheet at the tube outlets and are their ideal extension.

The probes were calibrated both in the pilot plant and on the condenser; the good agreement between the measured corrosion rates and weight losses is shown in Fig. 7.

In addition to the fluid aggressiveness the CISE probes are also capable of detecting the actual corrosion conditions of the tube. Fig. 8 shows, as an example, the corrosion rates of a probe set in an open tube and of another one set on a tube casually closed by a Taprogge ball: the abrasive effect of the carborundum Taprogge balls is evident on the first probe as well as the successive good oxidation during the treatment with ferrous salts; the second probe shows, instead, the anomalous condition of the tube on which it is set. As a matter of fact, the corrosion rate at the start-up is lower, probably because of the presence of a light passivation in the tube, whereas in the successive steps the corrosion rate rapidly increases, thus evidencing the effect of bad hydrodynamic and oxidation conditions. The first year of operation of the probes pointed out that the corrosion problems are ascribable to a bad initial oxidation of the tube. Some variations of Taprogge and ferrous salt treatments supplied satisfactory reduction in the corrosion rate. Fig. 9 reports, for instance in the case of copper–nickel, two different water treatments: it is evident that the latter treatment reduces the oxidation times to a half and, therefore, the metal thinnings diminish from 20 to 7 μm during the start-up.

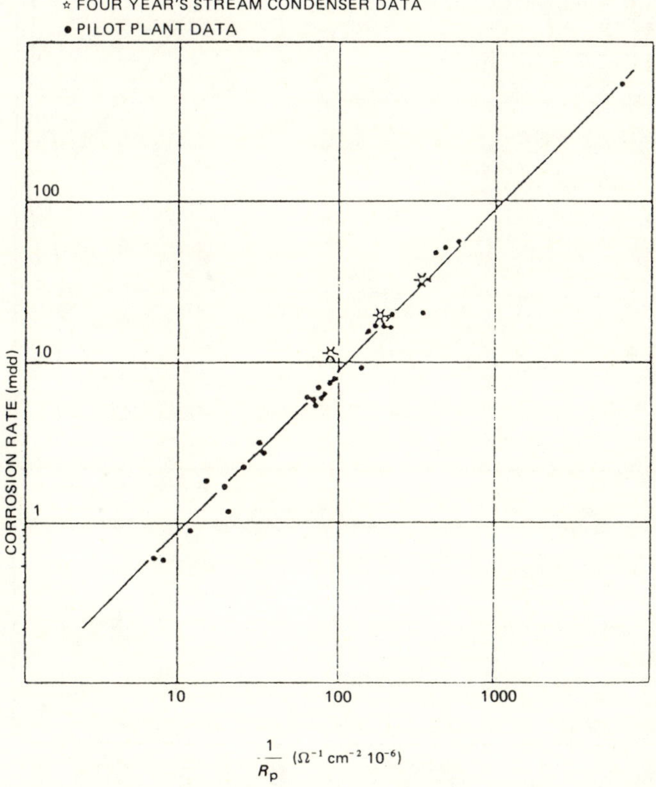

Fig. 7 – Relation between aluminium–brass corrosion rates measured by weight loss and by polarisation resistance method.

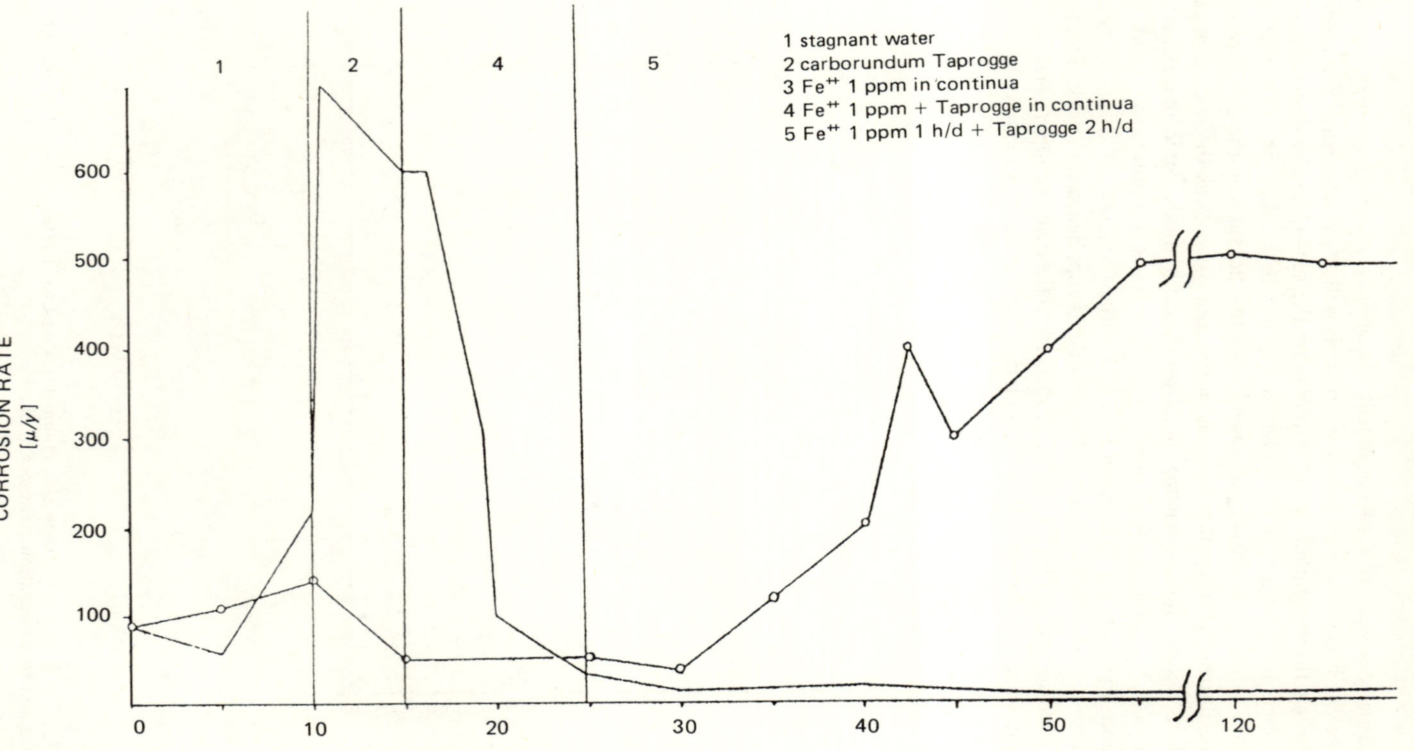

Fig. 8 – different responses of copper–nickel 70–30 corrosion probes mounted on a occluded by Taprogge ball tube and on a 'free' tube.

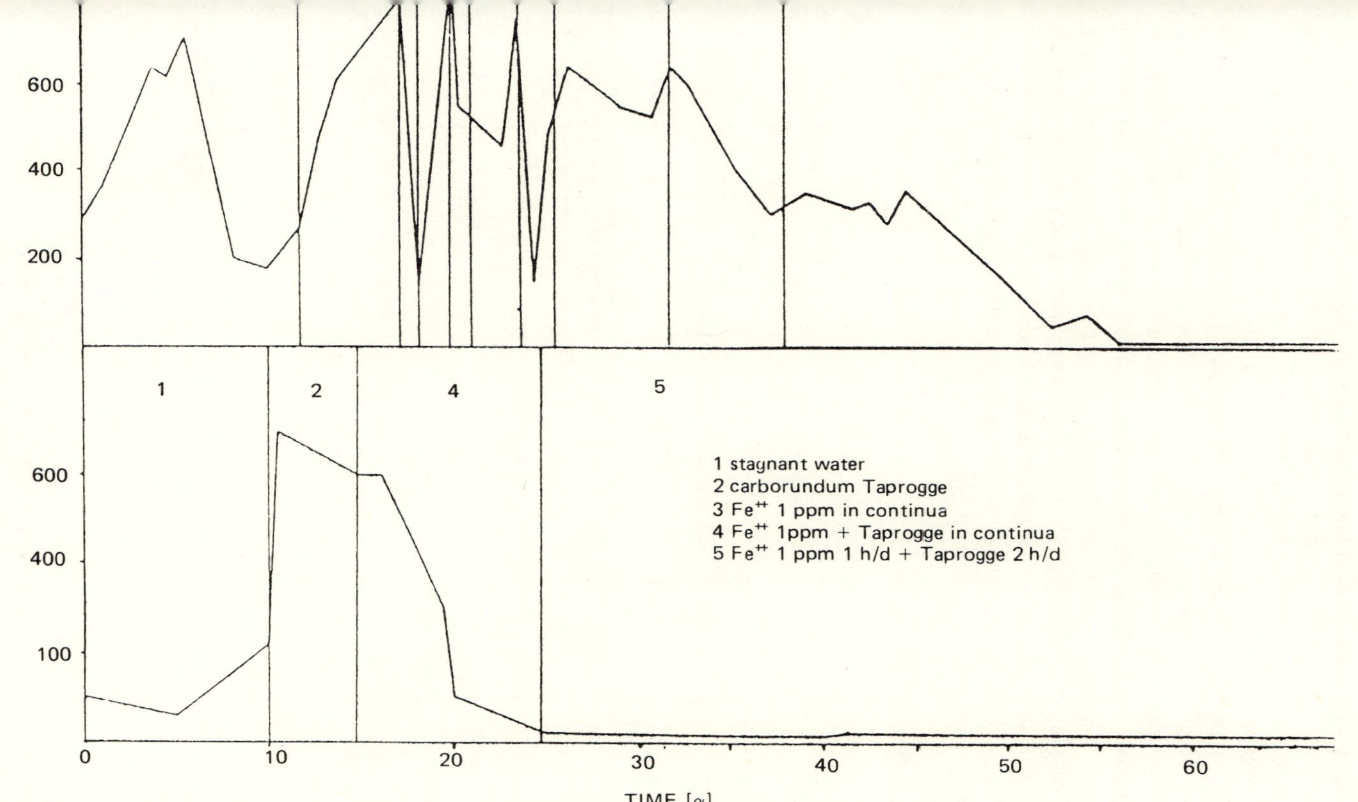

Fig. 9 – Copper–nickel 70–30 corrosion probes responses for two different 'starting time' treatments of sea water.

3.2 High purity water in high temperature and pressure conditions

The use of CISE probes for the measurement of corrosion parameters in very pure water has been carried out up to now on experimental loops working in high temperature and pressure conditions; the application on power plants of different type is being planned. Such experimental loops are exploited at CISE in the physico-chemical conditions typical of steam–water loops of nuclear and conventional power plants; moreover, they worked even in very difficult conditions of temperature and pressure. It was possible in this way to assess and characterise different materials, carbon steel in particular, and to deepen the understanding of the effect of the most significant chemical and physico-chemical parameters. Interesting information on the good behaviour of structural materials and on the fluid aggressiveness had already been obtained by corrosion potential and redox potential measurement [4].

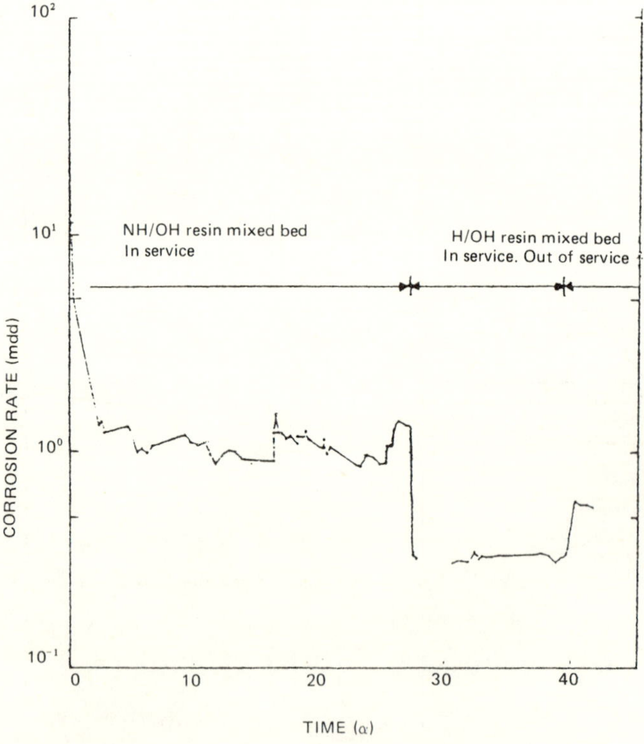

Fig. 10 – Corrosion rate of the carbon steel during a corrosion run in the REBO loop.

Much more important turned out to be the possibility of direct measurement of the corrosion rate of the materials in different parts of the loop [5, 6]. As an example Fig. 10 shows the behaviour of the corrosion rate measured during a corrosion run of carbon steel in basic reducing conditions at intermediate temperature. The figure clearly shows the initial transient due to the steel oxidation process and the further decrease of the corrosion rate following a chemical improvement after 28 days.

The monitoring system proved its capability of prompt detection of the possible difficulties in the passivation process and the effect of water purity change, as is well shown in Fig. 11.

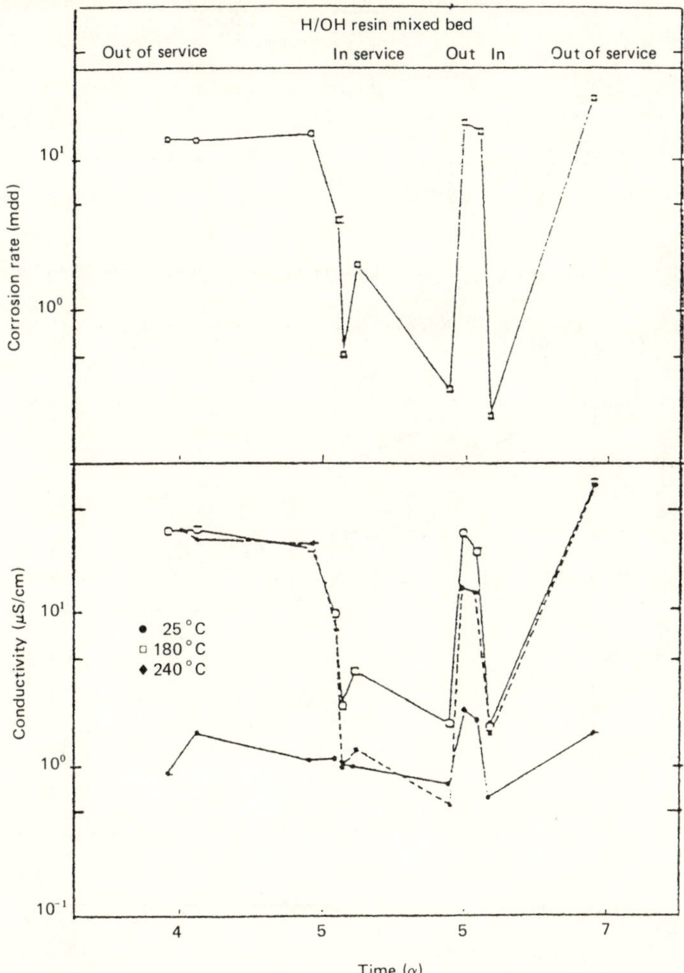

Fig. 11 – Corrosion rate transients during a corrosion run as function of conductivity.

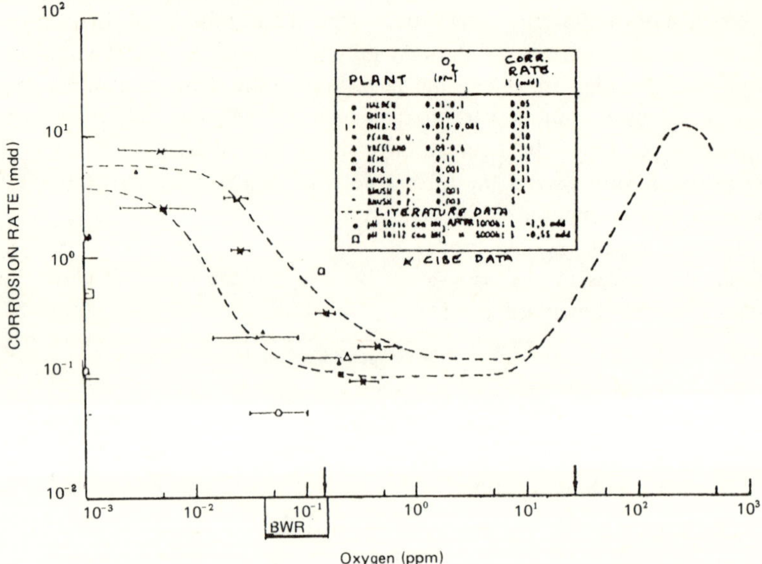

Fig. 12 – Corrosion rate of the carbon steel in different operative conditions.

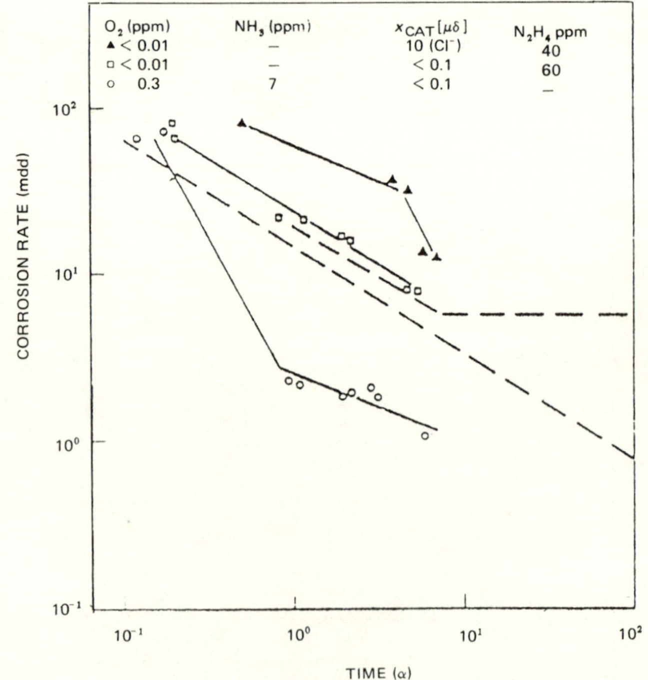

Fig. 13 – Corrosion rate of the carbon steel, during different corrosion runs in REBO loop at 240°C, compared with literature data.

The most interesting results, relevant to the corrosion rate of carbon steel measured in different operating conditions, are compared in Fig. 12 with the literature data; taking into account the relative experimental simplicity of these measurements the agreement can be considered good. In fact, significant results are obtained from few short corrosion runs in comparison with those published by other authors, which require several tests of different duration to obtain the corrosion rates from the weight losses (see Fig. 13).

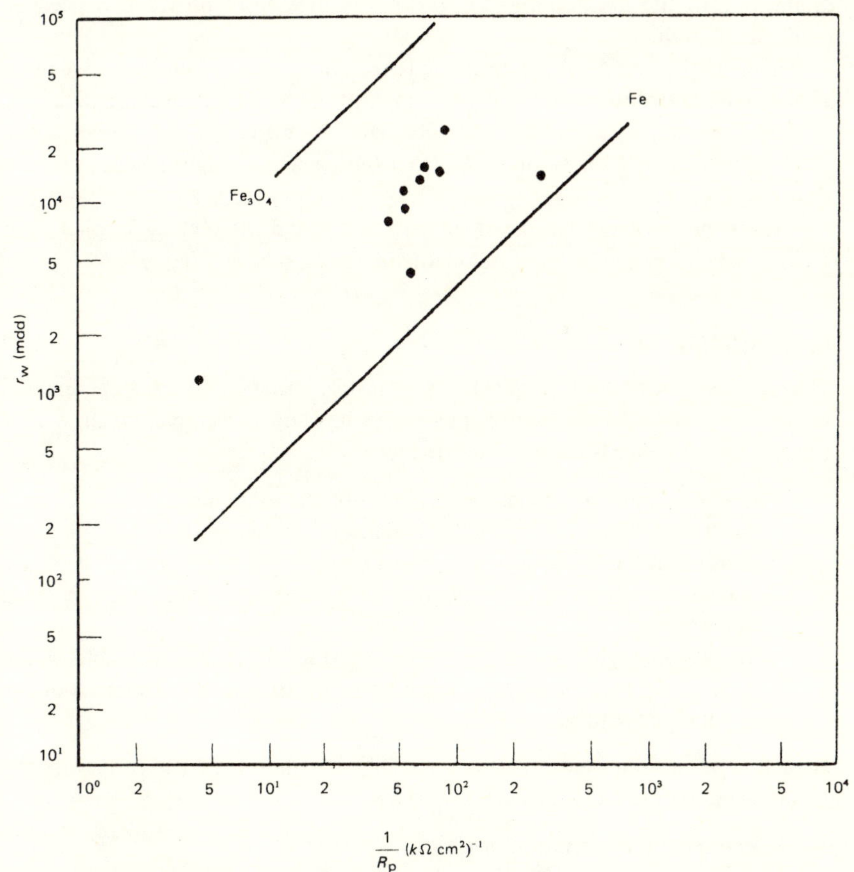

Fig. 14 – Comparison at high temperature between linear polarisation resistances and corrosion rates obtained from the weight losses of the oxidised specimens. Fe or Fe_3O_4, straight lines are relevant to integrated corrosion rates obtained using the average Stern and Geary constants for non-oxidised carbon steel or for magnetite electrode.

3.3 Other applications

On the basis of the theoretical and experimental data relevant to the application previously mentioned, probes for the monitoring of other systems were developed, such as: cooling water circuits both in the presence and in the absence of corrosion inhibitors or other additives to reduce the water aggressiveness; acid cleaning of circuits; decontamination processes, etc. In this last case, in particular, a probe was designed and constructed that permits the carrying out of electrochemical measurements during the treatment with concentrated solutions of mineral acids [7, 8], generally used in decommissioning purposes. The latest results, already partially presented at other conferences or to be presented in the near future, show the usefulness of the electrochemical measurements for the monitoring of the decontamination process in the aim of providing a selective oxide dissolution.

In fact, the trend of the linear polarization resistance can be a parameter of the oxide dissolution rate and consequently of the circuit decontamination. Also in these applications, the corrosion rates measured by the probes are in good agreement with those obtained from the specimen weight losses, as clearly shown in Fig. 14.

These probes were also used in both direct and indirect cycle plants for geothermal energy generation (use of volatile fluids, such as Freon).

4. CONCLUSIONS

The studies carried out at CISE Laboratories using complex experimental plants and thanks to the multiannual know-how on power plants, show that the corrosion rate measurements are true when:

- the measurement probes are made up of the same materials and operate in the same thermohydraulic conditions as the part of the plant of interest to the monitoring;
- a wide experimental understanding of the corrosion phenomena involved is available;
- there is enough information on the electrochemical processes and on the sources of possible electric noise or difficulties in the detection of the small signals utilised.

If such conditions are met, it will be possible to exploit the CISE probes for the real-time evaluation of:

- the trend of passivation processes;
- resistance of the protective oxide to pollutant aggressiveness;
- effectiveness of additives etc.

It therefore becomes possible to undertake the necessary interventions, set up treatments, schedule maintenance breaks, etc.

REFERENCES

[1] M. Stern and A. L. Geary, Electrochemical polarization. A theoretical analysis of the shape of polarization curves. *J. of the Electrochemical Soc.*, **104** (1), 56 (1957).

[2] G. Buzzanca, A. Cella, E. Diacci and C. Ronchetti, Ottimizzazione di celle per la misura della velocità di corrosione con il metodo della resistenza di polarizzazione. CISE-R-1575 (1980).

[3] G. Buzzanca, E. Diacci and C. Ronchetti, Misura continua in situ della corrosione dei condensator di vapore della centrale di Monfalcone. CISE-NT-81.123 (1981).

[4] G. Buzzanca, E. Diacci, R. Rizzi and C. Ronchetti, Misure in continua della corrosione di acciai al carbonio e basso legati in condizioni di riferimento per il prototipo CIRENE. CISE-NT-81.023 (1981).

[5] P. A. Borroni, E. Diacci, R. Rizzi and C. Ronchetti, Corrosion of carbon steel at a high temperature in reducing chemical conditions. 8th Int. Cong. on Metallic Corrosion, Mainz, West Germany, September 6–11, 1981.

[6] R. Rizzi and C. Ronchetti, La corrosione degli acciai al carbonio e basso legati in condizioni di riferimento del prototipo CIRENE. CISE-NT-82.070 (1982).

[7] M. Pavarotti, R. Rizzi and C. Ronchetti, Basic studies on carbon steel decontamination. Int. Conf. on Decontamination of Nuclear Facilities, Niagara Falls, NY, USA, September 19–22, 1982.

[8] A. Agostinelli, F. Bregani, R. Pascali and C. Ronchetti, Vigorous decontaminations tests of steel samples. 1982 International Decommissioning Symposium, Seattle, Washington (USA), October 10–14, 1982.

CHAPTER 15

On-line surveillance and monitoring developments for application in power plants

S. Ghia, V. Regis, G. Possa and F. Tonolini

1. FOREWORD

Plant surveillance should not be considered an isolated technical problem. More precisely, it is related to several problems, such as plant construction and operation, plant maintenance, adequate inspection interventions and the setting-up of the necessary technical systems.

The concept of surveillance by itself has to be implemented where severe safety requirements and maximum plant availability are looked for.

The design and construction of large power plants according to well tested codes and standards guarantee high quality levels, compliance of the materials with the necessary specifications, and an installation without defects, and assure an overall structural integrity. These guidelines find with the quality assurance organisation a mandatory application in the nuclear industry.

In the same way attention is paid to the optimisation of operative modalities and procedures, in particular during plant transient conditions (fully automatised in most cases), and to the observance of the operational parameters set up to widely assure mechanical and structural integrity of the plant.

Scheduled interventions are generally planned for each component and their periodicity will be determined by means of reliability studies and statistical considerations based on a filing of failure events.

Satisfactory surveillance and diagnostics in such a system should not therefore be limited to plant controls by means of non-destructive techniques, identification and location of defective areas or spot assessment of mechanical and structural integrity of components. It should rather rely on the behaviour of the whole system to be controlled, under the assumption that it is possible to assess its conditions by selected operational parameters.

Real-time monitoring of plant conditions and on-line detection of deviations from expected behaviour are the ambitious goals of the development of surveillance and diagnostics techniques.

Important applied research activity is under way and, on the basis of the positive results already available in laboratory and in field work, their technological transfer to the plant is being assessed in areas such as mechanical stress monitoring, mechanical and structural control of components, leak detection.

2. INTRODUCTION

It is known that the mechanical design of a plant is based on the definition of the stress–strain conditions calculated for the different assumed operative conditions (including the accidental ones) and on the sizing and anchorage of the different components so as to guarantee the safeguarding of the maximum stresses resulting from the material characteristics (yield stress, mechanical ultimate strength, fatigue, time dependent creep and fatigue creep).

However, little account is taken, at present, of the material internal stresses and of mechanical and microstructural dishomogeneities. Stress is obviously the key parameter for the determination of behaviour and microstructural integrity of the plant, since it determines plasticity, aging, creep, defect nucleation and crack-growth, and contributes heavily to corrosion phenomena. Stress is given with good approximation (linear elastic or even elastoplastic) by the design but its determination may be difficult in geometrical and structural singularities zones. Anyway, this stress is not systematically checked and measured. In addition this parameter is strictly correlated with plant operation parameters (pressure, temperature, etc.) and is therefore fundamental for surveillance purposes. At present, time-reliable measurements of this parameter over long periods using strain-gauges, even at high temperature, can be assured. The calibration of the experimental values with those supplied by stress analysis has given satisfactory results that, in any case, prove the maturity of the on-line strain/stress monitoring technique. Another evaluation of the deformation conditions components and the evolution of their metallurgical and structural integrity can be achieved by the detection of acoustic emission; the phenomenon being associated, as is known, with the local discrete and/or continuous release of stress waves emitted during irreversible processes occurring in the material (plastic deformation, thermal fatigue, microstructural evolution, etc.). Testings are under way in the laboratory on specimens, on experimental rigs and on mock-ups as well as in full-scale components. Optimisation of the signal/noise ratio, and a wide knowledge of the specific metallurgic phenomena contributing to stress wave emission, have to be obtained. It can also be expected that a generalised industrial application of acoustic emission techniques for metallurgical and structural integrity surveillance will occur within ten years.

It is now necessary to point out that the incorrect operation and alignment of a whole component can be monitored by specific transducers and by vibration analysis. This technique has turned out to be a powerful tool, in particular in the applications to rotating machines, for design checks, assembling optimisations and early monitoring of deviations due to malfunctions, geometrical and structural changes, and damage. It is already widely used for industrial applications in plants after accurate specific calibrations.

A further consideration should now be added. A plant is composed of a series of components (pipes, valves, pumps, exchangers, vessels, etc.) whose primary function is to contain and circulate the process fluids. Some material damage may generate important leaks, which generally precede the catastrophic rupture of the component, following the concept of 'leak before break' assumed at least for the ductile steels used in the construction of pressurised components.

The on-line detection of leaks, just at the beginning of their occurrence, associated with the information, even approximate, of their location is thought to be the last useful level available for any intervention before severe and irreversible damage to structures and plant. Many technical difficulties should be listed in the practical application of leak detection systems. They have to be highly reliable and provide elements to speed up and improve the necessary detailed checks by current non-destructive methods.

Enhanced results have been obtained on this topic, using both traditional leak-detection devices and high frequency acoustic emission techniques. The first are already currently monitoring preheaters in large power plants; and a wide development of acoustic emission techniques, that are at present undergoing in-plant qualification, can be foreseen.

3. CONTINUOUS MEASUREMENTS OF STRESS–STRAIN CONDITIONS

Strain-gauge techniques for the measurement of stress–strain fields in structures have been used for a long time now and give very good assurance of accuracy and reliability, particularly for dynamic or short-time measurements. The use of conventional resistance strain gauge for on-line continuous surveillance of components in thermal plants has, however, met with remarkable difficulties owing to the decay of the gauge performances due to high temperature effects for unavoidable long periods. The availability of capacitive strain gauges, specifically developed in the last ten years for long-period measurements in severe environmental conditions, has supplied a new and powerful tool of investigation, suitable for the structural monitoring of the pressurised components working at high temperature [1, 2].

In 1975 ENEL started an experimental programme for the development of high temperature strain gauges through laboratory and in-plant tests, referring to the encouraging results and know-how accumulated by British research laboratories [3].

An extensive characterisation of the CERL-PLANER capacitive strain gauge types C4 and C5 (for ferritic and austenitic steels) had been therefore carried out to assess their potentialities and limits [4]. Some of the obtained results (drift: $-0.05\ \mu\epsilon$/h for 10,000 hours at 540°C; thermal hysteresis: -10–$-15\ \mu\epsilon$/cycle 20 + 600 + 20°C) in good agreement with those of other authors [5, 6] have promoted the successive experimental step with in-plant testing. Nevertheless, the study of their behaviour under particular conditions (neutron irradation, severe thermal shock) is at present under way in cooperation with ENEA at La Casaccia Laboratories.

For the experimental in-field testing, a measurement and data acquisition system has been assembled (Fig. 1) with HP and ASL instruments, and controlled by a minicomputer that usually provides a first data processing, that, relying on simplified computer codes, can supply the on-line stress–strain range of the instrumental areas. The system accepts up to 100 capacitive strain gauge channels and 40 analogue channels. During the set of measurements carried out up to now, the system has shown high reliability and it has been found extremely easy to operate.

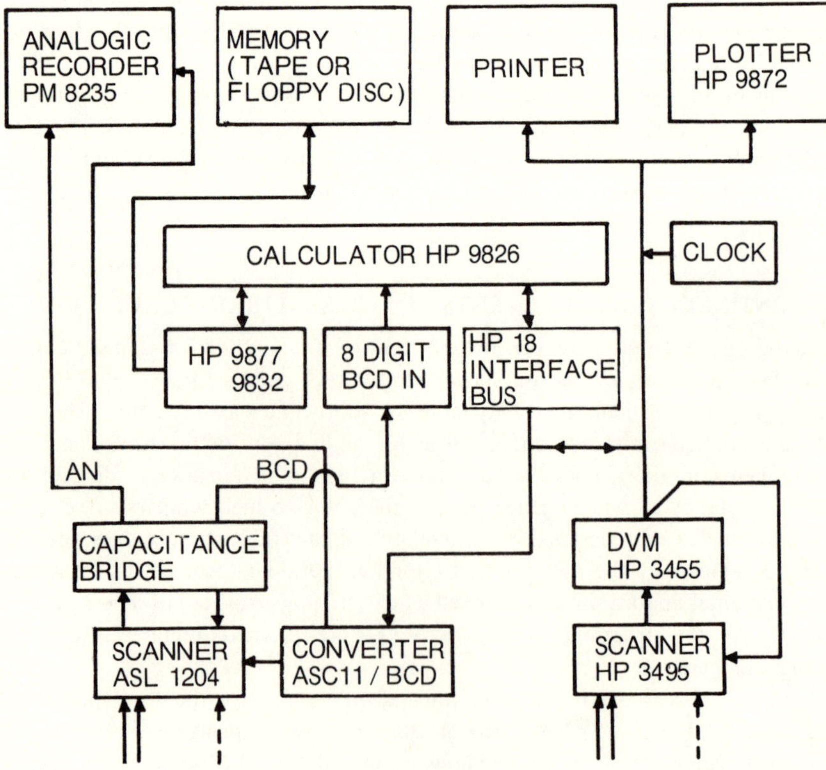

Fig. 1 – Data acquisition system.

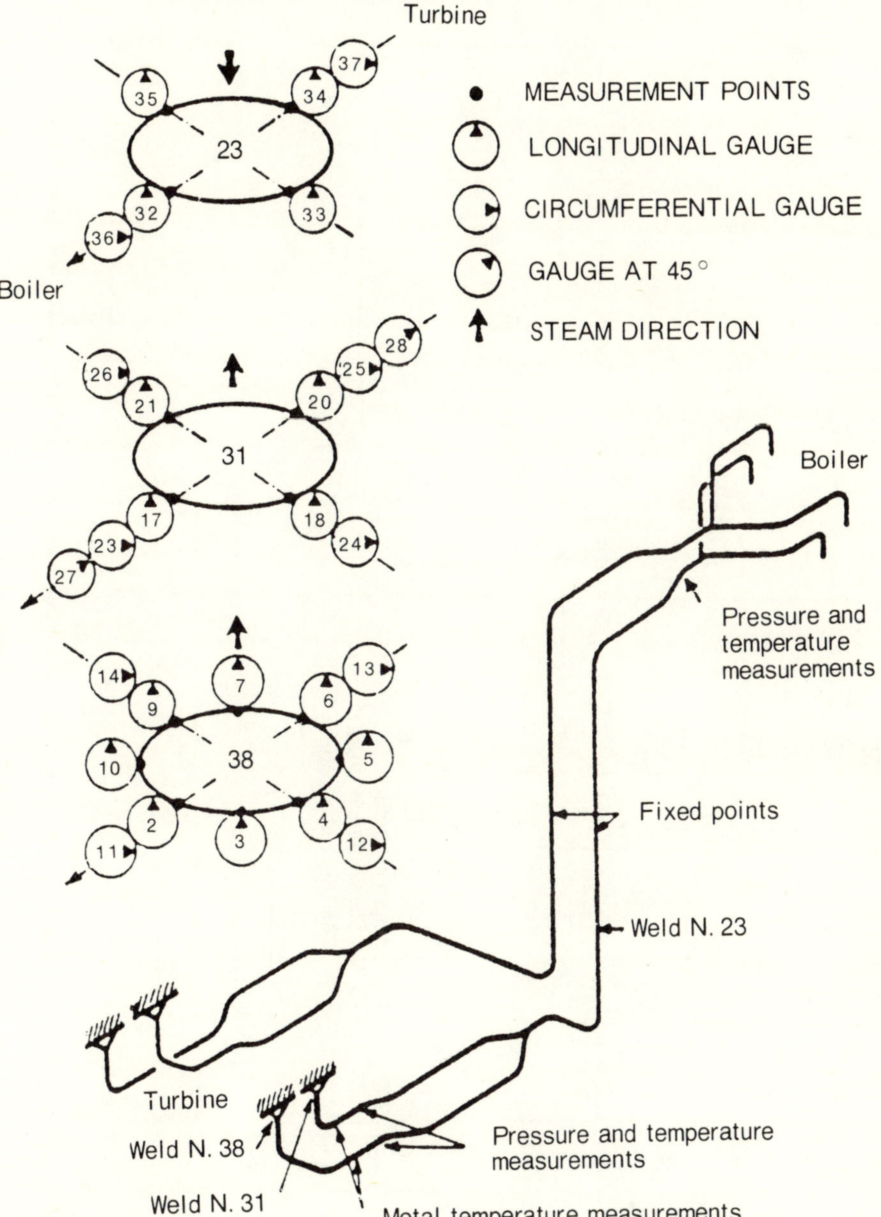

Fig. 2 – Sketch of main steam pipes in 600 MW unit and strain-guage positions near welds nos. 23, 31 and 38.

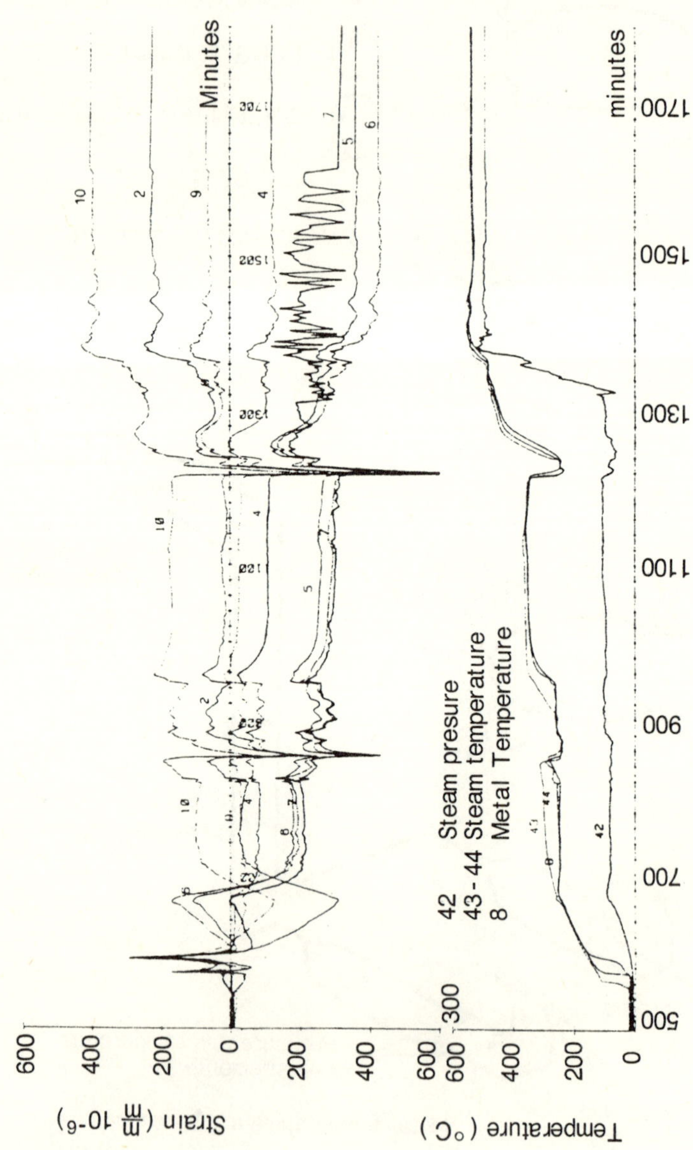

Fig. 3 – Typical strain outputs during a monitored cold start-up (main steam pipe of a 600 MW power unit – zone no. 38).

The experimental monitoring of components of thermoelectric power plant was started in 1977 with the installation of 17 strain gauges, type C4, on the pipes of the main and reheat steam in a 320 MV unit with a view to check the transducers stability and reliability in the real operational conditions and to evaluate their effectiveness in monitoring hot creep effects.

The strain gauges have been operating at present for 37,000 hours, with only two records of malfunction or rupture, and have supplied data in good agreement [7] with the periodic mechanical measurements carried out during scheduled shut-downs.

The monitoring of the stresses, which take place in the main steam pipe during cold start-ups and transients, has been carried out on a 600 MW unit of La Spezia, by equipping three regions of a steam line with a total of 29 strain gauges (Fig. 2). Typical strain trends during start-up are reported in Fig. 3. The stress data obtained from the test and the results of the thermomechanical and sectional flexibility analyses show a substantial consistency and have been used, for instance, to assess the line behaviour under different in-service conditions [8, 9].

The whole of in-plant and on-rig circuit tests (Table 1) has proved the good intrinsic reliability of the capacitive strain gauges, even for in-service time intervals larger than 20,000 hours (Fig. 4) and with acceptable stability even in load-transient conditions. The estimated sensitivity has always been of the order of 10 $\mu\epsilon$, and, in addition, good performances of the detection and measurement system for in-field applications, even after 30,000 hours, with particular reference to drift (-0.03 $\mu\epsilon$/h at 540°C) and calibration repeatability ($+15$ $\mu\epsilon$), have been noticed.

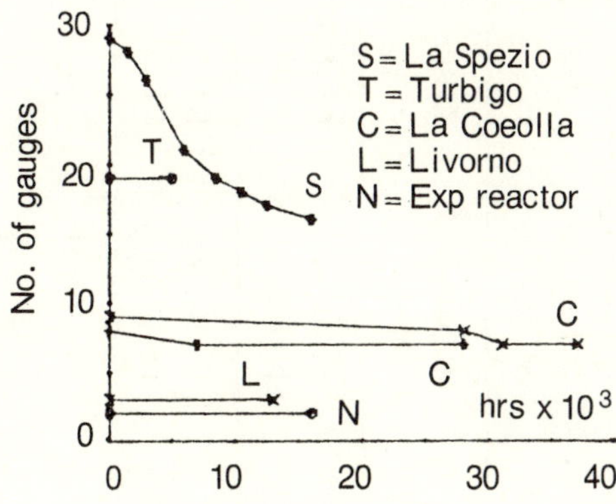

Fig. 4 – Guages operating life-time.

Table 1 – Capacitive gauges installed on plant components.

Unit	Monitored component	Steel	Temp. (°C)	No. of gauges installed/damaged	Operating time (hrs)	Scope	Remarks on tests
Casella 2 320 MW	RH pipes SH pipes	A335 P22	540 540	9/2 8/1	37,000 29,000	creep rate evaluation	in progress
Spezia 4 600 MW	SH steam piping	AISI 316	540	8/3 29/13	4,000 16,000	stress measurement	completed
Livorno 1 160 MW	SH steam piping	Cr–Mo–V	538	3/0	13,000	crack detection	completed
Turbigo 2 320 MW	feedwater heater	SA F12	400	19/0	5,000	stress detection	in progress
Triga	exp. reactor	316	70	2/0	15,000	neutron damage on gauge	completed
AST 2	thermal shocks rig	316	650	2	–	ratcheting	scheduled

4. ACOUSTIC EMISSION MONITORING

Many microstructural processes occurring in stressed materials, such as plastic deformation, phase transformations, slug disbonding and fracture, crack formation and growth, produce sudden and well localised vibrational elastic energy releases named acoustic emission (AE) events. These energy pulses propagate in the medium and may be detected and located by special sensors placed on the surface of the material.

The dynamic peculiarity of this process makes AE an appropriate and attractive method for the continuous monitoring of the structural integrity of plant components during operation.

A lot of problems, however, have to be solved for an AE full application to the plant surveillance: the complete characterisation of the AE events, the background noise rejection or filtering, the appropriate and reliable implementation of instrumentation systems having a good capability of operation in hostile ambient conditions.

The AE characteristics are strongly dependent on the process in progress (i.e. local yielding, stress corrosion, thermal and mechanical fatigue, ductile subcritical crack growth) and load-ambient conditions. For this reason in connection with the development under way abroad [10–13], an important research effort has been devoted by CISE-ENEL to AE on steel specimens measurements in tensile and fracture mechanics tests [14, 15], experimental rigs and plant components in order to find a correlation among AE, applied stress, mechanical properties and microstructural processes.

A first significant check of these AE phenomenology in real and full-scale situations has been obtained performing AE examinations during several pressure vessel hydrotests [16, 17].

Recent important research in progress concerns AE monitoring and evaluation produced by intergranular stress corrosion cracking of butt welds of tubes in AISI 304 material during measurements performed on experimental rigs.

The background noise level at the measurement point during plant operation is a decisive factor for the feasibility of a correct AE monitoring in continuous surveillance.

In the frequency range normally employed this noise is usually continuous and stationary, having a typical strongly decreasing frequency spectrum and being variable with the operating plant conditions (fluid turbulence, due to high flow-rate and the throttling effect of valves). In other cases AE background pulses may be present (fretting on pipes walls caused by vibrations and impacts, machinery, mechanical noise, etc.).

In the presence of high level noise it may be convenient to use high frequency AE sensors (the signal to noise ratio increases vs. frequency), but the number of the sensors has to be consequently increased because of the strong attenuation of AE signals at high frequency.

Generally a compromise has to be reached. Accumulated experience in noise measurements on test circuits or plant lines has suggested for each case to use special filtering, tuning and rejection techniques in order to improve the signal to noise ratio.

Special attention has to be devoted to environmental conditions imposed on the instrumentation chain which has to be placed in a hostile ambient (high temperature, high radiation level, humidity). The fastening of the sensors may be difficult. The instrumentation system has to work correctly for long periods without maintenance.

Moreover, the cable part between sensor and preamplifier may be rather long, requiring an appropriate impedance matching and an accurate design and installation of the signal transmission line in order to reduce the pick-up of electromagnetic spikes.

The above-mentioned problems were faced and solved, assuring therefore a correct AE monitoring.

The system design and electronic devices have to be implemented in accordance with low cost requirements and special on-site problems, at the same time obtaining meaningful and reliable results.

The main functions performed by the apparatus are: amplification and amplitude discrimination of AE signals, reflection of spurious signals and of AE pulses extraneous to the monitored area, recording of the event description data (the differences, ΔT, among the arrival times at various sensors according to the location of the sources, the peak amplitude, the pulse duration, the event-appearance instant, the typical plant parameters, etc.) and their transfer to the computer memory.

All these data have to be processed and analysed in real time or out-line depending on the operator demand. In order to satisfy the above requirements CISE, on ENEL behalf, manufactured and qualified by means of a relevant experimental activity of two main versions of the AE systems:

- a compact and reliable equipment (2–8 AE channels) for applications in laboratory experiments (stress-corrosion in pipe welds, fatigue phenomena on structure models, etc.) and in the surveillance of limited plant zones (nozzles, pipe and weld portions, valves, etc.) (see Figs. 5 and 6);

- a multichannel system (12–16–24 expandable channels), according to the standards recommendations, for the application to pressure vessel hydrotests and huge structures monitoring. The system performs, by means of microprocessor units, all the functions previously mentioned and is equipped with high performance software: acquisition speed up to 100 events/sec, efficient rejection of spurious signals, sensors set-up flexibility, location capability in different situations of geometry and thickness, operation capability up to 12 metres spacing between various sensors, high interactivity between systems and operator, data analysis and display and hard-copy representation (Fig. 7).

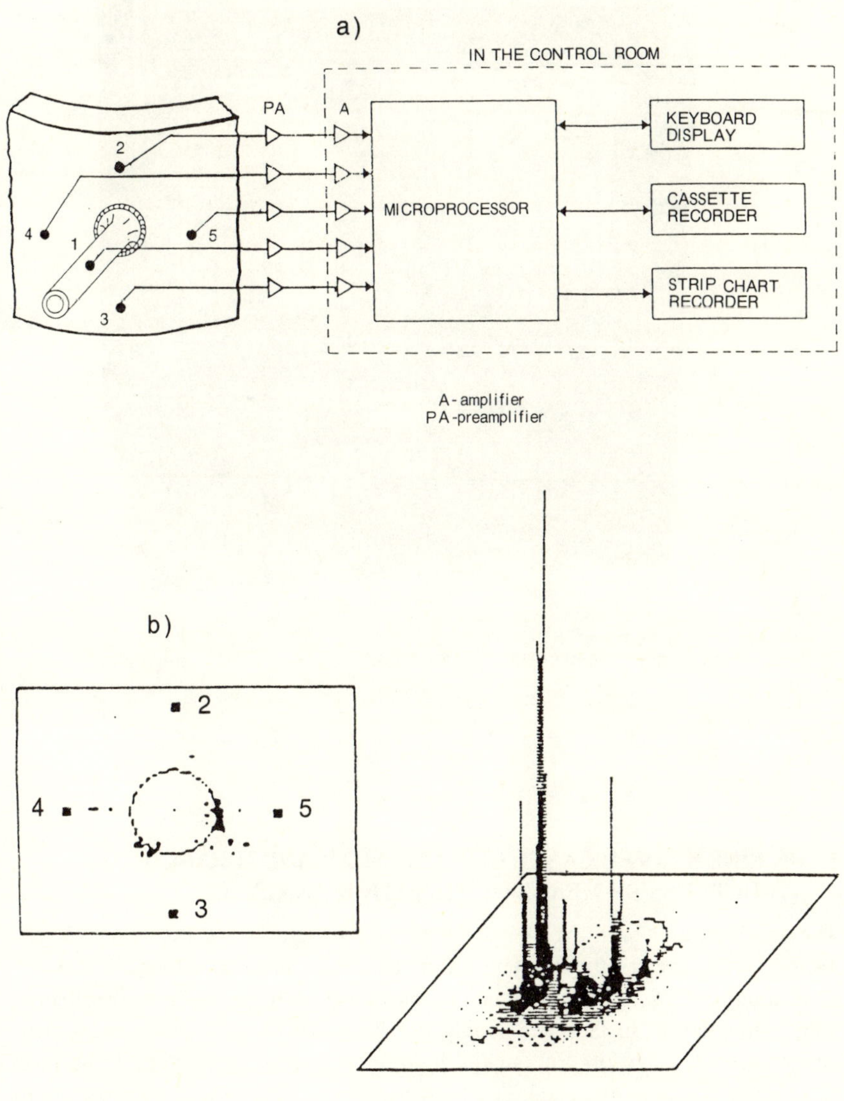

Fig. 5 – (a) block scheme of AE continuous surveillance system; (b) location map of AE sources along a weld of a nozzle; (c) tridimensional view of AE located sources.

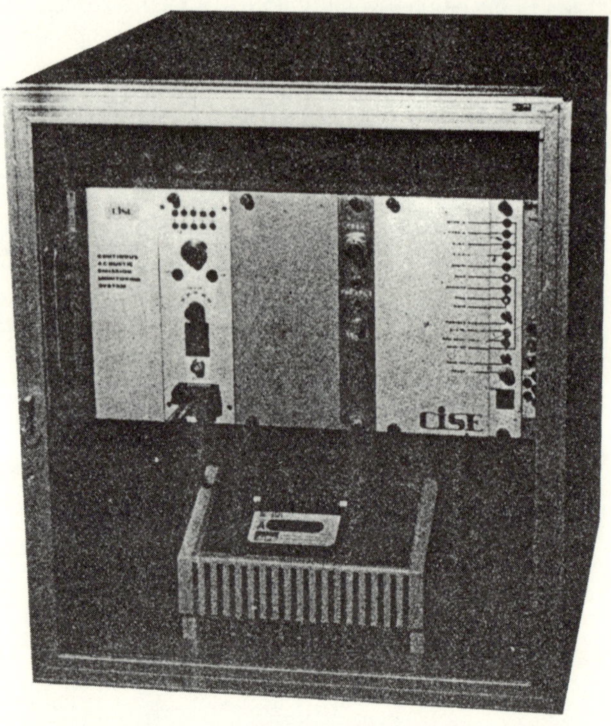

Fig. 6 – Acoustic emission continuous surveillance equipment.

5. ON-LINE SURVEILLANCE OF MACHINERY AND MECHANICAL STRUCTURES BY MEANS OF VIBRATION ANALYSIS

The mechanical vibrations of machinery and structures are the 'dynamic response' to the stresses they undergo. For this reason they can be regarded as sensible monitors of operation 'quality', particularly fit for on-line surveillance. Vibration analysis, in fact, can supply timely information on many phenomena active in the degradation processes of the structural integrity, such as: imbalances of rotating components, crack development, mechanical shocks, rubbing, joints becoming loose, fluid dynamics instabilities, etc.

The vibration analysis technique has been applied by CISE and ENEL in particular to rotating machinery surveillance, with very satisfactory results (see [18] for pump and blowers, and [19] for large turbogenerators).

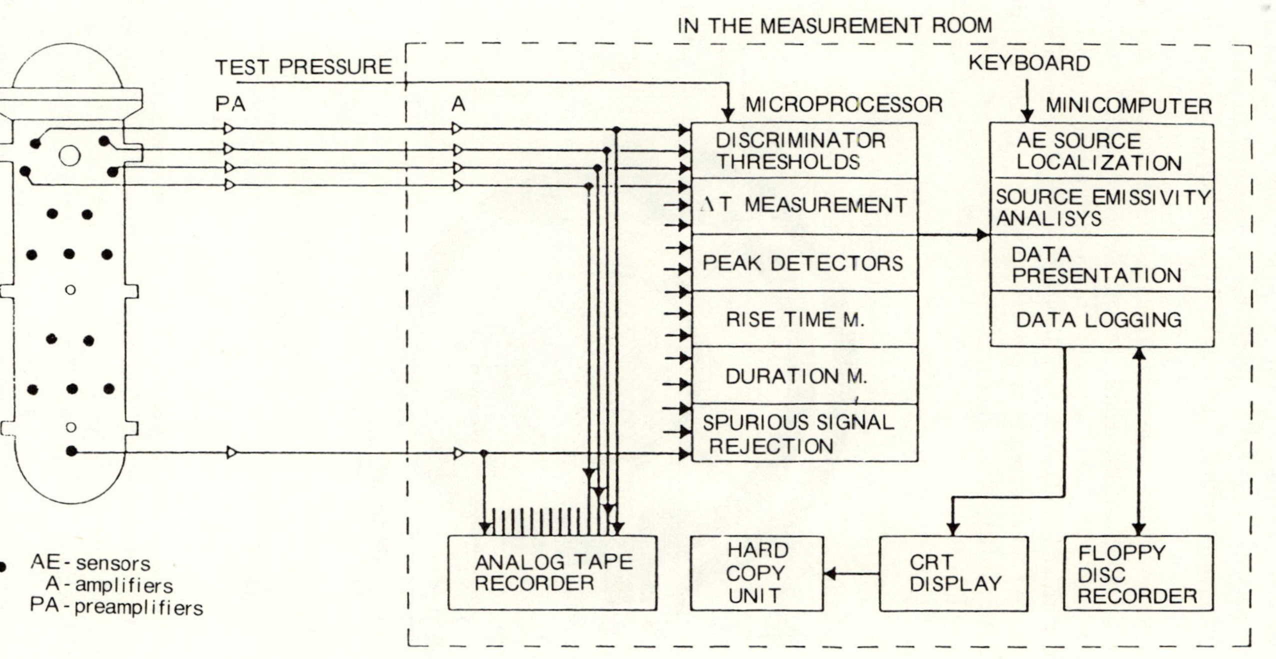

Fig. 7 – Block scheme of typical instrumentation system for AE measurement in pressure component hydrotest.

For large turbogenerators, a specific instrumentation system, termed VIBRO, has been developed. The VIBRO surveillance system uses as primary sensors proximity transducers (mounted in couples on the fixed parts, corresponding to each bearing, as shown in Fig. 8, measuring the shaft movement), oil pressure transducers (measuring the oil pressure in the lubricating meatus of each bearing) and precise transducers of the relative vertical position of each bearing [2]. The signals from these transducers are collected on-line, together with the main plant parameters, and are conveniently analysed to obtain the mean values and the vibrating components values at the rotation frequency and at other frequencies of interest [21].

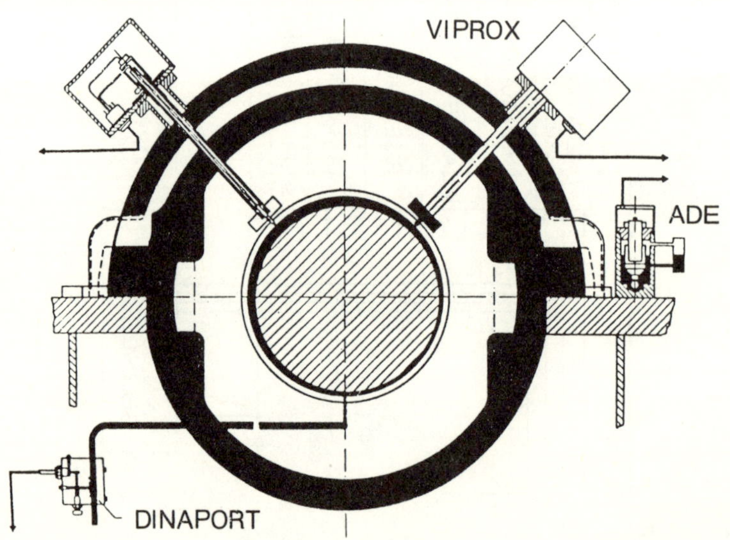

Fig. 8 – Vibro system.

The VIBRO system has been tested on several turboalternators in a large variety of operating conditions. These tests have allowed [20, 22] the formulation of diagnosis procedures for the prompt malfunction identification on the basis of the detected 'symptoms'. Typical examples of correlations between vibration condition and malfunctions are given in Table 2. Several different types of diagnostic intervention on a turboalternator can now be performed, e.g. on the first start-up, on a start-up after a major overhaul, on a turboalternator having vibration problems.

Table 2 – Correlation between shaft vibration and turboalternator malfunction.

Type of shaft vibration observed	Probable malfunction
High shaft vibration at rotation frequency, with marked peaks at critical speeds	Rotor imbalance
Shaft vibration at constant rotation frequency, steadily increasing with time	Rubbing of rotating parts on fixed parts
Marked alteration of the oil film pressure in a bearing	Bearing damage or misalignment of the bearing or an adjacent bearing
Relevant shaft vibration components at frequencies lower than rotation frequency	Oil-whirl instability in an adjacent bearing (if highly loaded)
Occurrence of abnormal vibration components at critical speeds (when rotating at full speed)	Crack development in the shaft

6. LEAK DETECTION

In many industrial plants prompt leak detection is a vital necessity: to limit the quantity of fluid losses, to avoid structural damages, to prevent major failures. One interesting possibility for achieving this prompt leak detection is continuous acoustic monitoring. In fact, the turbulent fluid outflow through cracks, leaking seals, etc., produces a remarkable acoustic emission in a broad frequency band (from some hundred hertz to some hundred kilohertz and even up to megahertz). The acoustic power generated propagates in the whole volume surrounding the leak position according to sound propagation laws. In industrial plants, owing to various favourable circumstances such as the existence of many effective acoustic wave guides (e.g. the pipe walls) and the low intensity of the high frequency spurious noise sources, it is often possible to perform an adequate on-line surveillance on plant components by installing only few acoustic sensors in suitable positions.

A first example of the acoustic leak detection technique relates to the high pressure preheaters of fossil power plants. The technique, called SIBILO, (developed by CISE under ENEL sponsorship) is aimed at the prompt detection of tube through failures. The leak detection is obtained by detecting the pressure

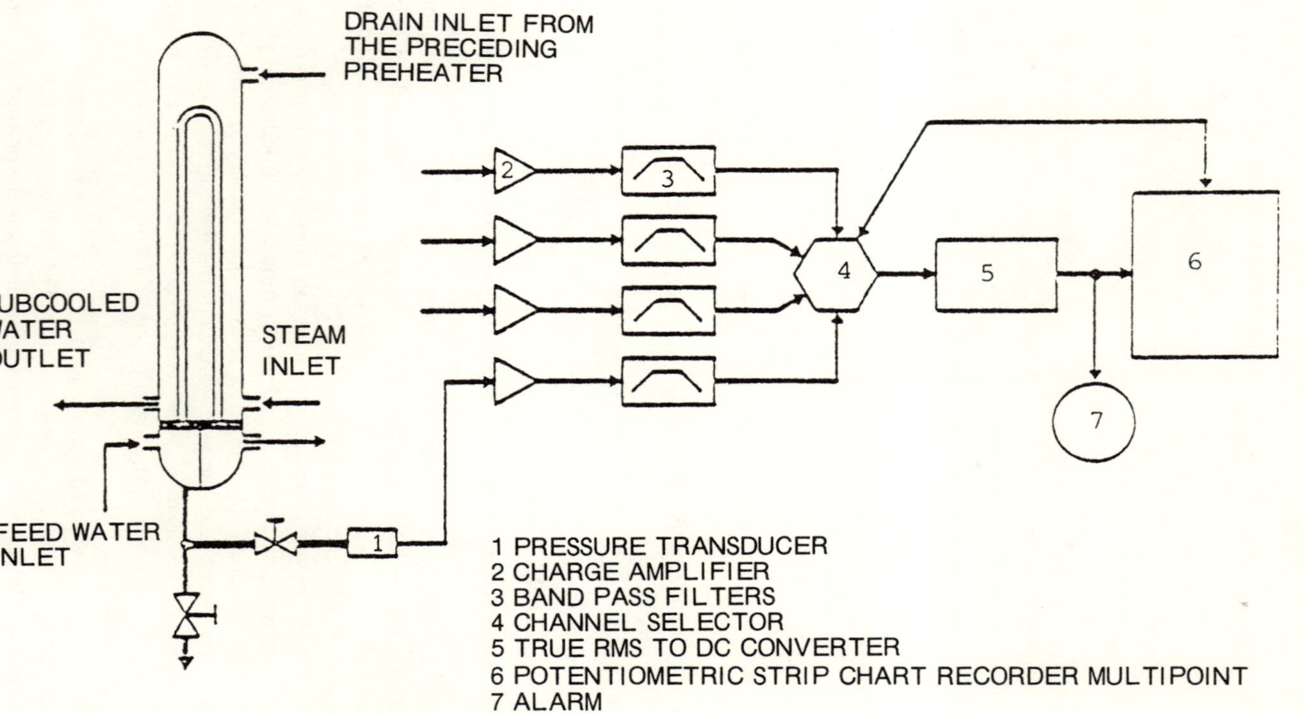

Fig. 9 – Typical block scheme of instrumentation for acoustic surveillance of HP feedwater preheater.

noise generated by the leak in the feedwater volume of the preheater in a suitable frequency band (e.g. 3–15 kHz) (Fig. 9). As surveillance primary sensors, piezoelectric pressure transducers have been chosen, owing to their well known characteristics of ruggedness, sensitivity and dynamic response. These transducers are located as close as possible to the preheaters (one transducer for each preheater), in easily accessible positions. The signal characterisation after amplification and frequency filtering consists simply in the measurement of the rms noise value, then its being compared with an alarm threshold level (Fig. 10).

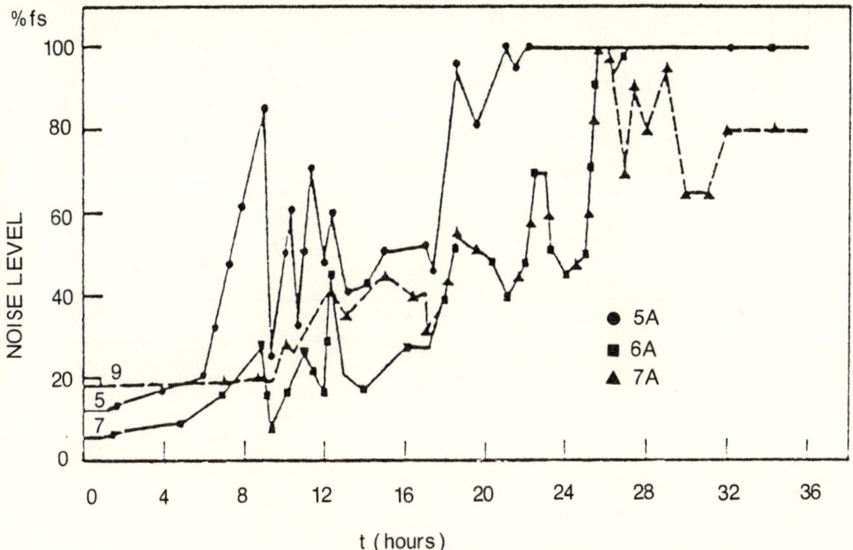

Fig. 10 – Example of signal trends in case of preheater leak occurrence.

The results obtained with this technique, which has been widely adopted by ENEL on its plants, are satisfactory [23, 24] with a remarkable reduction in tube damage and an increase in plant availability and reliability.

A second example of the acoustic leak detection technique relates to boilers of fossil power plants. The technique, also called SIBILO, similar to the one just presented, is aimed at the prompt detection of boiler tube through failures. The leak detection is obtained by detecting the pressure noise alteration in the boiler volume due to the leak development in the frequency band chosen for the auscultation (e.g. 1.5–10 kHz). The primary sensors are piezoelectric pressure transducers installed near boiler windows (Fig. 11). Typically 10–15 pressure transducers are needed to monitor the whole boiler volume (Fig. 12). The only signal characterisation, after amplification and frequency filtering, is the determination of the rms noise value, then comparison with an alarm threshold level. The results obtained with this technique are satisfactory [24].

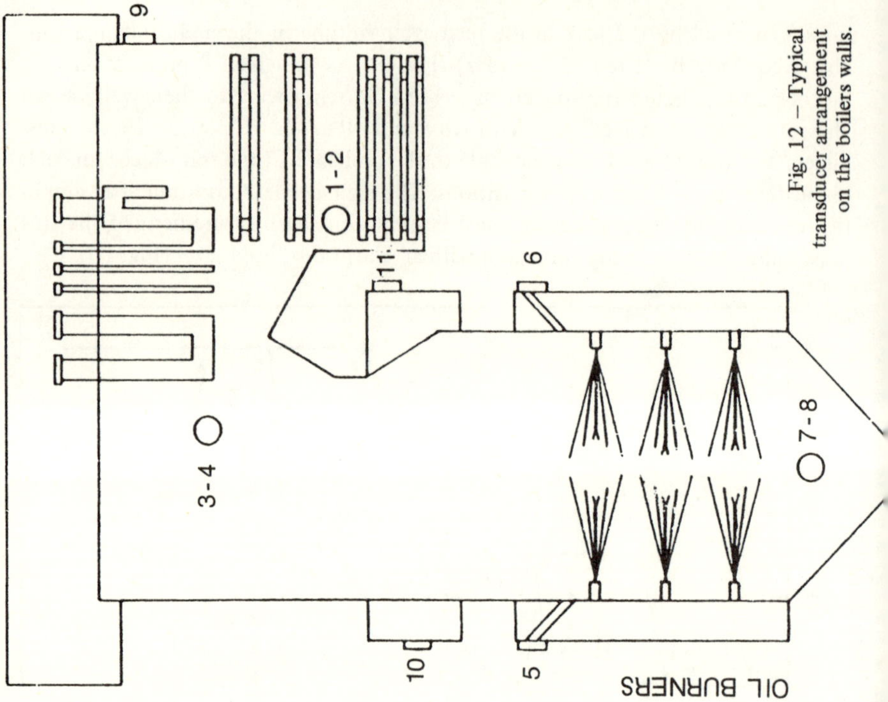

Fig. 12 – Typical transducer arrangement on the boilers walls.

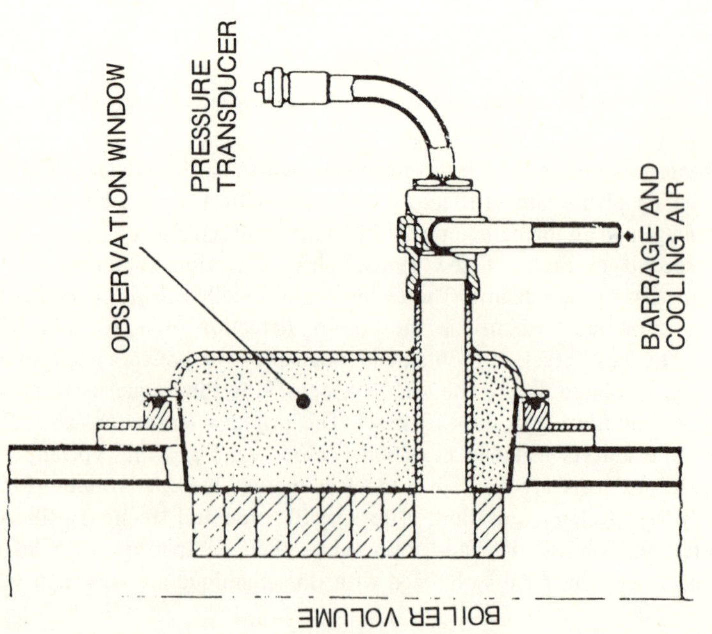

Fig. 11 – Pressure transducer installation on boiler.

A typical indication corresponding to a leak is shown in Fig. 13. The technique has been widely adopted by ENEL in its plants.

To give an idea of the large variety of situations where acoustic leak detection techniques can be applied, it is interesting to mention the steam leak detection from a tube in a sodium steam generator of a fast nuclear reactor [25]. Such a leak is particularly dangerous owing to the explosive sodium–water chemical reaction. On this subject CISE and ENEL have carried out several sets of measurements in collaboration with EdF (at the 'Les Renardières' 50 MW sodium facility, in France). The measurements have evidenced the possibility of leak diagnosis at low frequency (using the noise generated by the sodium–water chemical reaction) and at high frequency (using the noise produced by the steam outflow).

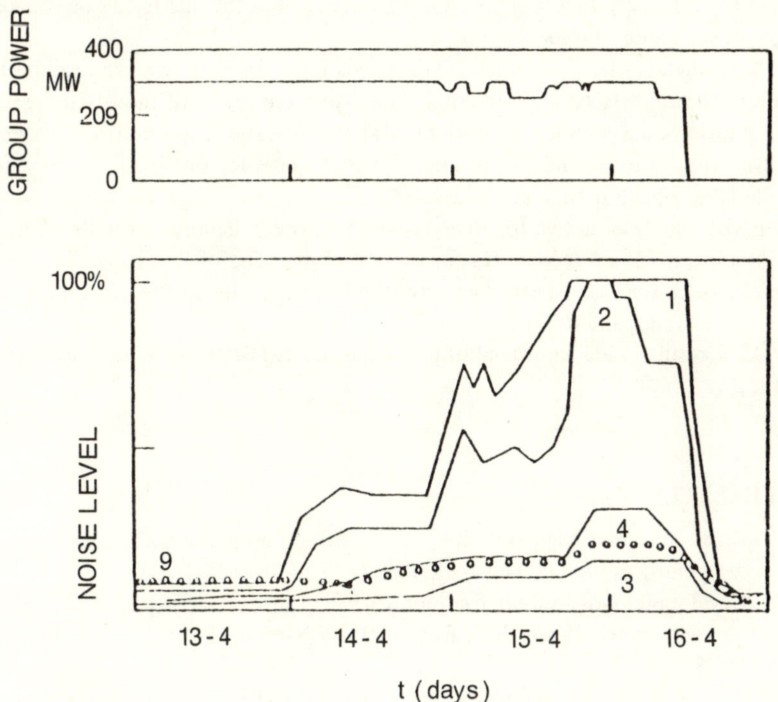

Fig. 13 – Example of signal trends in case of boiler tube leak occurrence.

7. CONCLUDING REMARKS

The optimisation of the management and utilisation of industrial plant constitutes a primary aim at the present time, as a consequence of enhanced safety and reliability requirements.

It is thought that the main tool for attaining this aim may be identified in the extensive adoption of modern on-line surveillance techniques, as a support to the current methods and control procedures, mainly based on non-destructive tests and generalised quality assurance.

Local control of the stress–strain field using strain gauge techniques, early detection of material damages at the microstructural level and surveillance of the defect evolution using acoustic emission, vibration analysis monitoring of the mechanical and service behaviour of the components, prompt identification of through defects or imperfect seal with acoustic leak detectors, all constitute important topics where surveillance techniques have already reached the development level required for wide industrial application.

The availability of powerful and flexible systems based on on-line computers (of relatively low cost), that can be installed in the plants, has been a key factor in such a development.

Nevertheless plant surveillance still needs R&D for improving and validating theoretical supports to experimental data. Also required are industrial system configuration developments, in-field test qualification, promotion of new specific applications, and collection of wider statistics on which to rely for detailed cost–benefit analysis.

ENEL has been active for many years, in close collaboration with CISE, in this area of advanced techniques for on-line monitoring of its plants. Extensive research programs have been sponsored and carried out, in the laboratory, on experimental rigs, and in plants.

As a result, wide industrial applications are expected with significant economical return.

REFERENCES

[1] B. E. Noltingk, High stability capacitance strain gauge for use at extreme temperatures, *IEE Proceedings,* **119,** (1972).

[2] Development of system for monitoring in-service strains in power plant piping, Electric Power Research Institute NP 186, Project 218-1, January 1976.

[3] P. C. O'Neil, The application of CERL PLANER high temperature strain gauges, CEGB/CERL RD/L/N 182/75, November 1975.

[4] S. Ghia, Estensimetri capacitivi per relievi a lungo termine su strutture operanti ad alta temperatura, 5th Congresso AIAS, Bari, September 1977.

[5] M. Schulz, Einsatz kapazitiver Dehnungsmessstriefen für statische Messungen bei hohen Temperaturen, VDI Bericte No. 313, SESA, Munchen, 1978.

[6] Downe *et al.,* Performance and application of the Cert Planer strain transducer, BSSM/SESA Conference, Edinburgh, UK, 1981.

[7] S. Ghia, Reliability of long-term strain-gauges monitoring in high temperature pressurised components, 8th Congress on Material Testing, Budapest, 1982).

[8] F. Cesari, S. Ghia, A. Macchi, S. Menghini and V. Regis, Theoretical and experimental analysis of steam piping for supercritical power plant, 6th SMIRT Conf. Paper L13/4, Paris, 1981.

[9] S. Ghia, D. D'Angelo and V. Regis, Monitoring of high temperature steam piping with capacitance strain-gauges, Int. Conf. on Engineering aspects of creep; paper C193/80, Sheffield, 1980.

[10] H. N. G. Nadley, C. B. Scruby and J. H. Speake, Acoustic emission for physical examination of metals, *International Metal Review*, n. 2, p. 41–64, (1980).

[11] Development of an acoustic emission zone monitor and recorder for BWR pipe cracking detection, Electric Power Research Institute, NP 1407, May 1980.

[12] Vibration signature analysis and acoustic emission monitoring at Brayton Point, Electric Power Research Institute CS-1938, July 1981.

[13] Emission analysis during corrosion, stress corrosion cracking and corrosion fatigue process on type 304 stainless steel, Central Research Institute of Electric Power Institute and Tokyo University, 5th International Acoustic Emission Symposium, Tokyo, November 1980.

[14] G. Airoldi, A533 B and A 508 steels: deformation processes evolution with temperature by means of acoustic emission, 8th European Working Group on Acoustic Emission Meeting, Dortmund, 3–5 October, 1979.

[15] C. Rinaldi and F. Tonolini, A contribution to the problem of detection of subcritical crack growth of structural steels, 10th European Working Group on Acoustic Emission Meeting, Winterthur, 14–16 October, 1981.

[16] C. Panzani, G. Possa and F. Tonolini, CISE experience on acoustic emission monitoring in pressure vessel hydrotests, Symposium Die Shallemission, Bad Nauheim, (FRG), 26–27 April 1979).

[17] E. Fontana, C. Panzani, F. Tonolini and G. Villa, Experience in acoustic emission examination of pressure vessels during hydrotests, 6th International Acoustic Emission Symposium, Susono City, Tokyo, Oct. 31–Nov. 3, 1982.

[18] E. Gadda and A. Vallini, Diagnostica del macchinario per mezzo di rilievi di vibrazioni, Congresso Annuale AEI, Bari, 1980.

[19] G. Lapini, T. Rossini, E. Gadda and A. Benanti, In-service measurements of turbogenerator bearing vertical misalignments, Electric Power Research Institute Workshop on Rotor Forgings for Turbines and Generations, Asilomar Conference Site, Valif., 22 September 1980.

[20] A. Clapis and T. Rossini, Il sistema strumentale ADE per la misura in continua delle variazioni di posizione verticale di punti caratteristici di strutture complesse, Giornate di Studio CISE su 'Nuove tecniche di

Sorveglianza di Apparecchiature e Processi in Impianti Industriali', Segrate, 4 April 1979.

[21] A. Clapis, G. Lapini and T. Rossini, Il sistema strumentalw VIBRO per il monitoraggio di macchinario rotante, Giornate di Studio CISE su 'Nuove Techniche di Sorveglianza di Apparecchiature e di Processi in Impianti Industriali', Segrate, 4 April 1979.

[22] A. Clapis, G. Lapini and T. Rossini, Instabilità del film d'olio in un cuscinetto di un turboalternators da 320 MW: diagnosi e analisi delle cause e soluziones del problema, A. I. Man., X Congresso Manutenzione e Management, Trieste, 10–12 November 1981.

[23] F. Dallavalle, G. Possa and F. Tonolini, Early leak detection in power plant feedwater preheaters by means of acoustic surveillance, Symposium On-line Surveillance and Monitoring of Process Plant, London, 26–28 September 1977.

[24] G. Parini and G. Possa, New acoustic techniques for leak detection in fossil fuel plant components, EPRI Workshop on Incipient Failure Detection for Fossil Power Plant Components, Hartford, Connecticut, 25–27 August 1982.

[25] A. Clapis, D. Scandolo, V. Regis and R. Rappini, Hig frequency noise measurements during CNEN/NIRA steam generator testing at Les Renardières, Topical Meeting on LMFBR Sream Generators, Genova, 30 November–2 December 1981.

CHAPTER 16

The use of a local on-line mathematical model to optimise the performance of a 1500 MW power station

K. Riley

1. INTRODUCTION

In 1981/82 almost 60% of the cost to the CEGB of generating electricity was attributable to the fuel bill (£3,797M) [1]. Thus it is clear that major savings can be made if thermal efficiencies can be raised by even a small amount. For a 500 MW generating unit, an increase in thermal efficiency of just 0.1 percentage points reduces fuel costs by £125K a year. Equally, the cost of replacing electricity lost due to an unplanned breakdown on a high efficiency 500 MW unit with that from a less efficient station is £30K per day, in addition to the repair costs. Hence, it is a continuing objective of the CEGB to minimise fuel burn through maintaining or increasing plant cycle efficiency whilst at the same time maintaining plant availability.

At the design stage of a station, full consideration is given to thermal efficiency for a notional design condition. However, in practice a generating unit frequently runs off its design point. Controlling the unit under these conditions requires continual operational compromises to be made in the light of changing constraints.

Such compromises are made using a great deal of skill and detailed plant knowledge, but are often made with incomplete information. As a result, the operation of the plant at any time is not necessarily optimal. Since small changes in thermal efficiency produce large savings in fuel costs, then optimising for maximum thermal efficiency within existing but varying constraints is essential. In addition, overall system efficiency is also affected by the availability of particular generating units. Hence maximising the availability of high merit plant is a further important consideration.

Reports have begun to appear in the literature recently of other work in this area. Sonnenschein [2], for example, discusses the application of a whole plant model to the thermodynamic design of a complete power station and shows the value of such a tool to the designer in investigating the optimum arrangement of plant within a system.

The application of a boiler model to the continuous monitoring of boiler performance is described by Heil *et al.* [3]. The use of the model to optimise sootblower usage illustrates the potential performance improvements which can result from the availability of an on-line plant model.

The purpose of this paper is to demonstrate how a local on-line mathematical model of a complete 500 MW generating unit can be used to optimise the day-by-day running of a complete generating unit and to monitor the performance of plant components against operational or design targets. It also indicates how the model can be used as a tool to influence the development of optimum maintenance and operational strategies. The development of a prototype system which integrates the plant model into a local on-line performance optimisation tool suitable for commercial use is broadly described, with additional details of its implementation where appropriate. However, this chapter does not attempt to show how effective improvements can be made through plant modifications, although the model has been successfully used in this respect.

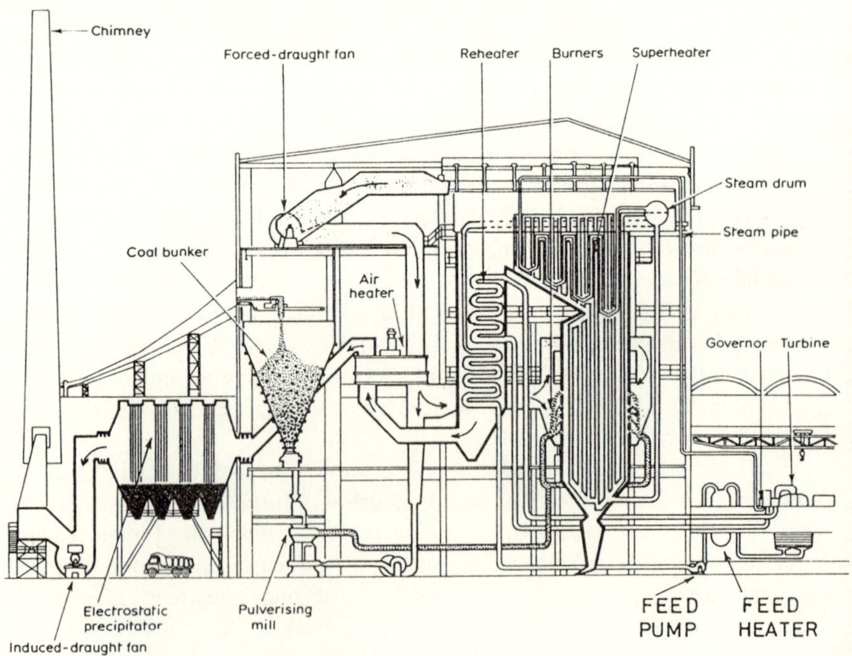

Fig. 1 – Simplified layout of generating uunit.

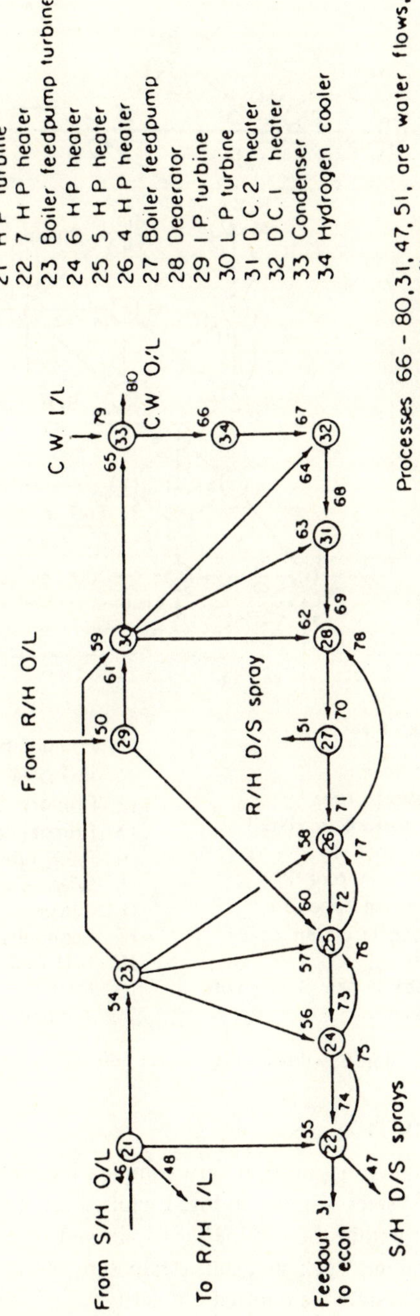

Fig. 2 – 500 MW turbine and feed train: process flow diagram.

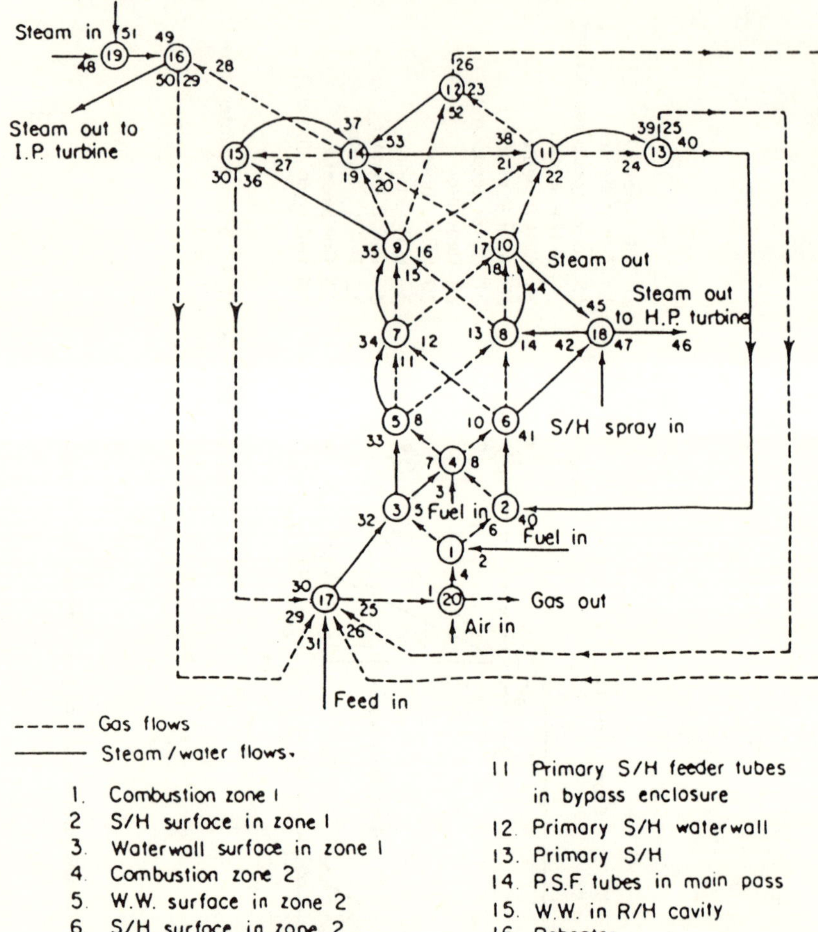

Fig. 3 – Process diagram for 500 MW boiler.

2. PLANT DESCRIPTION

A large modern generating unit is a very complex interactive system. A simplified picture of the system is shown in Fig. 1 whilst Figs. 2 and 3, which show the main process flows within the turbine/feed train and boiler of a 500 MW fossil-fired unit. Minor items such as gland steam supplies to the turbine and the feed heater cascade system are omitted for clarity.

Steam is raised in a sub-critical drum boiler with single reheat. Within the boiler are a number of heat transfer surfaces across which the hot furnace gases flow. A diagram of the disposition of the heat transfer surfaces relative to the main gas flow is shown schematically in Fig. 5.

The main turbine consists of HP (high pressure), IP and LP cylinders. Steam from the HP turbine exhaust is returned to the boiler for reheat. At various stages bled steam is fed to a set of high-pressure tubed heaters which raise the feed water temperature to within about eighty degrees of saturation.

The complete flow system is very complex and there is a considerable degree of interaction between constituent components. This in intself is an indication of the difficulty of identifying shortfalls in component performance in isolation and of the scope for performance optimisation.

It is clear that predictions of the behaviour of such complex plant needs to be done within a systematic framework. A mathematical model provides this framework, but if it is to provide realistic performance predictions it also needs to be on-line.

The need for the model to be on-line is demonstrated by the variability of the plant configuration due to the need to take defective plant out of service and by the varying nature of some of the major boundary conditions. Thus if one of the feed water heaters is out of service the temperature profile of the steam within the system will change, as well as the load. An important feature of a fossil-fired boiler is that the heat transfer surfaces become fouled by ash deposits. This fouling, which is both variable in extent and a function of the fuel supplied and the operational regime of the boiler, also brings about a change in the load and the steam temperature within the system.

Thus it is necessary for any plant model which aims to give realistic performance predictions to be on-line to the plant.

3. THE MODELLING SYSTEM

The mathematical model which has been developed for this work has been reported previously [4]. The model, KRAB1, is a one-dimensional steady state mathematical model. It can be run in either a predictive or a plant fitting mode and thus can follow changes in boundary conditions and plant configuration as they occur. It is written in BASIC and runs on a PDP-11 computer in 32K words of memory. Typical run times for a performance prediction are currently 30 secs, although the production version will run in about 3 secs. The language BASIC was chosen mainly for two reasons:

(1) it was accessible to plant engineers and did not require substantial software expertise to be understandable or usable, and
(2) it was compatible with a CEGB real-time plant logging and contrrol language, SWEPSPEED 2.

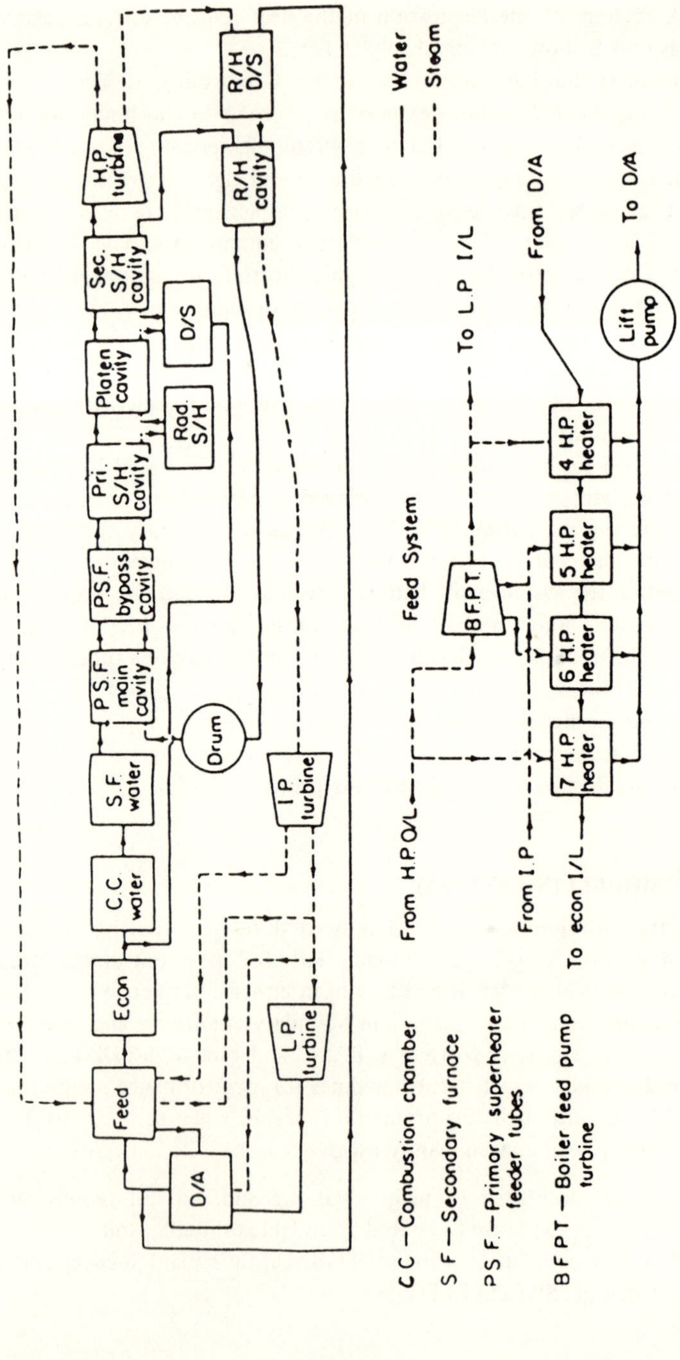

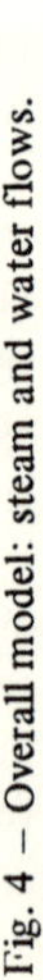

Fig. 4 – Overall model: steam and water flows.

It was felt to be important that the software should be easily maintained and developed by non-specialist site users. This would ease the burden of support on the central facility whilst at the same time making the tools available to the people with the most immediate knowledge of the sources of any performance shortfall and of the potential or performance improvement.

The component parts of the plant model are in the form of a series of plant modules, each of which represent mathematically one particular item or type of plant with appropriate input and output streams. The user provides a specification of the overall model required in the form of two files:

(1) A topology description; this is a mathematical translation of Figs. 2 and 3 and defines the number and type of connections between the individual components of the model, and
(2) A specification of data values for each of the plant modules (e.g. tube surface areas, number of turbine stages etc.).

From this problem description a set of software utilities generates a specific plant model with all the necessary stream interconnections. A major software design aim is that the generated model should be data driven. Hence if changes to the model structure or plant data items are needed, they can be made whilst the model is on-line. A possible translation of Figs. 2 and 3 into a model is shown in Figs. 4 and 5.

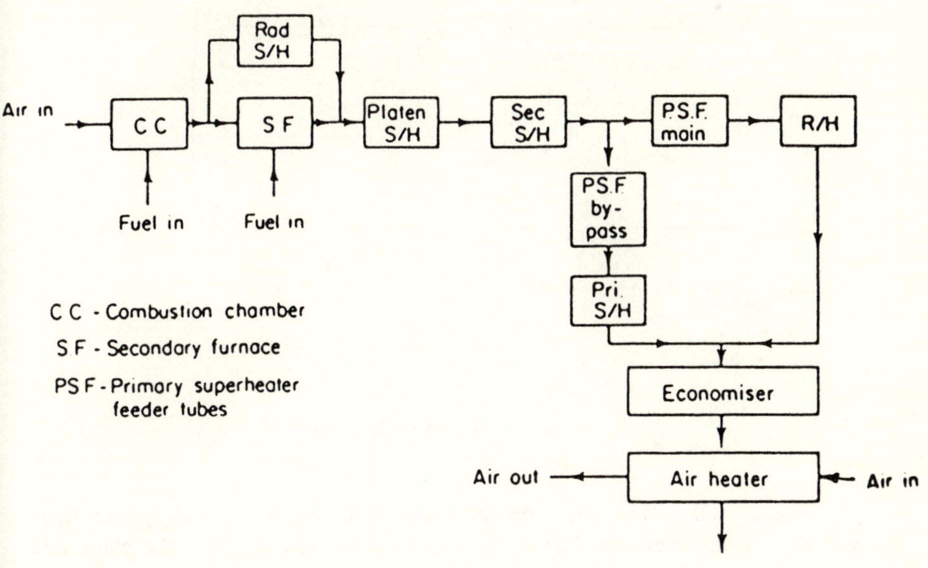

Fig. 5 – Boiler model: gas flows.

Whilst the plant model is responsible for carrying out the actual performance calculations, it is the job of the software support system to integrate this into an effective plant monitoring tool. Fig. 6 indicates how the overall plant modelling system can be arranged.

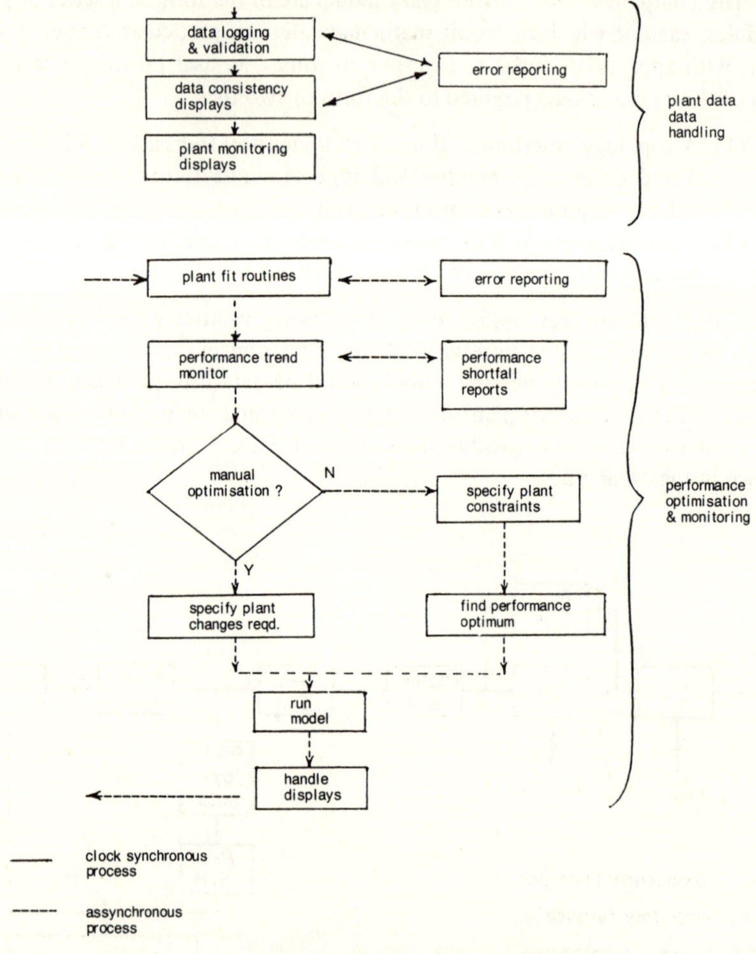

Fig. 6 – Plant modelling system.

The overall system must be capable of handling both clock-synchronous and asynchronous events. The clock-synchronous events, such as the plant data acquisition and handling routines, occur at regular clock-controlled intervals. The asynchronous events are mostly associated with the processes which control the search for a performaance optimum. These events occur at an essentially arbitrary rate. One of the system design decisions to be made is how these two types of events will be handled and whether they will be handled by one or more machines, since if the problem of contention for c.p.u. time is not handled correctly the resulting conflicting demands on the system can cause a system lock-up. In the prototype system this problem is resolved naturally since separate machines already exist to handle the plant data acquisition functions.

4. INSTRUMENTATION AND DATA LOGGING

All large power stations within the CEGB are equipped with a substantial coverage of monitoring and control instrumentation aimed principally at all the major process variables but including many of the subsidiary ones. For a plant modelling application the data requirements are not large, typically requiring less than 100 channels of data. Most of this data, in the form of pressures, temperatures and flows, is available on the Station system (600–1000 channels of data per unit). Where this data is not available extra instrumentation is usually installed to normal contractor standards.

A problem with a lot of the plant monitoring signals which are available is the presence of a high common-mode electrical noise level. This can have a serious effect on the quality of the plant data. In addition, instrumentation faults occur from time to time and it is the job of the logging software to detect these conditions and to carry out the first level of data validation as well as the data logging and conversion. As the quality of the model predictions directly reflects the quality of the plant data, significant effort is spent on the data validation software. This is effort well spent since the effect of the resulting poor predictions or a system crash due to bad data can be a severe loss of credibility of the system, resulting in a loss of the benefits that the system can bring.

5. INITIAL RESULTS FROM A FEASIBILITY STUDY

A study of the feasibility of installing and operating an on-line local plant monitoring system of the type described and the benefits to be obtained has been carried out on a 500 MW coal-fired unit. The system was installed on a PDP-11 but the necessary data was obtained from normal control room commercial instrumentation and entered into the machine through a keyboard. Despite the low accuracy of the instrumentation used and the frequent errors in the data which was entered into the machine, the system performed very well. A consistent fit with the running plant was invariably obtained. The resulting predictions made by the model closely matched the state of the plant after the relevant plant changes had been made.

A typical set of results obtained from one such test is shown in Tables 1 and 2 [4]. In this case the metal temperatures at the outlet of one of the superheater

Table 1 – Data required for model fit: initial and final values with predicted results.

Location	Initial value	Final value	Predicted value	Error Col. 4 – Col. 3
Pri s/h o/l steam tmp. (°C)	381	376	375	−1
Rad. s/h o/l steam temp. (°C)	387	381	378	−3
Platen s/h o/l steam temp. (°C)	511	497	493	−4
Sec s/h i/l steam temp. (°C)	450	464	460	−4
Sec s/h o/l steam temp. (°C)	568	568	568	0
Reheater i/l steam temp. (°C)	381	378	378	0
Reheater o/l steam temp. (°C)	567	542	535	−7
Econ. i/l feed temp. (°C)	206	206	208	+2
Econ. o/l feed temp. (°C)	290	279	275	−4
Econ. o/l gas temp. (°C)	301	288	288	0
Pri s/h o/l gas temp. (°C)	573	557	547	−10
Total air flow (kg/s)	403	363	363	0
Steam flow (kg/s)	252	252	259	+7
Reheater d/s spray flow (kg/s)	0	0	0	0
Drum pressure (bar)	167	171	171	0
S/h pressure drop (bar)	15	14	15	−1
Temp. of air to burners (°C)	280	272	275	+3
No. of oil burners	12	12	12	0
No. of P.F. burners	20	20	20	0
Boiler feed pump turbine i/s?	Yes	Yes	Yes	

Table 2 – Some additional results.

Location	Initial value	Final value	Predicted value
Load (MW)	298	301	305
S/h d/s spray flow (kg/s)	17.7	11	11.4
Excess air (%)	24[a]	–	9
R/h bypass flow (kg/s)	45[a]	–	45
Thermal efficiency (%)	35.0	35.3	35.6

[a]Calculated

sections (Platen) was unacceptably high and causing a load limitation. The data shown in Table 1 was fitted by the model which was then used to explore ways in which the particular temperature could be reduced. In the event, it was predicted that the temperature could best be reduced by simply reducing the total air flow which was supplied to the boiler. On carrying out this air flow reduction it was found that the predictions were substantially correct. In addition, the load increased (by 3 MW) as did the thermal efficiency (by 0.3%). (The mismatch is due to the fact that the plant was not allowed to reach its steady state owing to loading requirements.)

Table 3 – Further improvements to plant performance from moving R/H bypass dampers.

Location	Initial value	Final value	Optimised value	Change Col. 4 – Col. 2
Pri s/h o/l steam temp. (°C)	381	376	368	−23
Rad. s/h o/l steam temp. (°C)	387	381	369	−18
Platen s/h o/l steam temp. (°C)	511	497	470	−41
Sec s/h i/l steam temp. (°C)	450	464	460	+10
Sec s/h o/l steam temp. (°C)	568	568	568	0
Reheater i/l steam temp. (°C)	381	378	378	−3
Reheater o/l steam temp. (°C)	567	542	568	0
Econ. i/l feed temp. (°C)	206	206	207	+1
Econ. o/l feed temp. (°C)	290	279	274	−16
Econ. o/l gas temp. (°C)	301	288	287	−14
Pri s/h o/l gas temp. (°C)	573	557	484	−89
Total air flow (kg/s)	403	363	363	−37
Steam flow (kg/s)	252	252	255	+3
Reheater d/s spray flow (kg/s)	0	0	0	0
Drum pressure (bar)	167	171	171	+4
S/h pressure drop (bar)	15	14	15	0
Temp. of air to burners (°C)	280	272	275	−5
No. of oil burners	12	12	12	0
No. of P.F. burners	20	20	20	0
Boiler feed pump turbine i/s?	Yes	Yes	Yes	–
Load (MW)	298	301	306	+8
S/h d/s spray flow (kg/s)	17.7	11	3.7	−14
Excess air (%)	24[a]	–	9	−15
R/h bypass flow (kg/s)	45[a]	–	25	−25
Thermal efficiency (%)	35.0	35.3	35.7	+0.7

[a]Calculated.

The model indicated that more could be achieved by also adjusting the fraction of the gas flow which bypassed the main reheater gas pass. In doing this it was predicted that the particular temperature which was causing the problem could be reduced by about 40 degrees Celsius. In addition the load would further increase as would the thermal efficiency, with indicated savings equivalent to £0.5M per annum for that unit.

A substantial number of such tests were carried out over a period of six months. The results were consistently good, indicating that potentially substantial efficiency and operational improvements could be made and justifying the extension of the work to the implementation of a prototype commercial system.

6. DESCRIPTION OF THE PROTOTYPE SYSTEM

The prototype commercial system is being designed to be able to monitor the performance of two 500 MW units. The major design decisions to be made were:

(1) how to handle the synchronous data logging demands as well as the synchronous optimisation requirements, and
(2) how to communicate with the user effectively and efficiently considering the large quantity of information which is available.

The first problem was solved by deciding to separate the data logging functions from the modelling functions. The plant data handling (clock-synchronous) events are carried out on the two separate computer loggers which are already carrying out that function as part of the normal commercial instrumentation.

The modelling and control activities are resident in their own small machine (PDP-11/23). Inter-processor communications software allows the modelling computer to interrogate the plant logging computers when data is required. The man-machine interface is handled through the use of a touch-sensitive colour VDU. With this device, system software can define arbitrary areas of the screen as being touch-sensitive. The user can thus control the progress of the optimisation process and the display of plant data in an efficient manner and without the need for a keyboard by simply touching the relevant part of the screen. The data displayed will enable the user to discover the effects of potential optimisation schemes as well as the current state of the two units.

7. FUTURE DEVELOPMENTS

The short-term development of the system is aimed at securing the cost reductions obtainable from optimisation of thermal performance. The first phase of this development will aim at:

(1) the provision of the basic monitoring system with manual control of the optimisation process followed by,
(2) the installation of an automatic optimising routine.

However, other uses for a whole station model are also being pursued which it is expected will bring substantial additional benefits:

(1) Optimisation of maintenance strategies based upon the early warning of declining performance in specific plant components.
(2) The global optimisation of control system set-points. Control systems are usually designed to drive a target variable towards a predetermined set-point value on the basis of the values of a small number of varying parameters. The value of the set-point is determined only within the context of the local environment of the control action and not of the whole system. Thus although a particular control-system set-point may be optimal with respect to its local environment, it is not necessarily optimised with respect to the whole system.

    The up-dating of individual control-sytem set-point values from an on-line model of the complete plant should bring additional cost reductions through increased exploitation of the plant capabilities.
(3) The development of a management information system which will enable future plant development and operational strategies to be investigated.

In the medium term. the system will be developed to generate information which will form the basis of optimised maintenance schedules (although this can take place to some extent in parallel with the initial phases). By comparing the actual performance of plant components against design expectation any shortfalls can be identified at an early stage. This should substantially reduce availability losses as well as maintaining optimum performance.

The final phase would see the incorporation of additional factors, such as the trade-off between the benefits from additional short-term load generation and the longer-term costs of any damage resulting from the easing of load-limiting plant constraints, into the optimisation process. This would produce a very powerful management tool which could be used by senior management for the investigation of alternative overall operating strategies.

ACKNOWLEDGEMENTS

This paper is published with the permission of the Director General of the South-West Region of the Central Electricity Generating Board.

REFERENCES

[1] CEGB Annual Report & Accounts, 1981–82.

[2] H. Sonnenschein, A modular optimizing calculation method of power station energy balance and plant efficiency, *Trans. ASME, J. Eng. Power*, **104**, 255–259 (1982).

[3] T. C. Heil, R. M. Nethercutt and S. A. Scavuzzo, Boiler heat transfer model for operator diagnostic information, Joint ASME/IEEE Power Generation Conference, October 4–8, 1981, Paper 81-JPGC-PWR-14.

[4] K. Riley, Whole-station thermal performance prediction, Part 1, Modelling the boiler, Power Industry Research **1**, 201–218 (1981).

CHAPTER 17

The vibratory surveillance of the turbogenerator sets at Electricité de France

J. P. Fanton

1. THE DEVELOPMENT OF THE SURVEILLANCE METHODS

The vibratory surveillance methods for the turbogenerators are at present developed on a large scale at Electricité de France. The turbosets of all nuclear power plants are equipped with adequate devices and at the same time some of the older classical power plants are also equipped.

The reasons for such developments are twofold; they are linked as much to the machines' behaviour as to the progress of instrumentation techniques. We shall specify successively these two converging tendencies.

1.1 Reasons related to the machines

The turbogenerator sets which come into service today, whether they operate at 3000 rev/min or 1500 rev/min, are very large machines. Furthermore the trend to increase the size of such machines is still economically justified.

The notion of size being important only makes sense, obviously, by comparison with other sets of machinery that can be met throughout the industry, and also when the limits of the manufacturing technologies used are considered.

The available computational techniques for predicting the dynamical behaviour of the machines are capable of increasingly more (determination of stresses, fatigue, stability, imbalance response). They allow us to come still closer to the construction limits and to search for optimization.

The mechanical and thermal dimensioning of the turbosets being so accurate, their exploitation becomes more critical. On the other hand, the level of their unit power renders their outbreaks all the more unlooked-for.

These general reasons mean that the implementation of surveillance methods with a predicting capability may have a wide interest today.

Vibratory surveillance is, properly speaking, more a surveillance 'by the means of the vibrations' rather than a surveillance of the vibrations themselves. This means that what we try to detect are vibratory events, and that these vibratory events do not constitute by themselves infringements of the control limits assigned by the manufacturer. Vibratory events include slow evolutions, fast variations, alterations in a vibratory signature. These events can, however, provide a lot of information on the mechanical state of health of the machines [1].

1.2 Reasons related to technical evolution

The progress of technical and theoretical knowledge also leads to a more systematic surveillance of machines. First, the capability of the numerical models for the turbogenerator sets dynamics allows us to predict the expected efficiency of a surveillance criterion and in due course to optimize it; for instance in the case of a simulated transverse crack. Secondly, the modern technical means available in the field of instrumentation permit us to reach the required performance: analogue or numerical vibration analysers, large-scale microcomputer systems, etc.

2. OFF-LINE SURVEILLANCE OF TURBOGENERATOR SETS – THE PRESENT STATE

2.1 Basic principles

According to the complexity of the problem, and especially of the difficult interpretation of the vibratory data in terms of the mechanical condition of the machines, the surveillance of the turbosets of Electricité de France has, until recently, been organised for off-line exploitation.

It follows a constant organisation scheme, but in some power plants some technical elements may be slightly different.

The functional operation is assured regularly by the production services, and consists in the exploitation of a data acquisition system on one hand, and of data processing software, on the other [2].

The acquisition system operates permanently, while the treatment software is executed only periodically. The nature of the results is such that their interpretation, though made more easy, remains necessary. The remedial action of the surveillance process on the machine always passes through the decision of a human operator.

The general organisation is represented in Fig. 1.

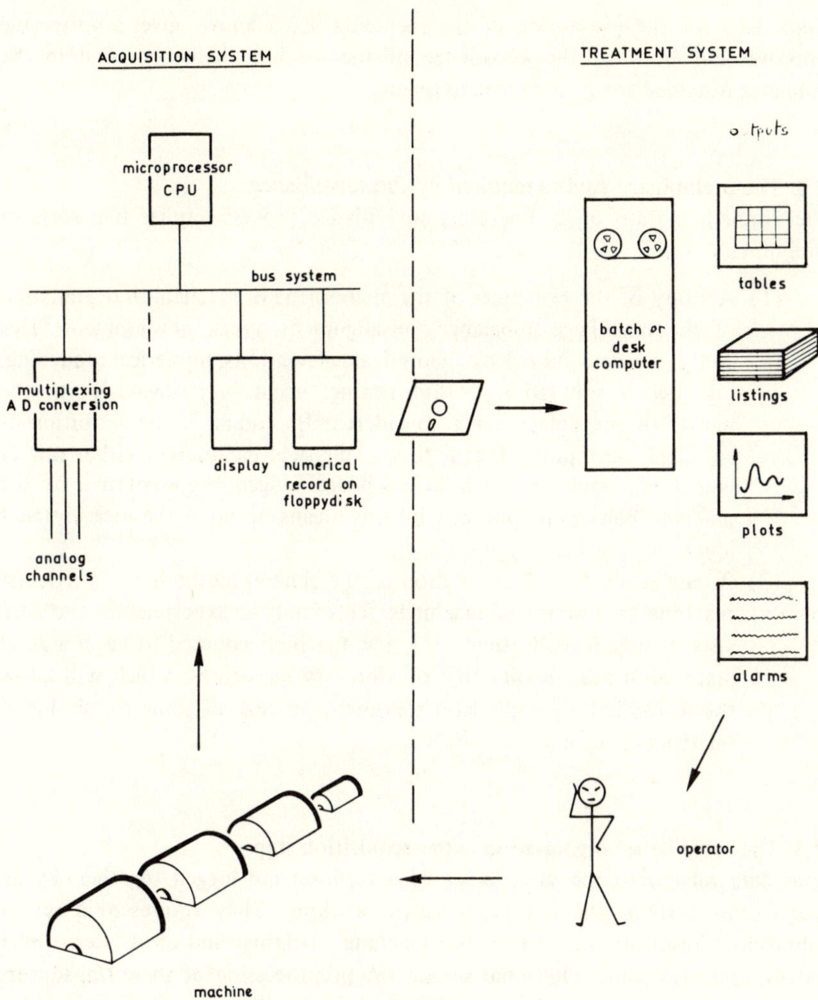

Fig. 1 – Surveillance process block diagram.

2.2 Recall of the first Electricité de France objectives

The first motivation for surveillance implementation at Electricité de France has been to meet requirements of nuclear safety rules. The same surveillance installed in classical power plants may lead to similar benefits.

The aim is mainly to prevent missile emission from the turbosets; this could result either from a machine overspeed caused by a defect of the steam inlet valves or from a crack in the shaft line [3]. In addition, the practical dispositions used for the prevention of the incidents listed above, gives a noticeable improvement both in the knowledge of the machine behaviour and in the guidance obtained for predictive maintenance.

2.3 The preliminary studies required by the surveillance

The implementation of an operating surveillance process implies two sorts of preliminary studies.

(1) A study of the principles of the phenomena of mechanical degradation of the machines; imbalance, misalignment, crack development. This study can be achieved by theoretical calculations, numerical modelling, test bench simulation or life testing, according, obviously, to the nature of the defect being considered. Its ending is the definition of so-called descriptors, that is to say, physical parameters, either raw or issued via signal processing, which are judged representative of the machine behaviour; one can by this means optimise the measurement locations, the analysis process, etc.

(2) A case study for the adaptation of the general methods to a particular machine or a series of machines. This study is experimental and consists mainly in collecting, on a new machine assumed to be in a good state of health, a quantity of vibratory signatures, which will allow the introduction, in the later treatment process, of some thresholds of relative evolution.

2.4 The surveillance organisation – the acquisition step

The data related to the surveillance of a turboset are logged together, by an acquisition system which is permanently working. They respresent a set of vibration transducers located on the machine's bearings and on its steam inlet valves, and also some additional sensors. A preprocessing of these transducers leads to the acquisition of about one hundred parameters.

The preprocessing for the shaft line sensors is a synchronous analysis: the main harmonic components of the signal are determined with respect to the rotary speed. Table 1 summarises the different parameters involved.

The data collected are recorded in numerical form on a floppy disk and through a data compression process. This results in a far superior autonomy compared to an analogue magnetic tape recording, with a quite similar observation accuracy.

The acquisition system is designed around a microprocessor CPU. This processor drives a set of peripheral cards through a logical bus system. This principle confers to the system a moderate cost and the capability of configuration adaptation to the different power plants.

Table 1 – Surveillance parameters selected as descriptors.

(1) *Steam valves*
- *overall vibration level* (rms) determined by an axial measurement accelerometer + charge amplifier,
- closure time of cut out valves,
- position of control valves.

(2) *Shaft line*
- overall vibration level (rms),
- synchronous components:

$$\left.\begin{matrix} X_{\Omega} \\ Y_{\Omega} \end{matrix}\right\} \left\{\begin{matrix} A_{\Omega} \\ \phi_{\Omega} \end{matrix}\right. \qquad \left.\begin{matrix} X_{2\Omega} \\ Y_{2\Omega} \end{matrix}\right\} \left\{\begin{matrix} A_{2\Omega} \\ \phi_{2\Omega} \end{matrix}\right.$$

$$\left.\begin{matrix} X_{3\Omega} \\ Y_{3\Omega} \end{matrix}\right\} \left\{\begin{matrix} A_{3\Omega} \\ \phi_{3\Omega} \end{matrix}\right. \qquad \left.\begin{matrix} A_{\Omega_n/2} \\ A_{4\Omega n} \end{matrix}\right\} \text{ scalars}$$

(3) *General condition*
- rotary speed Ω (nominal value Ω_n),
- active and reactive power,
- condenser pressure,
- rotor current.

2.5 Information processing

The data treatment is processed every week, either on a batch computer or on a dedicated computer.

The software organisation is based on independent modules, so that the experience can be incorporated. It includes the capability of constituting very long-period archives.

Each software module corresponds to the detection of a particular mechanical defect and results in adequate paper outputs: tables, diagrams, or alarm messages.

A conversational exploitation is possible to make easier for the control staff the access to all the information; a general graphics capability is also provided.

The treatment process separates observed periods into two main cases: the steady states (constant output power) and the variable ratings (rundowns, load variations); the detections operated depends on these possible situations; they are summarised in Table 2.

CENTRALE: ***	VIBRATIONS GTA LE: 2 octobre 1983 à: 23 heures 6 minutes 0 seconde	
TRANCHE: *	PUISSANCE ACTIVE BRUTE: 486 MWatts	REACTIVE: Pas de mess

	EDITION:
	Demandée
*	Etat hebdomadaire

HP MP BP1 BP2 BP3 ALT

0 1 2 3 4 5 6 7 8 9 10 11

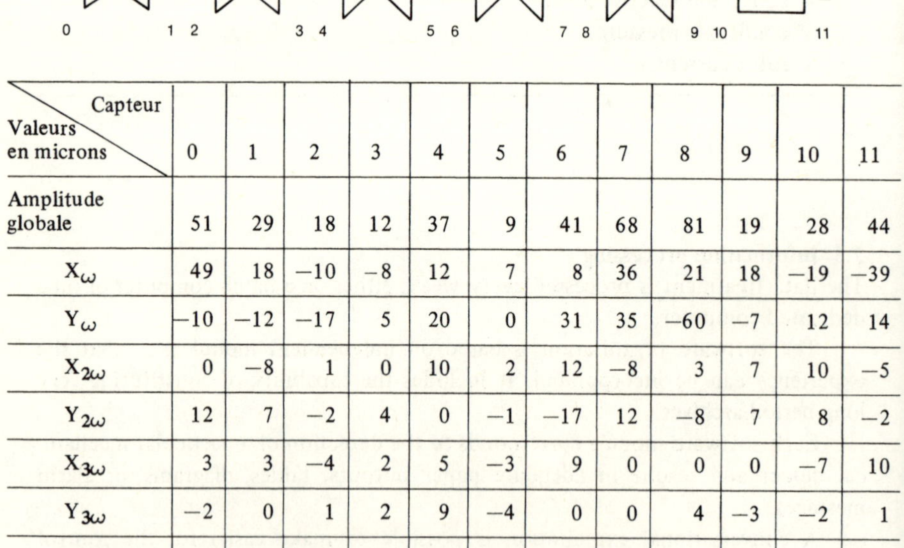

Capteur / Valeurs en microns	0	1	2	3	4	5	6	7	8	9	10	11
Amplitude globale	51	29	18	12	37	9	41	68	81	19	28	44
X_ω	49	18	−10	−8	12	7	8	36	21	18	−19	−39
Y_ω	−10	−12	−17	5	20	0	31	35	−60	−7	12	14
$X_{2\omega}$	0	−8	1	0	10	2	12	−8	3	7	10	−5
$Y_{2\omega}$	12	7	−2	4	0	−1	−17	12	−8	7	8	−2
$X_{3\omega}$	3	3	−4	2	5	−3	9	0	0	0	−7	10
$Y_{3\omega}$	−2	0	1	2	9	−4	0	0	4	−3	−2	1
$A(\omega/2)$	7	5	0	0	0	0	0	0	0	0	0	0
Température	69	81	57	61	70	75	63	68	67	73	84	59

Fig. 2 – Sample of surveillance process output for the shaft line's steady state.

Table 2

Case	Component		
Steady states	Shaft line	(H_1)	Vibration analysis (see Fig. 2).
		(H_2)	Trend computations.
		(H_{14})	Vibration discontinuity detection.
	Steam valves	(H_{10})	Excessive permanent vibration levels.
		(H_9)	Valve stem rupture.
		(H_5)	Reliability indicators computations number of manoeuvres – total operating durations.
Variable ratings	Shaft line	(H_3)	Rundown frequency analysis. Nyquist plane diagrams.
		(H_{13})	Computation of machines abnormal sensitivities (to power change, intensity, condenser pressure.)
	Steam valves	(H_{11})	Global untightness detection.
		(H_6)	Excessive transient vibration levels.
		(H_8)	rubbing detection by closure times evolutions.

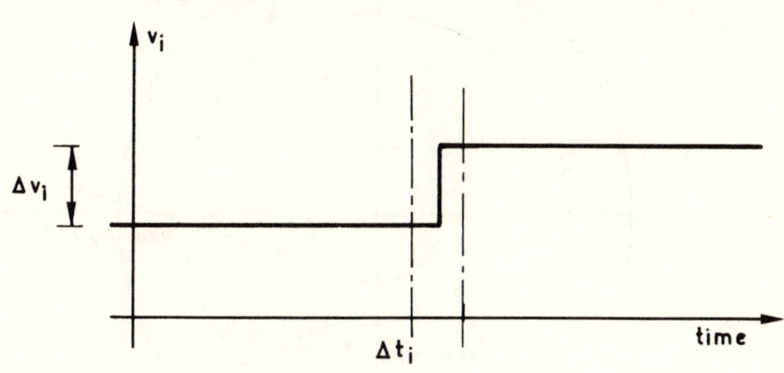

$$\begin{cases} \Delta t_i < \Delta t_{ref} \\ \Delta v_i > \Delta v_{ref} \end{cases}$$ any fast and noticeable variation (increase or decrease)

Detection applied to: $\begin{cases} X_{j\omega} \\ Y_{j\omega} \end{cases}$ $j = 1, 2, 3$

(ω revolution speed)

Fig. 3 – Abnormal changes in vibration, blade departure-like.

Three of the detection criteria listed above are more precisely described in Figs. 3, 4 and 5 which give examples of enhanced detection

- fast changes in vibration similar to blade departure,
- abnormal sensitivities of the machine to the control parameters,
- trend computations.

$$\text{To} \left\{ \begin{array}{l} \text{P : active power} \\ \text{Q : reactive power} \\ \text{I : rotor current} \\ \text{p : pressure to the condenser} \end{array} \right\} \begin{array}{c} \text{Exploitation} \\ \text{parameters} \\ p_{j} \end{array}$$

$$\left. \begin{array}{l} A_{\omega} \\ A_{2\omega} \\ A_{3\omega} \end{array} \right\} \text{vibration parameters } a_i$$

Systematic computation and plotting of: $\dfrac{\partial a_i}{\partial p_j} = s_{ij\,\text{sensitivity}}$

The s_{ij} are vector-plotted

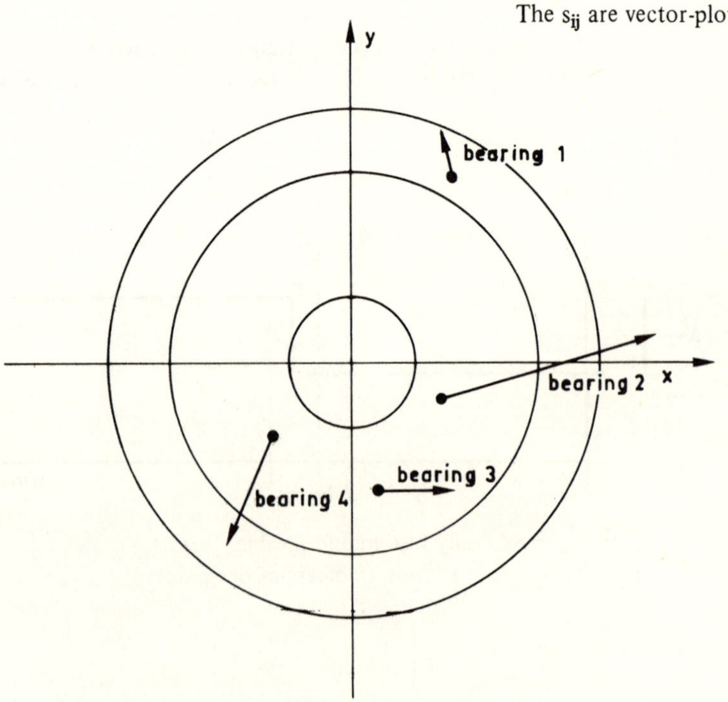

Fig. 4 – Machine sensitivity.

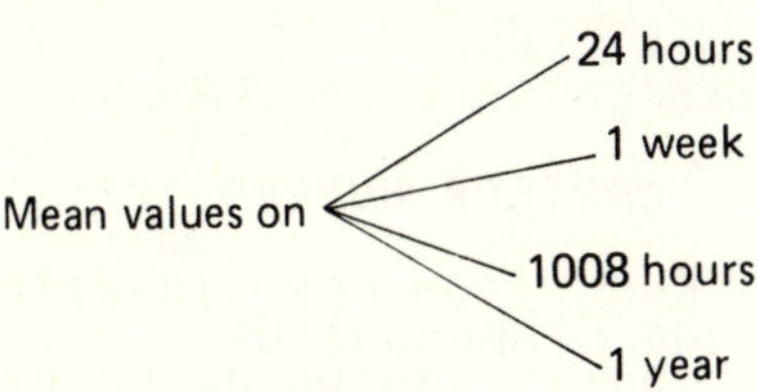

and elementary statistics:

– standard deviation
– histograms

under similar power conditions.

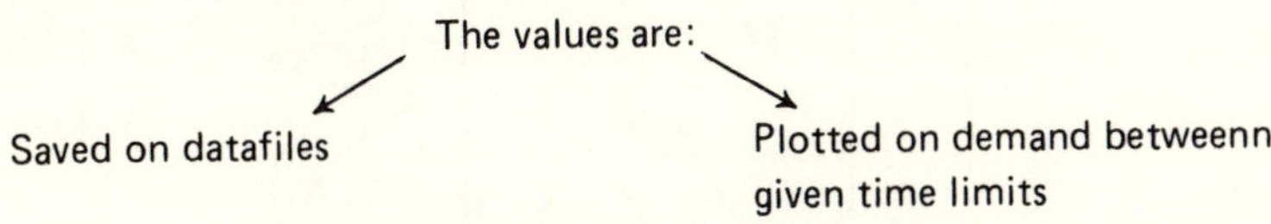

Fig. 5 – The trend computations.

3. ON-LINE SURVEILLANCE OF TURBOGENERATOR SETS

On-line surveillance has been applied to some of the most recent Electricité de France turbosets. This evolution of the process is the result both from the already-acquired experience with periodical turbosets surveillance, and from the better knowledge of the time-evolution of some mechanical defects.

A greater degree of confidence has been obtained in the surveillance criteria; this may authorise a more immediate use of the information, and above all, it becomes sometimes necessary to provide, at very short notice, adequate information to modify or to correct the operation ratings of the machine.

We must here make clear by a practical description the particular meaning given to the term: on-line surveillance.

The functional surveillance system has the same constitution as for off-line processing, except that it includes an additional element. This element is a small desk-top computer connected to the acquisition system via a serial transmission line. The minicomputer receives, and processes every minute, information on a part of the scanned parameters. It works in parallel with the data logging of the measurement channels on the magnetic support. Thus only a part of the surveillance may be described as on-line. In fact it concerns only the detection of some fast evolutions.

The parameters involved are the components of the shaft line vibrations filtered at once and twice the revolution speed. They are measured through additional non-contacting sensors located in orthogonal positions on the bearings.

** CENTRALE DE ***

CAPTEUR NUMERO 7 : P+RV

EVOLUTION DES VIBRATIONS
DE L'HARMONIQUE 1
SUR 1 SEMAINE DE ROTATION A
1500 tr/mn
Dernier Jour: 38 a 16 H 1 mn

Rayon Gd Cercle 40 Mic CC
R Cercle d'ALARME 20 Mic CC

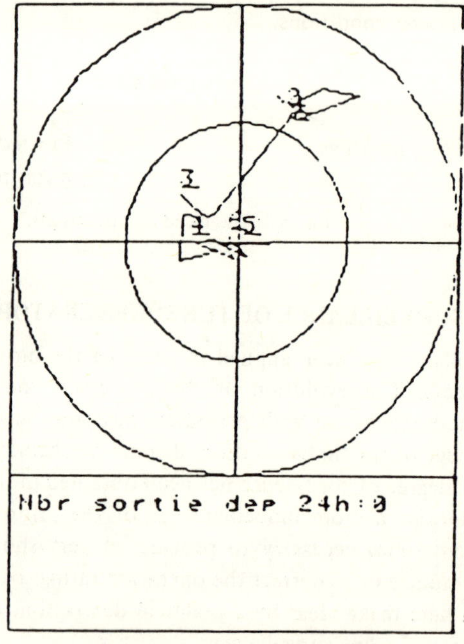

heures des SORTIES
du CERCLE D'ALARME
depuis 24 HEURES

JOUR 37 15 he 59 mn
JOUR 37 19 he 59 mn
JOUR 38 0 he 0 mn
JOUR 38 3 he 59 mn
JOUR 38 7 he 59 mn
JOUR 38 11 he 59 mn
JOUR 38 16 he 1 mn

Fig. 6

The values derived from these sensors, after a convenient averaging process (one hour time constant) should constitute constant vectors when the machine is in a steady state. The surveillance principle is to display a polar diagram of the vector's *deviations* around their mean values. A favourable circumstance for the detection of evolution is encountered when there is a limited sensitivity of the machine to the exploitation parameters such as active and reactive power, etc. (Fig. 6).

Basically we assume that the progressive evolution of a vector in its plane outside given circles with an existing anomaly on the machine [4].

According to the preceding, this method can be applied quite automatically by the control staff. Control orders are in effect so that the control staff must make sure every eight hours that the circles are not reached, that is to say, at every control team change. There are two thresholds; the alarm circle, and the imperative shut-down circle. A practical example of the radius of the circles is given in Table 3. When the corresponding thresholds have been reached, remedial action must be taken: either power decrease or shutdown.

Table 3

		Alarm	Stop	
1 X rev. speed	Horizontal direction of vibration	1.6	3.2	Peak to peak displacement values in mils
	Vertical direction of vibration	0.8	1.6	
2 X rev. speed	Hor.	.3	.6	
	Vert.	.3	.6	

3.1 Further development of on-line surveillance

The work that we are at present doing in the field should lead to a step by step generalization of on-line surveillance, that is its extension to every vibration surveillance criterion. The evolution of the hardware and measurement tools leads obviously towards an increased speed and power of treatment.

Such a transformation should be of wide interest:

- One will achieve the integration of the acquisition and of the treatment in the same process. Hence the hardware system will become easier to design and to maintain.

– The delivery of the information will be faster and, above all, more automatic. The human intervention will be less required, the turbogenerator sets will be more effectively surveyed.

REFERENCES

[1] R. M. Steward, Vibration analysis as an aid to the detection and diagnosis of faults in rotating machinery. Institution of Mechanical Engineers, September 1976.

[2] J. P. Fanton, A rapid survey of the research and devlopment activitu at EDF in the vibrations surveillance and monitoring. Rapport EDF No. P43-175.

[3] J. C. Sol and A. Jaudet, Monitoring steam valves of PWR 900 MW turbogenerator. MFPG 28th meeting, July 1979, Aan Antonio, Texas.

[4] J. Morel, Proposition pour une détection précoce des comportements vibratoires anormaux. Rapport EDF No. P34-166.

Part 5
SYSTEM MONITORING – GENERAL

CHAPTER 18

The detection of mechanical seal failure due to face vaporisation in laboratory tests

D. Harrison and R. Watkins

1. INTRODUCTION

In late-1978 severe operational problems were encountered due to failure of the mechanical seals on Forties Main Oil Line pumps. These failures were believed to be due to unstable vaporisation, between the seal faces, of the film of pumped fluid which normally lubricates the faces.

Seal face vaporisation is a common mode of mechanical seal failure when seals operate at temperatures above the atmospheric boiling point of the pumped liquid. Seal failure is characterised by the emission of bursts of vapour from the seal, often referred to as puffing or popping, together with clacking noises from the seal as the faces, which are temporarily forced apart, regain contact.

A seal testing programme was initiated to evaluate the performance of proprietary seals under conditions closely simulating field operating conditions using live Forties crude, mixtures of crude oil and formation water, and water as test fluids.

2. MECHANICAL SEALS

In any conventional rotary pump it is necessary to form a seal between the rotating drive shaft and the pump casing. Several sealing arrangements may be used, e.g. packed glands, labyrinth seals, lip seals and mechanical seals. Mechanical seals, owing to their ability to run at high pressure and speeds with low leak rates are the most common method of sealing high performance centrifugal pumps. A typical sealing arrangement is shown in Fig. 1.

A mechanical seal consists essentially of two faces one rotating with the shaft and the other stationary relative to the pump body. The faces are loaded by a combination of spring and fluid forces . The thin film of pumped fluid between the seal faces serves as a lubricant. Fluid is circulated from the high pressure side of the pump, through the seal flush to cool the seal in operation.

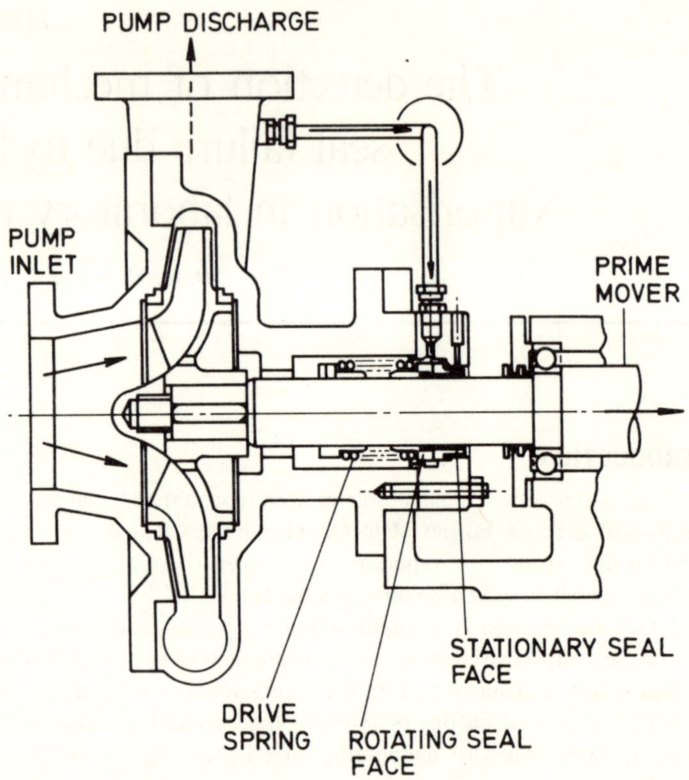

Fig. 1 – Typical centrifugal pump sealing arrangement.

3. TEST APPARATUS

A general view of the test rig is shown in Fig. 2. Two seals are tested simultaneously in a back-to-back arrangement to eliminate end thrust. The test fluid is circulated at pressures of up to 100 bar through steam heat exchangers to the seal flush, exiting through the top hat units which simulate the pump end of the seal.

The test fluid is stored in an accumulator system under nitrogen pressure and circulated to the seals using positive displacement pumps. The seals can be driven at variable speeds of up to 4200 rpm. Seal testing was generally carried out at 3600 rpm, seal face sliding speed being 21 m/sec. Seal drive torque, flush flow rate, inlet and outlet seal flush temperatures and operating pressure are logged during testing. The rig uses a microprocessor based system for data logging and shutdown. In addition operating parameters are logged on a multichannel quick response recorder.

Fig. 2 – General view of test rig.

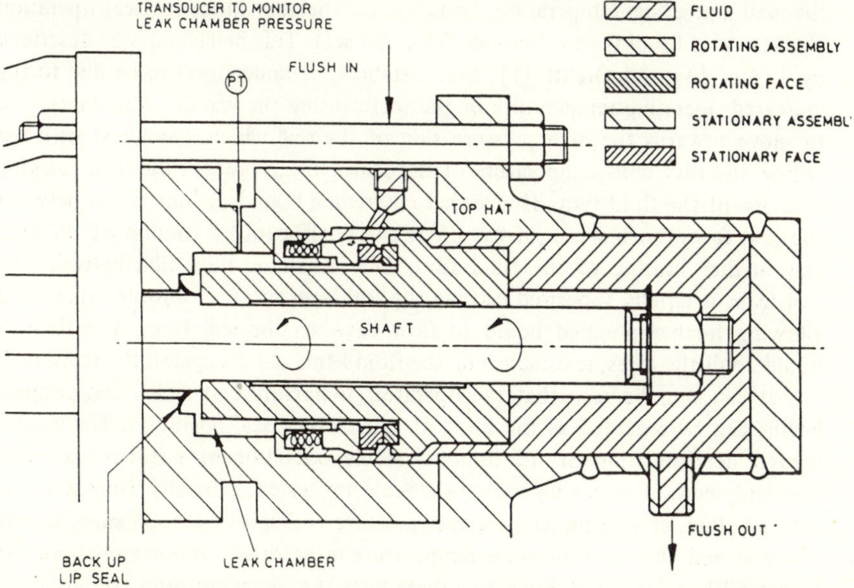

Fig. 3 – Schematic mechanical seal assembly. Seal assemblies Fig. 3 shown in foreground.

A schematic diagram of a seal installed on the rig is shown in Fig. 3. Fluid leaking from the seal passes into the leak chamber from where it is piped to a vented leak pot. Fluid leakage to the bearing assembly is prevented by a dry running secondary lip seal which acts as a back-up seal in the event of failure of the primary mechanical seal.

4. SEAL FAILURE DUE TO SEAL FACE VAPORISATION

In normal operation the faces of a mechanical seal are separated and lubricated by a thin film of the pumped fluid. Over limited areas of the seal, boundary lubrication occurs, with some asperity contact of the faces. Fluid film pressures are maintained by a combination of radial outward leakage, giving a hydrostatic pressure distribution between the seal faces, and the development of hydrodynamic pressures between converging areas of the seal face.

Experimental and theoretical work [1–3] has shown that in many cases a two-phase fluid film exists between the seal faces. Owing to the power dissipated between the sliding faces the fluid between the faces is heated and vaporises towards the low pressure side of the seal. This is believed to be a common and stable mode of seal operation, the fluid volume changes associated with vaporisation enhancing the load-carrying capacity of the seal. If, however, the seal operating temperature is increased further, as has been shown experimentally by Orcutt [1] and theoretically by Lebeck [3], the zone of vaporisation widens, moving towards the high pressure side of the seal. Further increase of the seal operating temperature leads to an unstable mode of seal operation characterised by puffing of vapour from the seal. This behaviour was described by Lymer [4] and Orcutt [1]. Seal instability is understood to be due to the increased operating temperature of the seal causing the vapour–liquid interface to move towards the high pressure side of the seal which, combined with the falling viscosity with temperature of the sealed fluid, reduces the load carrying capacity of the fluid film. This leads to increased boundary lubrication between the seal faces, an increase in the operating coefficient of friction of the seal, and further heating of the fluid film. The remaining fluid film between the seal faces is rapidly vaporised causing the seal faces to part, emitting vapour and allowing further pumped liquid to flow between the seal faces. The flow of liquid cools the faces, re-establishing the fluid film, and the cycle is then repeated.

It can thus be seen that the operating temperature at which seal popping begins can be regarded as the limit of satisfactory seal operation. The experimental determination of the limit of stable operation of a seal is used as a standard method of testing seals [4]. Seals are tested, generally using water as the test fluid, at varying temperature, pressure and speed. Thus, if speed is held constant and at a given pressure temperature is increased, an operating limit can be established for a seal. Repeating these tests at varying pressures establishes an operating envelope for the seal (Fig. 4). Vaporisation occurs at a temperature

ΔT below the sealed fluid's boiling point at the test pressure. The operating ΔT of a seal at a given pressure can be used as a criterion of seal performance, seals with low ΔT values having superior perfomance being able to operate closer to SVP conditions without suffering unstable vaporisation.

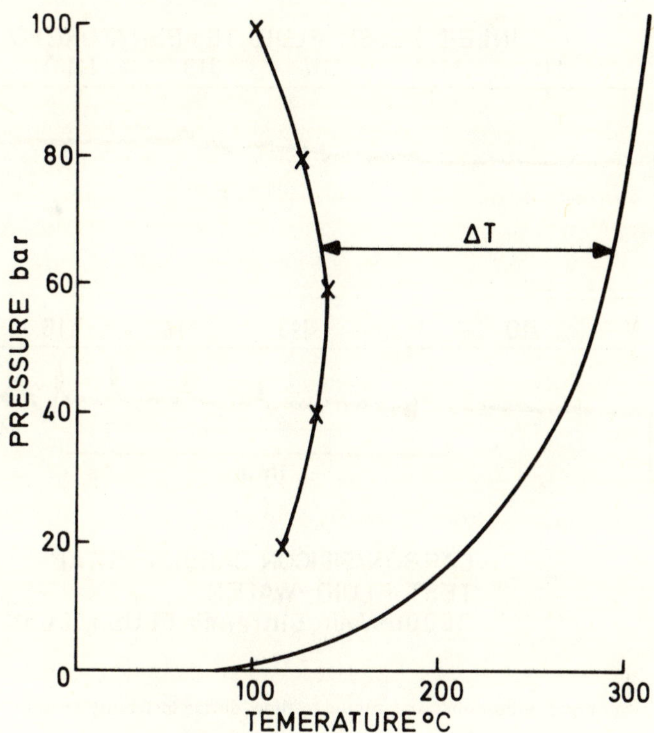

Fig. 4 – Typical seal operating envelope on water.

5. EXPERIMENTAL PROCEDURE

5.1 Evaluation of seal performance on water

The performance of the seals was initially evaluated on water. The seal was operated at constant pressure and speed and the fluid temperature progressively increased until instability was observed. The limit of seal stability is characterised by a progressive rise in the operating torque level from the normal operating level followed, at slightly higher temperatures, by large fluctuations in torque. If a seal is tested with the low pressure side open to the atmosphere the torque fluctuations are associated with 'clacking' noises from the seal as the faces lose and regain contact and the emission of puffs of steam from the seal.

A typical torque/time trace from a seal is shown in Fig. 5. The test shown was terminated by automatic rig shutdown as high leak chamber pressures were detected. On starting subsequently the seal leaked badly and the face showed comet trail damage of the carbon face which is typical of seals where vaporisation has occurred.

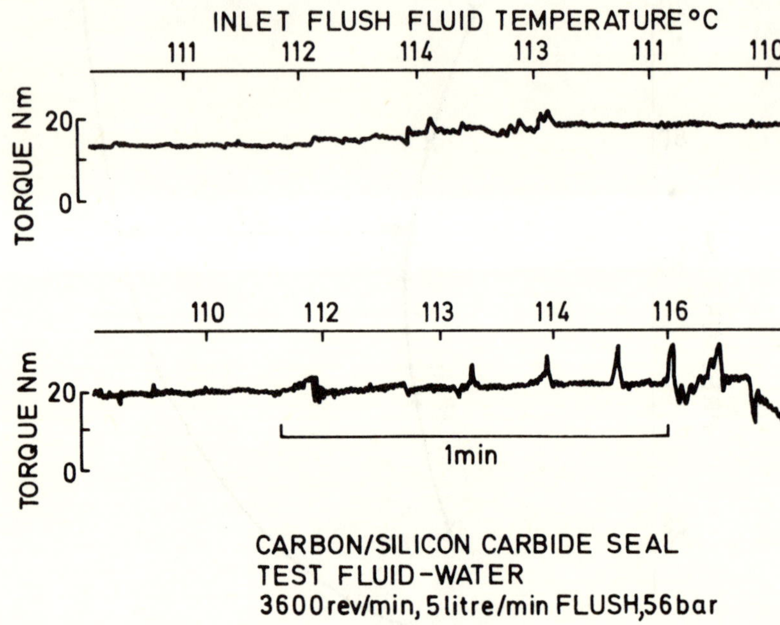

Fig. 5 – Seal operating torque readings during instability testing.

To investigate seal vaporisation in more detail and detect seal puffing, which owing to the enclosed design of the test rig was not visible, seal leak chamber pressure was monitored using a sensitive pressure transducer. Fig. 6 shows a trace and seal leak chamber pressure obtained during an envelope test. Typically, corresponding to the start of the rising torque characteristic, a series of rapid pressure pulses at a frequency of 1.5 Hz can be seen in the leak chamber pressure trace. These pressure pulses can thus be seen to indicate, at an early stage, the onset of seal instability. Further runs were carried out to determine the seal operating envelope at varying pressures. Similar seal behaviour was observed, a series of pressure pulses indicating the onset of seal instability. On dismantling the seals the faces were found to be in good condition with no physical indications of vaporisation damage.

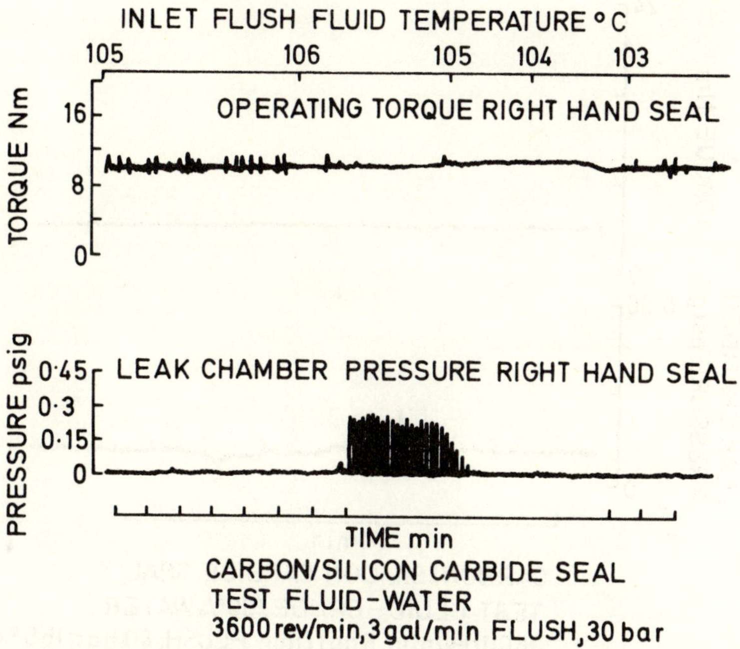

Fig 6 – Seal operating torque and leak chamber pressure during envelope test.

5.2 Evaluation of seal performance on crude oil

The performance of seals was subsequently evaluated on crude oil and crude oil/water mixtures. The same experimental technique was adopted, seal operating temperature being progressively increased at constant operating pressure.

As the vapour pressure of crude oil/water mixtures is higher than that of water, seal vaporisation was expected to occur at lower operating temperatures. Seal operating behaviour on crude oil and crude oil/water mixtures was found to be significantly different from that on water. Increasing operating temperature did not give a clear indication of seal vaporisation, either from the torque transducers or by detection of pressure pulsations from the atmospheric side of the seals. When seals were operating at high temperatures with significant quantities of water (30%) present, isolated pressure pulses of small amplitude were observed from the seals (see Fig. 7). During endurance runs at lower temperatures a constant low level of leak chamber pressure activity with intermittent pressure pulses of low amplitude was observed (see Fig. 8), variations in seal leak chamber pressure were associated with variations in the seal operating torque. This observed pattern of slight intermittent face vaporisation correlated with observed seal face damage. Whereas on water seal face damage and failure was rapid, seal damage on oil/water mixtures was progressive, cumulative damage over a period of 40 to 100 hours causing significant deterioration of the seal faces.

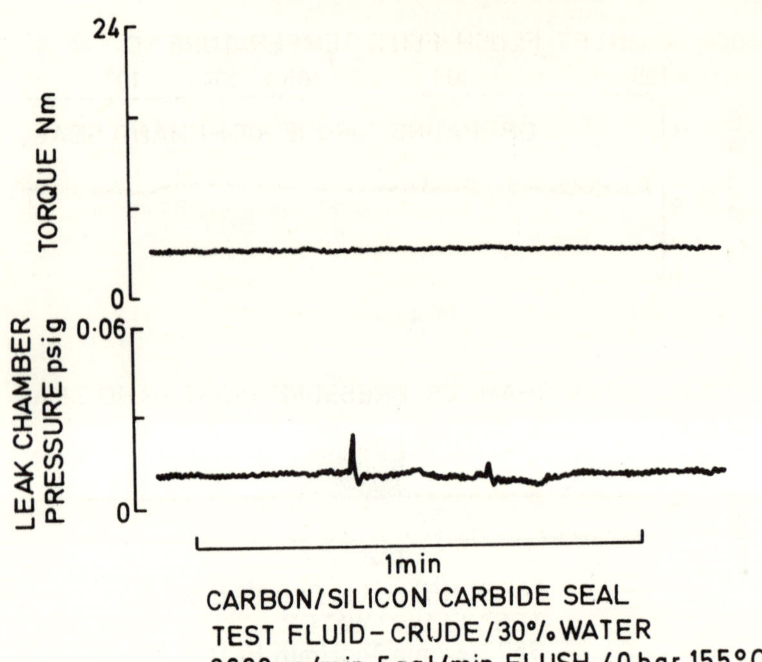

Fig. 7 – Seal operating torque and leak chamber pressure during high temperature operation.

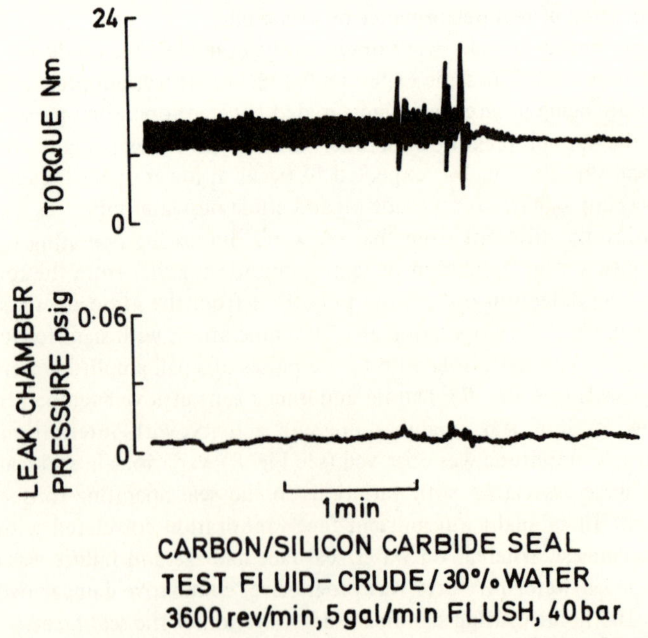

Fig. 8 – Seal operating torque and leak chamber pressure during endurance test at 105°C.

6. COMPARISON OF VAPORISATION CHARACTERISTICS OF WATER AND CRUDE OIL/WATER MIXTURES

Explanations for the different behaviour of the seals when operating on water and crude oil/water mixtures were sought. As has been discussed the vapour pressure of the sealed fluid is a key parameter in determining the operating envelope of a mechanical seal. Comparative vapour pressure/temperature curves are shown in Fig. 9 for water, crude oil, and crude oil/water mixtures. The vapour pressure curve for crude oil/water mixtures can be seen to be considerably higher than that for water alone, thus seal failure would be expected to occur at lower temperatures when a seal is operating on a crude oil/water mixture.

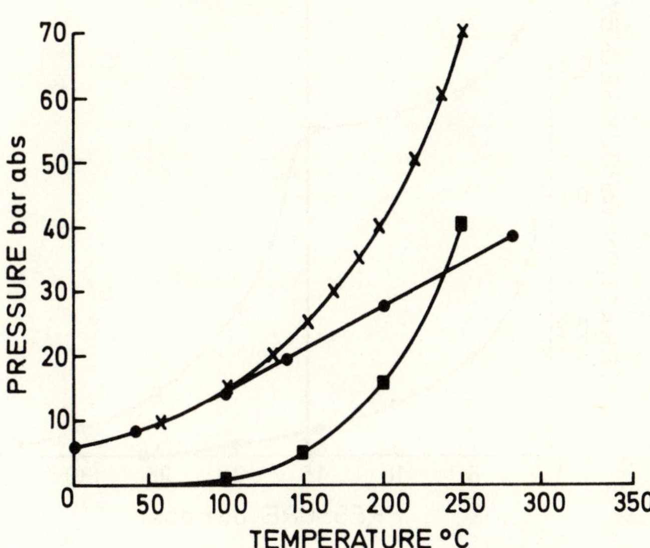

Fig. 9 – Bubble point curves.

As the expected behaviour was not observed a theoretical study was carried out to compare the vaporisation characteristics of crude oil and crude oil/water mixtures. The study showed that at a given pressure a crude oil/water mixture, being a multi-component mixture of compounds with widely varying boiling points, vaporises progressively over a wide temperature range, unlike water which vaporises at a single temperature. Furthermore smaller volumes of vapour are evolved from crude oil/water mixtures compared to water, under similar conditions. Comparative flush curves for water and a crude/30% water mixture are shown in Fig. 10.

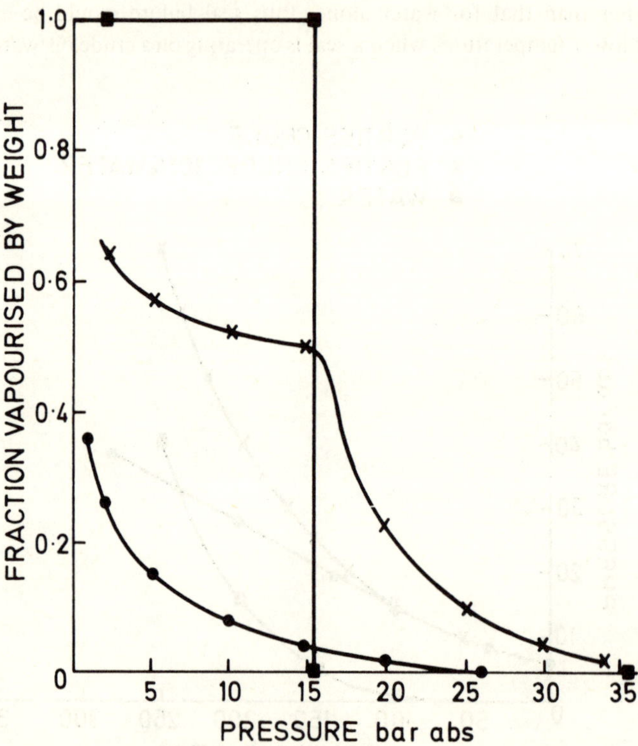

Fig. 10 – Flash curves at 200°C.

The reduced level of detected vaporisation and the associated reduced level of seal damage is consistent with the vaporisation behaviour of the test fluid.

7. MONITORING OF OPERATIONAL SEALS

A seal operating stably has a low and relatively constant rate of vapour emission indicated by a low, slowly varying seal leak chamber pressure, increase in seal leakage being indicated by a rise in leak chamber pressure. When operating under marginal conditions, fluctuations in leak chamber pressure are indicative of non-steady seal face conditions. Thus it is suggested that monitoring of seal leak pressure may offer a means of monitoring, on line, seal conditions in operational pumps, giving early warning of seal distress.

Further experimental work is required to characterise leak chamber pressure signals obtained from pumps handling a range of fluids, over a range of typical operating conditions and utilising seals employing different seal face combinations. Leak chamber pressure signals obtained from seals failing for reasons other than fluid vaporisation at the seal faces should be investigated. At present the technique appears most useful for use on pumps handling pure liquids.

8. CONCLUSIONS

During seal testing on water, monitoring of seal leak chamber pressure proved a reliable method of detecting the onset of instability due to face vaporisation, enabling seal performance to be rapidly evaluated.

When seals were operated on a liquid which vaporised over a range of temperatures seal instability could not be detected, at a particular temperature, by monitoring the leak chamber pressure. Intermittent pressure pulses indicated a progressive mode of seal failure.

It is suggested that monitoring of seal leak chamber pressure should be considered as a method of monitoring on line seal condition in operational pumps.

ACKNOWLEDGEMENT

Permission to publish this chapter has been given by the British Petroleum Company plc.

REFERENCES

[1] F. K. Orcutt, An investigation of the operation and failure of mechanical face seals, *Journal of Lubrication Technology, TRANS ASME,* Series F, **91,** No. 4 (1969), pp. 713–725.

[2] W. F. Hughes, N. S. Winowich, M. J. Birchak and W. C. Kennedy, Phase change in liquid seals, *Journal of Lubrication Technology, TRANS ASME,* **100** (1978), pp. 74–80.

[3] A. O. Lebeck, A mixed friction hydrostatic face seal model with phase change, *Journal of Lubrication Technology, TRANS ASME,* **102** (1980), pp. 113–138.

[4] A. Lymer, A engineering approach to the selection and application of mechanical seals, presented at the Fourth International Conference on Fluid Sealing, held in conjunction with the 24th ASLE Annual Meeting, Philadelphia, May 1969, FICFS Preprint No. 22.

CHAPTER 19

A systematic approach towards automating the management of machinery complexes

R. M. Stewart

1. INTRODUCTION

The purpose of this chapter is to convey some idea of developments currently taking place in the field of machinery management. Current practice is largely 'manual' and so what this chapter considers is how much of the 'process' we can automate.

To begin with the aim must be to improve the effectiveness of machinery managers. As will be discussed later in the paper, the removal of professional manpower presupposes the existence of some equivalent 'knowledge base' of experience, judgement etc., and this is patently not yet available, even with the advent of expert computing technology. Whether or not an increase in effectiveness would allow the plant to get by on fewer engineers or technicians is a matter only the plant can decide, and obviously this would be a powerful selling point.

As with many such developments, and notably those sweeping through manufacturing industry, the driving force is a desire to capitalise on opportunities created by the ubiquitous microcomputer. Of particular note in the machinery management field are arrangements of those that implement the technologies of artificial intelligence (or expert computing), system modelling, signal processing, measurement, database management and control.

The chapter is organised to appeal in particular to the technical manager. First to be presented is a small amount of general information on the methods being developed to introduce the new ideas, followed then by a much larger amount of information on the problems currently being tackled.

Central to any concept of machinery management is the goal of keeping machines running, sometimes at almost any cost. Engineers who have tried to automate the process will therefore appreciate that the most crucial problem computerised systems face is when to hit the stop button; to press it at the first sign of trouble is certainly to court disaster. Systems based on the measurement of vibration and debris are particularly prone to problems of this type.

From a technical standpoint the crucial thing is to build up a picture of the machine from a diverse set of measurements and experiences, so long as it is clearly understood why the various measurements are being made and therefore, by implication, what each can reasonably be expected to show.

2. ORGANISATION AND METHOD

First, let us look at the general approach. Shown in Fig. 1 is the layout of a generalised methodology currently being used by the author's company to transfer machinery management technology. The recipient (generally a large organisation) begins the transfer process at the top and works downwards through the various steps in a controlled and tested way. In particular, the information gathered during each phase is given a bias towards establishing (a) the things that must be measured, (b) the costs involved, and (c) the false alarm rates likely to be experienced. For installations of any appreciable size, sooner or later a capital budget would have to be approved, so placing the onus on the engineer to demonstrate actual financial performance.

In simple terms the steps are, however, as follows:

(i) *Design audit,* involving a thorough investigation of the machine from the point of view of its construction, known failure record, economics of availability/unreliability etc., detectable faults and environment (access, safety requirements, ambient temperatures, sensor cable runs etc.).

The final step in this procedure is then the scheming out of a management system that defines what technology can be applied to health monitoring, alarm/threat evaluation, prosecution of overall plant management goals, machine control and maintenance.

(ii) *Technology proving,* which in the majority of cases involves the engineering of some pilot system, perhaps an element of development and a great deal of discussion between the various parties involved on the matter of the system's integration into existing management structures (if they will take it).

The primary objective of this step is therefore to demonstrate the practicality of the technology at minimum cost and prepare the case for any large capital expenditure.

(iii) *Technology implementation,* which would take place only after successful completion of the proving step and generally involve a phased expansion of the proven technology.

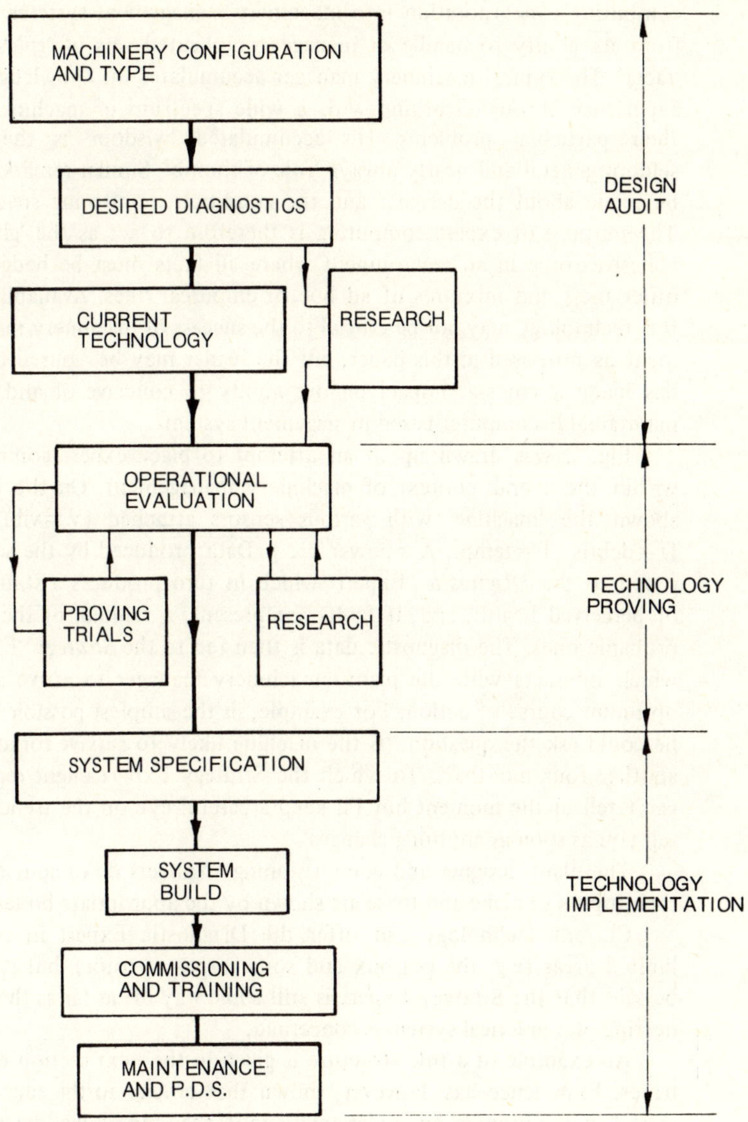

Fig. 1 – The systematic approach.

So much for the broad concepts; now what about the technology itself? The bones of this are as follows:

(i) *Expert computing*, which in the case of Stewart Hughes technology is based on the language known as PROLOG. The main reason for expert computing's incorporation into machinery management systems stems from its ability to handle or manage complicated sets of 'rules' and 'facts'. The typical machinery manager accumulates his knowledge and experience through working with a wide spectrum of machines and their particular problems. His accumulated 'wisdom' is therefore seldom general and nearly always 'rule of thumb'. Similar remarks may be made about the designer and the condition monitoring specialist. The purpose of expert computing is therefore to act as the 'glue' or cohesive force in an environment where all facts must be hedged by other facts and mixtures of ad hoc or empirical rules. Availability of this technology may not be crucial to the success of machinery management as proposed in this paper, but the reader may be assured that it has made a colossal impact on our ability to conceive of and write maintainable computer-based management systems.

Fig. 2 was drawn up in an attempt to place expert computing within the overall context of machinery management. On the left is shown the 'machine' with various sensors attached (V = vibration, D = debris, T = temp., A = power etc.). Data produced by the sensors is fed to the *Diagnostic* 'Expert' which in turn produces a statement of perceived health, and, if faults are present, a ranking of the most probable ones. The diagnostic data is then fed to the *Strategy* 'Expert' which interacts with the plant's machinery manager to arrive at the optimum course of action. For example, in the simplest possible terms he could ask the question, 'Is the machine likely to survive for at least another four months?'. To which the Strategy Expert might reply, 'I can't tell at the moment but I'll keep a careful eye on the trends and tell you as soon as anything changes'.

The plant designer and commissioning engineers do of course have their inputs to make and these are shown by the appropriate boxes.

Current technology can offer the Diagnostic Expert in certain limited areas (e.g. the gearbox and some types of rotor), but it must be said that the Strategy Expert is still some way off as far as the engineering of a practical system is concerned.

An example of a rule structure is given in the next section of this paper. Experience has, however, shown that crucial to the success of expert computing is an understanding of the underlying diagnostic processes, which is where System Modelling enters the picture.

(ii) *Sytem modelling,* which for Stewart Hughes Ltd means a mixture of machine dynamics, thermodynamics, gas dynamics, process control and management structure. The first three are examples in later parts of the chapter.

The art of modelling is knowing just exactly what to include, and therefore also what to ignore. Analytical competence is crucial and a

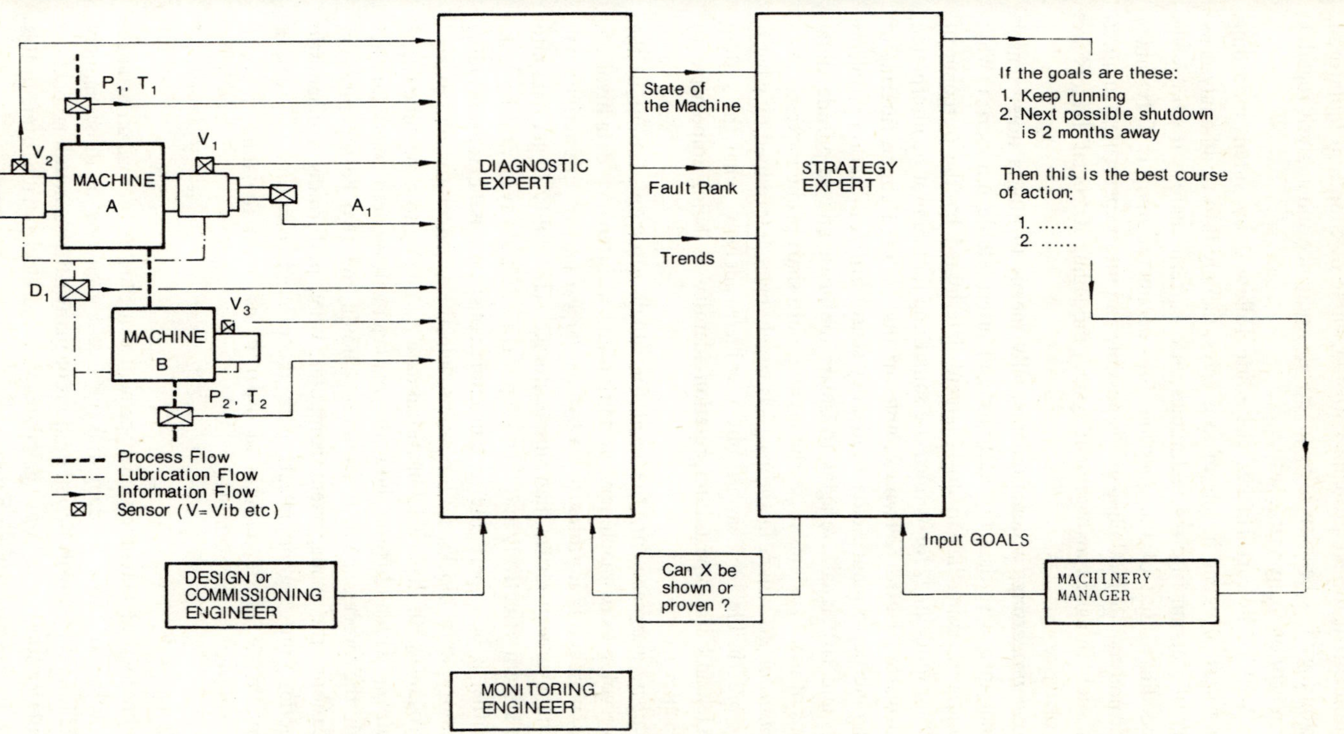

Fig. 2 – Information processing within a machinery management system.

significant historical problem with the whole of condition monitoring technology has been to bring practical experience and analysis together in an economically viable way.

Within Stewart Hughes the route followed by sytem modelling has invariably gone in one of four ways, namely into trend analysis by way of Kalman filtering techniques, into pattern analysis if wholescale modelling is out of the question, into parameter analysis if dynamic/thermodynamic modelling etc. is possible and into expert rule structures for such things as management system modelling if 'social' factors are involved.

(iii) *Signal processing,* which fundamentally means turning a noisy sensor signal into a statistically accurate and interpretable number(s). What signal processing has to do is largely determined by the outcome of system modelling, for to measure something that cannot be interpreted is a waste of effort. Two examples of this are (a) Kalman filtering of machine performance data to identify sensor drift/calibration problems etc. and (b) signal averaging to isolate the phase and amplitude vibrations of rotor systems in gearboxes and multi-shaft prime movers.

(iv) *Measurement,* which in the past has largely been determined by what it was practicable to measure, but now increasingly by what the system model and its expert interpretation identify as being imperative to measure.

(v) *Database management,* the need for which arises on account of the large volumes of machinery, condition and maintenance data involved. If experience is anything to go by, the largest part of this activity will be data upkeep rather than simple storage. The ability to get data into the system effectively and then keep it current is always a priority task, whether it be expert rule sets, machine constructional details or fading memory data blocks set up to provide the engineer (or his 'expert' surrogate) with the crucial historical data leading up to a machine trip.

(vi) *Control,* which dictates how the management system must interact with the process control computer should the two be separate. On occasions, the management computer, when performing a diagnostic function, must communicate with the process computer to create a step change in the process in order to ascertain dynamic response characteristics, for example.

The technology of machinery management can thus be seen as an arrangement of at least five more fundamental ones, and even one of those, namely system modelling, may involve combinations of many others, including machine dynamics, lubrication, metallurgy, gas dynamics and electromagnetism. In this way it is no different from modern design.

3. ROTOR DIAGNOSTICS

Most machinery engineers will be familiar with rotor proximity probe technology and this is generally fitted under the heading of condition monitoring. In the vast majority of installations, however, little more is done with the data other than simple thresholding. The management loop is seldom, if ever, closed and the author knows of only a very small number of cases in which the vibration data was a crucial factor in the decision to shut down a major machinery train. In the true sense of the word it cannot therefore be said that industrial rotors are often managed in the way implied by this paper. Furthermore, the author would argue that one of the main reasons for this is lack of diagnostic ability on the part of the monitoring systems.

A machine for which very substantial advances have, however, been made is the helicopter. There is considerable advantage to be gained from management of the rotor head system (a) to tune the vibration produced for the various roles (hover, high-speed cruise, ferrying, etc.), (b) eliminate wasteful component changing when the rotor becomes rough and (c) guard against component failure or degradation (e.g. water seepage into blade pockets).

A look at helicopter rotor head management technology therefore leads one to see many of the primary technologies mentioned in Section 1 at work.

Fig. 3 shows a picture of a fairly typical helicopter rotor head. Though not obviously complicated, it is in fact extremely so and this is the root cause of it being so difficult to 'manage' by conventional means (i.e. the mechanic in the field). The important parts of the mechanism are the blades (1), the aerodynamic tabs, the control linkages (2), dampers, control bearings (3) and flexures (4). Faults in these components produce some very complicated forcing functions indeed as the system is essentially aero-elastic and totally three-dimensional.

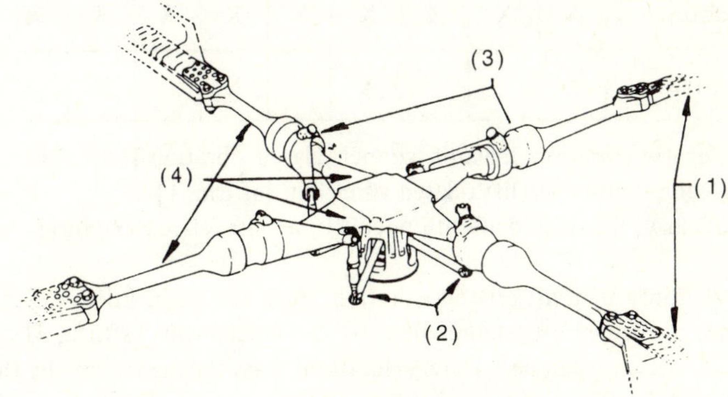

Fig. 3 – The Westland Lynx rotor head.

A list of typical faults and their associated symptoms is given in Table 1. In practice, the faults seldom if ever exist alone. For example, once-per-rev vibration (one-R), which most engineers associate with imbalance, arises not only as a fundamental fault in its own right but also as a secondary consequence of almost every other fault. to therefore re-balance a rotor head purely on the grounds of high one-R is a most dangerous move and therefore usually forbidden.

Table 1 – The fault/symptom matrix for a helicopter rotor head.

	SYMPTOMS or MEASURED PARAMETERS									
	Near zero-lift col. pitch					Positive col. pitch				
ROTOR HEAD IRREGULARITY	1R in-plane vibration	1R vertical vibration	lag error	blade tip track error	blade pitch mom. error	1R in-plane vibration	1R vertical vibration	lag error	blade tip track error	blade pitch mom. error
Mass imbalance on the blade flexural axis	X		X			X		X	X	
Mass imbalance not on flexural axis	X		X			X	X	X	X	X
Track rod error	X	X	X	X		X	X	X	X	
Blade tab error	X	X	X	X	X	X	X	X	X	X
Lag damper fault	X			X		X		X		

Rule (1, category, vibration, mass or damper related vibration [false, 1]).
Rule (2, category, vibration, lift related vibration, [affirm, 1]).
Rule (4, category, lift related vibration, confirmed lift related vibration, [affirm, 2]).
Rule (5, leaf, confirmed, lift related vibration, track rod error, [affirm, 3]).
Rule (6, leaf, confirmed lift related vibration, blade tab error, [affirm, 4]).
Rule (7, leaf, mass or damper related vibration, mass imbalance on the flexural axis, [false, 5, affirm, 6]).
Rule (8, leaf, mass or damper related vibration, mass imbalance not on flexural axis, [affirm, 5, affirm, 7, affirm, 8, affirm, 6]).
Rule (9, leaf, mass or damper related vibration, lag damper fault, [false, 5, false, 6, affirm, 9]).

The technical approach adopted by the author's company in the search for a management solution to the helicopter rotor head problem involved a very careful analysis of the system's dynamics. These are, however, far too complicated to solve in any complete way (even with the largest computers) and therefore the model is used (very effectively as it turns out) to indicate and define the patterns of vibration, blade track spread and velocity that should emerge for all combinations of faults.

In common with nearly all other efficient rotor diagnostic systems, the helicopter one requires measurement of at least two parameters (ideally three) and the conduction of a transient test.

The basic process of moving from the complete simulation to the groups of equations that produce the useful patterns is shown in Fig. 4. Starting with helicopters in general (and all the possible variations introduced by manufacturing tolerances, externally attached stores, fuel load, etc.), assumptions that permit effective simplifications are gradually made. The basic analysis is carried out in the rotating frame of reference so the first step is to effect the transformations that will allow interpretation of measurements made in the fixed frame. Next the blade is assumed to be rigid and hinged at the hub. Finally, given a set of equations that are simple enough to yield practical insights, the effects of first single faults and then combinations of faults are investigated by inserting their parametric effect (e.g. reduced damping) or state (e.g. blade will fly high) into the equations and solving them by such ploys as ignoring higher order terms etc.

Also shown in Table 1, beneath the irregularity/symptom matrix, is the PROLOG rule set for the track and balance management of a rigid rotor helicopter. The details of both are not really relevant to this discussion and they have been introduced solely to show what is produced in the way of intermediate diagnostic data (in the sense that one would not present the field engineer with anything as complicated as this). In practice the various pigeonholes within the fault matrix are filled with actual measured values and weights or probability measures that derive from both the model and aircraft experience; however, to have shown those would have introduced too much detail. Obviously, interpretation of the PROLOG rule set requires knowledge of PROLOG, just as to have presented a FORTRAN program would have presupposed a knowledge of FORTRAN.

It would appear that an efficient helicopter rotor head management package can be built up from a system analysis that condenses to four or five simple fault models, around 50 rules and data from five sensors signal processed to yield some 100 separate measurements.

In practice the rotor head sensor kit includes three accelerometers, a blade tracking sensor and a tachometer. To conduct a test the aircraft must be put through a set of manouevres in either the ground run, hover or flight regime (preferably the former) and the sensor data recorded for subsequent analysis and interpretation on the ground.

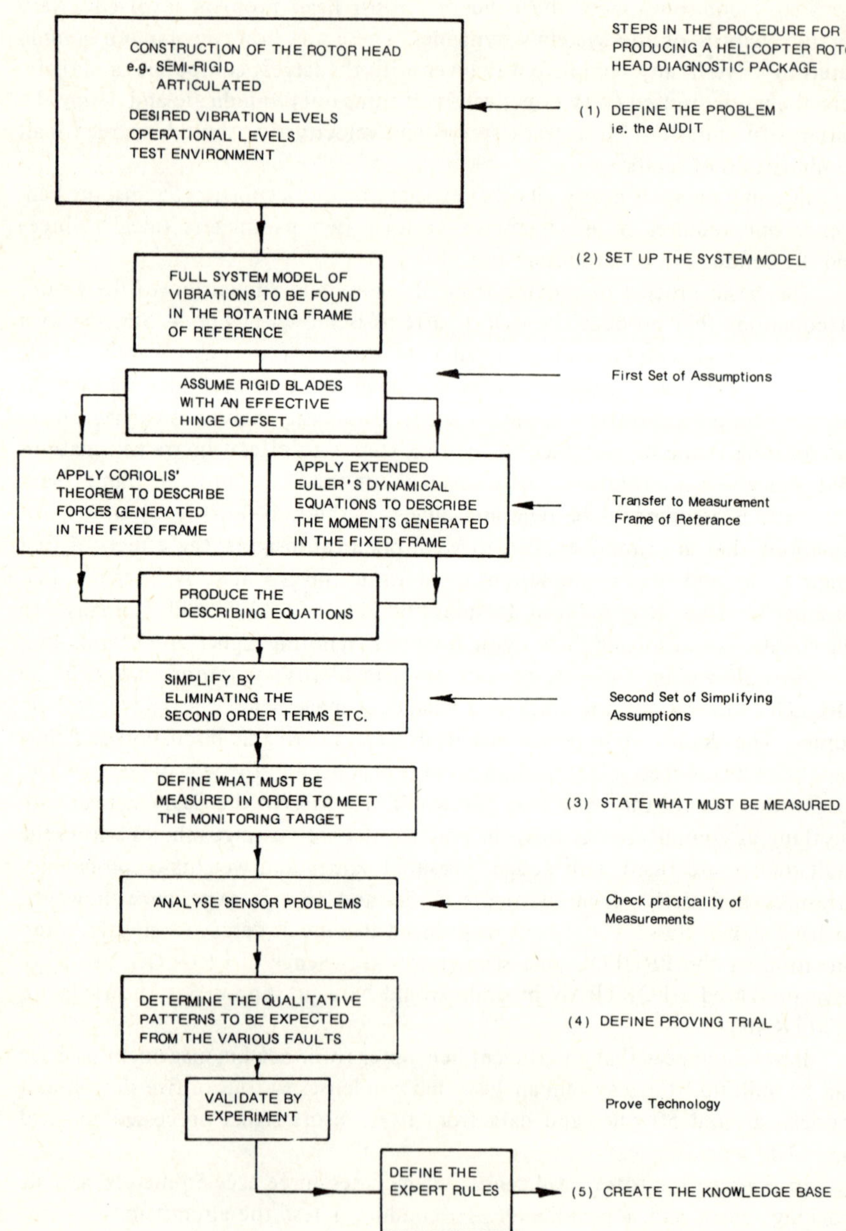

Fig. 4 – Development of a diagnostic package.

4. GEARBOXES

The imperative to manage gearboxes in a way that leaves them undisturbed for as long as possible is great indeed. For some reason gearboxes have acquired a reputation for unreliability and many machinery engineers therefore begin to worry at the first sign of a change in the vibration.

As is the case with many other mechanical components (e.g. alternator, bearing), the gear has many possible failure modes and a fair-sized armoury of diagnostic tools must generally be on hand to treat all the practical problems of epicyclic unit complexity, low-speed drives found in coal mines, etc.

From a management point of view, however, the typical gearbox possesses several features that make it relatively easy to monitor. Vibration-wise, for example, the 'normal' faults tend to produce quite distinctive patterns which are predictable in terms of simple kinematic and dynamic models [1], and can therefore make the basis of quite efficient diagnostic tests. The art of monitoring the gearbox therefore lies in prediction of these patterns and the structuring of what is known as the 'monitoring policy' to obtain a reliable diagnostic.

A fundamental problem with the vibration analysis of gearboxes is the high level produced by normal meshing action and the small changes that tend to accompany significant faults. Gearbox monitoring (for bearings as well as gears) in the limited sense of measuring simple vibration level therefore fails more often than not, especially for high-speed units.

The other major technique appropriate to gearboxes is oil system debris monitoring. However, in the majority of industrial situations this is even more difficult to implement than vibration monitoring. The gains in diagnostic efficiency that accompany the combined application of vibration and debris are nevertheless very great and one important reason why it is so desirable to begin the technology transfer process outlined in section 2 at the earliest possible time. Instrumentation-wise, debris analysis is probably marginally simpler to apply than vibration (given situations where both are judged to be feasible), yet from a mechanical point of view it nearly always has to be designed in from the beginning. Diagnostic systems known to the author that have combined vibration and debris monitoring have achieved some quite remarkable performances, the primary reason being the ease with which expert rule structures may be written to screen out false alarms and promote accurate diagnosis.

The kernel of the Stewart Hughes vibration monitoring system for gearboxes is the so-called figure-of-merit pattern group [2]. Starting with signal averages for each gear in the system, each one is reduced to a group of some four to ten members (the exact number depending on many constructional features of the unit). Perhaps two of these measure localised gear tooth damage, another two misalignment and the remainder miscellaneous faults such as shaft/housing fracture and parametric excitation. Over the past ten years a vast body of experience has been built up around their application, with the result that several management systems now exist into which they have been incorporated.

At this point a case study of their use would be interesting. Shown in Fig. 5 is the layout diagram of a gear box operated by the Elastomers Division of Esso Chemicals (Fawley) and it can be seen to consist of an electric motor, gearbox and centrifugal compressor (drive power approx. 5 MW). The system was being managed by the normal complement of sensors on the compressor rotor (thrust, bearing temp., clearance, etc.) plus three accelerometers on the gearbox and one on the motor.

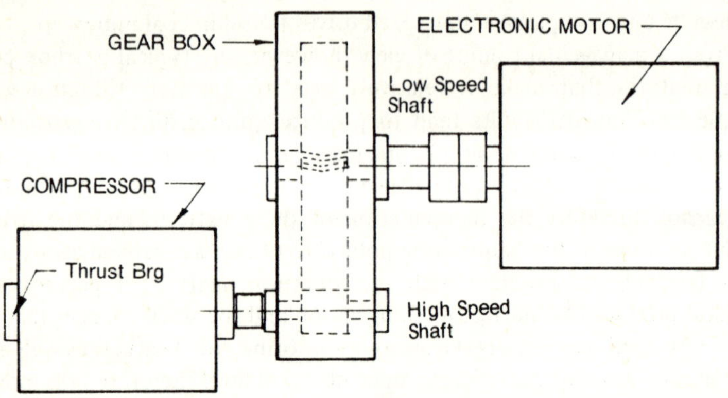

Fig. 5 – Compressor-drive gear box.

After six years of relatively trouble-free operation the gearbox began to radiate a knocking noise, attributed by the vibration analysis system to two localised spots of damage on the low-speed gear. Reference to trends of vibration also spotted a slowly rising level of vibration on the high-speed side of the gearbox (but at a very low level). The plant operators therefore began to make contingency plans for shutdown of the plant and replacement of the gears. Plant management on the other hand required that the plant be kept going as long as possible and ideally until the next scheduled downtime (four months hence) as production was vital.

A log of events over the next four months follows:

	Plant operator	*Management system*
Nov. 10	Filter pressure drop alarm. Bright metal flakes found in filter.	Management system begins to look for possible source of the debris. Gearbox vibration level seen to be rising.
Nov. 22	First alarm by operators regarding the knock.	Management system says that it is acceptable to continue. Analysis of the vibration indiccates the possibility of very minor damage to the low-speed gear. High-speed pinion OK.

Dec. 15		Analysis of debris says not from gearbox.
Dec. 20		Three pieces of evidence begin to point to the lube pump as the source of debris: (i) material, (ii) ferrography shows signs of cavitation, (iii) plant records show a blockage of the suction filter 2 years earlier.
Jan. 3		No change in gear vibration. High-speed shaft vibration still rising but not consistently.
Jan. 17	Plant operators becoming more concerned about the gearbox knock.	Monitoring system says that gear box still OK. Management decides to press on without stopping so that a vital production contract can be met.
Jan. 25	Temp. alarm on thrust bearing of the compressor (up to 150 degC)	Calculation based on this measurement and other indicate that nearly all white metal has been lost (if the sensors can be trusted). The site of attention now swings round to the compressor.
Jan. 26		Decision taken to monitor the system even more closely and plan for early shutdown to replace the compressor's thrust bearing.

The 'Expert output at this point was as follows:

No obvious connection between the gearbox and compressor problems.
The gearbox fault not propagating, damage local.
Six possible causes of the compressor problem:

(i) HS coupling lock (20% probability)
(low because no problem on G/B)
(ii) Fouling in compressor (40%)
(known history of fouling)
(iii) Balance piston labyrinth failed
(pressure check puts probability to zero)
(iv) Gas balance line blocked
(pressure check puts probability to zero)
(v) Process upset
(check of history puts probability to zero)
(vi) Bearing corrosion (40% probability)
(chlorides found in the oil)

	Plant set to produce maximum amount of intermediate product.
Jan. 29	Plant shutdown, thrust bearing opened up, pads on active side wiped, black deposits noticed on the bearing (tin dioxide corrosion), loss of rotor float in the inactive direction, coupling apparently free.
	Plant restarted after 16 h. Contribution loss minimised by the planned nature of the shutdown (accurate prediction of the fault) and the decision to build up intermediate stocks and retain inventories.
Feb. 24	Plant scheduled shutdown. Gearbox opened up: very minor damage to wheel found plus some corrosion and fouling of the pinion bearings. Gears considered to be quite serviceable, but replaced because of a planned five-year run. Rotor rear face found to be fouled.

Following this exercise it was concluded that the management system had saved the company a large amount of money simply by telling it at an early stage that a gear box had not sustained pinion damage and that the problem it did have was not immediately serious. When the compressor problem therefore did arise maximum attention could be given to it.

The quality of information produced led to a fairly certain conclusion that all that was needed in the short term was replacement of the compressor's thrust bearing, which in turn kept the production loss down to the absolute minimum.

The machinery manager in charge of the operation was exceptionally qualified to handle the situation and this undoubtedly tipped the balance. Nevertheless, such experiences form the basis upon which the whole technology advances.

5. GAS TURBINES

The rotor and gearbox cases were presented as examples of mechanical systems that could be treated in relative isolation. In this respect the gas turbine represents one step higher up the system tree in that it incorporates a rotor, but in addition to that a compressor, a turbine, a combustion chamber, a lubrication system and often a gearbox. It therefore provides an excellent vehicle for discussion of monitoring system design and integration.

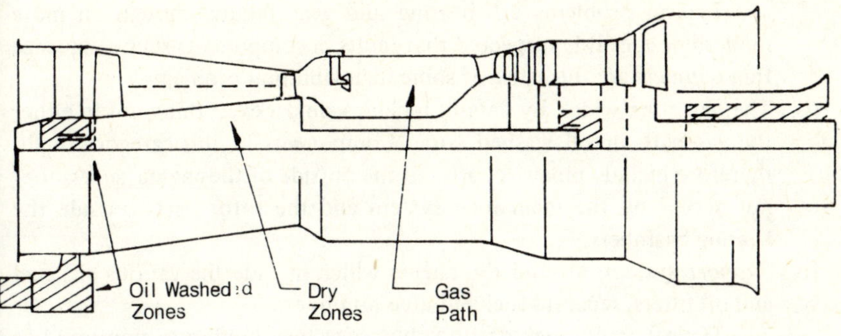

	PERFORM.	DEBRIS Oil/Gas	VIBRATION	VISUAL	USAGE
GAS PATH					
OIL WASHED					
DRY ZONES					
SUPPORT					

In-fill gives indication of potential coverage

Fig. 6 – Integrated gas turbine diagnostics.

The 'generalised' gas turbine is shown in Fig. 6. For monitoring purposes it can be broken down into three or perhaps four distinct zones, namely

(i) The *gas path*, which is generally the part of the machine most susceptible to damage and wear, and therefore the part which generally receives most attention. Typical faults are hot spots in the combustor or turbine leading to premature failure of these parts and corrosion in all parts due to ingestion of 'impure' air or fuel. Normal wear and tear on the other hand extends to blade creep and low cycle fatigue.

Typical signs of distress are rough running (i.e. high vibration), performance fall-off and evidence of metallic debris particles adhering to internal surfaces of the exhaust pipe.

The diagnostic problem is that in themselves none of the above are generally sufficient to confirm the existence of a fault.

(ii) The *oil washed zone* which includes principally the bearings and the accessory drives.

Typical problems are bearing and gear failure, though on more modern designs it is suspected that faults in components such as squeeze film dampers are the cause of some high vibration problems.

(iii) The *dry zone* which by default includes almost everything not in either the gas path or oil washed zone. Components in this category might therefore include pipework around the outside of the engine, accessories not served by the main lube system and the rotor parts outside the bearing chambers.

(iv) *Support systems* around the engine which include the various air, fuel and oil filters, separate fuel and lube pumps etc.

Typical faults include filter blockage (nearly always monitored as part of the engine's control system) and provision of fuel not conducive to long life.

Having already discussed in previous sections the problems of rotors and bearings, it is appropriate at this point to look at some of the technology that has been developed to handle those of the gas path.

The crux of the gas path problem is quite simply related to the huge number of structurally non-redundant parts that make it up, i.e. the compressor and turbine blades, failure of any one of which can lead to horrendous secondary damage. Unfortunately for the monitoring system, however, the impact on the performance of the system of one blade undergoing failure may be very small if not negligible, and so in the past detection of the fault has had to wait until either the blade has fallen off (detected perhaps by the step jump in vibration) or there occurs a serious deterioration in performance due to the ensuing secondary damage.

In recent years however the situation has changed for the better. For one thing, performance monitoring methods have become much more sensitive (largely due to improved system modelling techniques) and for another we can now see the real possibility of detecting blade/combustor distress at a much earlier stage in its development (long before the blade falls off). An example of the latter is electrical detection of distress-created metal particle clouds exiting the engine by way of the exhaust pipe. Somewhat further off are 'productionised' versions of vibration-based techniques currently being used by engine developers to monitor the amplitudes, natural frequencies and damping factors of all blades in specific blade rows. The important thing about these new techniques is the simplicity and ruggedness of their sensors, obviously key factors in their acceptance for field use.

The most important aspect of putting together a viable management system for gas turbines is therefore how to achieve a reasonable detection efficiency whilst at the same time holding false alarm rates down to an acceptable level. This problem is far more severe for gas turbines than on any other type of plant.

Nevertheless it is interesting to note that gas turbines were among the first machines to be operated 'on-condition'. Two bits of technology, namely the endoscope and magnetic chip detector were largely responsible for this.

The gas turbine can therefore be used to illustrate one of the fundamental concepts behind automated machinery management, namely that of gaining 'strength from diversity'. A way of putting this important point across is to consider the block diagram at the foot of Fig. 6, which considers the *diagnostic* performances of the various techniques. No one technique has anything like a 100% performance.

The point therefore being made by Fig. 6 is that to achieve any satisfactory level of performance for gas path *diagnostics* one must be prepared to look at several indicators (e.g. performance, vibration, debris in the jet pipe) and analyse these in ways that screen out problems of sensor failure, drift and straightforward technical ambiguity. Thus if a particular fault can be identified from two quite separate directions (better still three) the end situation in terms of fault ranking is likely to be much better. Expert rule structures for gas path diagnostics are thought to be approaching several hundred for the basic diagnostics, and perhaps a further fifty for the sensors.

Diesel engines can be treated in much the same way.

6. HARDWARE

It is not an axiom of machinery management as outlined in this paper that health monitoring is a crucial part, but it is very nearly so. In the vast majority of cases the technology of machinery management is inextricably linked to that of the hardware and software available for gathering and processing data. For example, vibration analysis would not really be effective without special-purpose equipment to process its signals, and the expert computing element would fail were there no access to a computer, albeit just a conventional one.

Hardware systems specifically designed to handle the machinery management task have therefore evolved and one of them is shown in Fig. 7 (shown in its naval ship form but first developed for use in a refinery).

At the heart of the system lies the central computer, denoted MSDA-2 (short for 'Mechanical Systems Diagnostic Analyser), and its database back-up MSDA-3. From this central facility radiate the main data highways (analogue and digital) to first the non-spared machines, then the major auxiliaries and finally the general purpose pumps and motors that form the general infrastructure of most plants. Other data highways travel to the process control computers and workstations provided for use by the mechanical test/efficiency engineers and maintenance staff.

The PMS system is unique in the sense that it was designed from the start to handle the outputs from a whole family of low-cost sensors (vibration, tem-

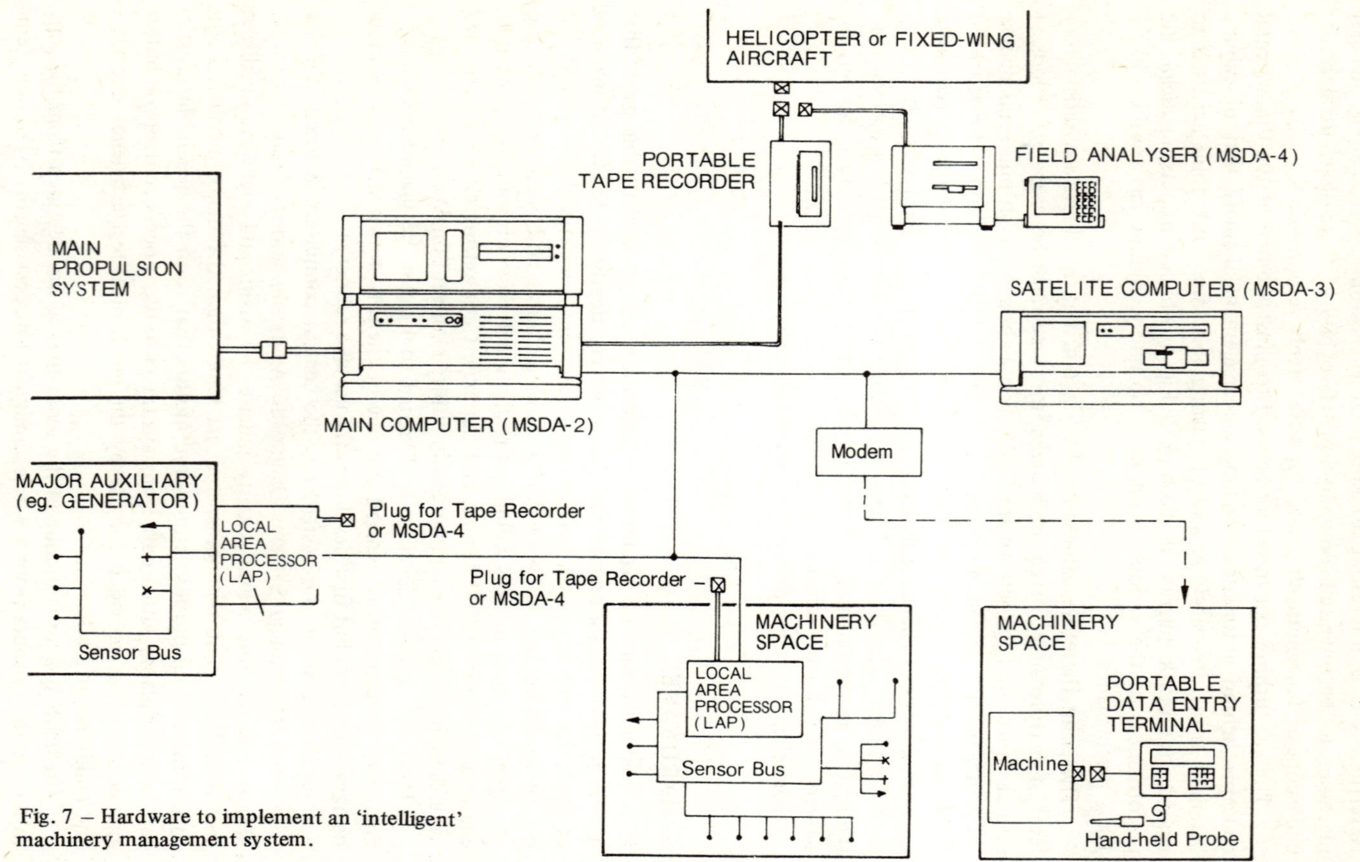

Fig. 7 – Hardware to implement an 'intelligent' machinery management system.

perature, debris, proximity, wear, power, hydrocarbon gas, corrosion, etc.). In addition, these may be coupled together on a low cost local area network data bus.

Needless to say, the PMS system's high level software takes care of all decisions regarding the details of signal processing, expert computing, etc. What it does, however, allow and encourage is trace back through decision making procedures (e.g. why it reached a certain conclusion regarding the health of a machine or what could have caused a machine to trip out).

7. THE ECONOMICS

Finally, we come to the most important section, namely how to make an economic case for investment in this technology. On the assumption that plant management will judge proposals for new technology on the same grounds as any other capital investment, the points to be borne in mind are as follows:

Profitability – making a profit over a period of time means that more resources are being created than consumed. Here, of course, market conditions must be taken into account. For example, consider the situation of a plant with two similar machines producing one product, and which in times of recession finds both machines under-utilised by a combined 40%. The choice open to management is therefore either to produce more product at a marginal saving in cost (using of course both machines) or reduce the amount of product by 10% and shut one machine down. The second would in most instances be the more profitable. The risks of both actions must, however, also be considered; for the first action it could well be that a further fall in the price would take place and lead to much higher losses. For the second the danger could be machinery breakdown leading to no production at all and loss of long-term business etc. An investment in machinery management technology might therefore be argued on the basis that it offers the possibility of operating non-spared machinery in situations where the marginal gains in profitability are greatest but the engineering risks too high for conventional management procedures to handle.

Return-on-investment (RoI) – anything that requires an investment must generally be justified in terms of the predicted return. The investment is therefore the sum of all the costs of introduction and the return is the accumulation of profits. The mechanics of computing RoI are much more complicated than just indicated, but the result is a compound interest rate which allows a comparison to be made with any other form of investment. Values of 20 and upwards are generally looked for in medium-risk technology.

Payback period – this is obviously just the time it takes to pay back the investment and it often determines the method of funding. In these volatile times payback periods have tended to become very short indeed.

It is therefore orchestration of the technology to produce the right numbers of profitability, RoI and payback that is likely to be the hallmark of the successful engineer. As far as the technology outlined in this paper is concerned, the key feature is probably the design audit for it is there that the areas of highest return and risk should be identified.

8. SUMMING UP

At the risk of introducing too many new concepts at one time the author has tried to show how a combination of basic technologies may be put together in order to create a 'machinery manager'. The essential elements of the package were Expert Computing, Machinery Modelling, Signal Processing, Measurement, Database Management and Control.

Implementations of the package are currently working in various parts of industry and defence and the experience to date has been very encouraging.

There would appear to be no clearly optimum set of circumstances for application of the technology. Plants with highly skilled resident engineers can use it as well as plants without. The change in efficiency promised by the technology is therefore thought to have a 'DC' component as well as one that rises in line with the degree of resident skill. The main risk associated with the technology is that a badly designed and installed system could easily lower machinery management efficiency owing to high false alarm rates.

The technology requires an enlightened attitude on the part of plant management towards instrumentation of machinery. The expert interpreter needs accurate data in order to compete with the conventional manager and his much greater ability to absorb background information of many forms. The combination of both ought, however, to be unbeatable.

The essential starting point is the design audit.

REFERENCES

[1] R. M. Stewart, The specification and development of a standard for gear box monitoring, *Vibrations in Rotating Machinery*, I. Mech. Engrs, Cambridge, 1980.

[2] Gear monitoring manual, Stewart Hughes Ltd, Report No. SHL/18/81, 1981.

CHAPTER 20

Mobile laboratory for plant surveillance

A. Péter and P. Pellionisz

In Hungary, acoustic emission investigations for different applications have been performed since 1976. Together with the experimental and measurement program, instrumental development has also progressed. The most recent result – a mobile laboratory comprising a computer-based 32-channel, acoustic emission measuring and data processing system as well as the additional equipment – is briefly described. The design philosophy is dealt with and some of the most interesting measurement solutions are described.

1. INTRODUCTION

After not much more than 15 years since its first industrial application, acoustic emission testing can be regarded as an established inspection technology, a valuable addition to the numerous, non-destructive testing methods [1]. With these measurements acoustic waves of wide frequency band give information on the state of important engineering structures or on technological processes like corrosion, leakage, fatigue, etc. Acoustic emission is in worldwide use to monitor pressure or loading tests, to detect propagating material flaws and the deformation of microstructures, to signal leakage or other anomalies.

Although acoustic emission is a mature inspection technique, many problems have not yet been eliminated. In order to study these problems and to introduce this surveillance method to Hungary, investigations were commenced in 1976 at the Central Research Institute for Physics of the Hungarian Academy of Sciences.

Acoustic emission experiments, laboratory and industrial measurements were accompanied in the Institute by our own instrumental development program [2, 3]. During this series of measurements, very different targets and problems were deliberately selected in order to test the application possibilities of this new method in different inspection fields and for different purposes. We would mention here, metallurgy [fracture mechanical tests of welding materials], stress corrosion measurements on heavy machine parts, leakage detection at underground pipelines or in hydraulic valves of aircraft, checking of composite helicoptor rotor blades, laboratory tests with geological materials, etc. [2, 3].

Plant surveillance applications were also considered. Pressure tests of power plant boilers, spherical gas reservoirs, etc. needed methods and instrumentation appropriate to heavy, industrial conditions, requiring not just a single detector but a network of them; these tests demonstrated the special difficulties and requirements of plant surveillance techniques.

For illustration, in Fig. 1 a part of the arrangement of one of our spherical gas reservoir measurements is shown where a welding with flaws – discovered by ultrasonics – had to be checked. Here an array of four detectors was used with the aid of which the geometrical location of the sound-emitting sources could be calculated. The positions of the detectors and the welding, as well as the localised sources, are shown in Fig. 2. In Fig. 3, our home-developed four-channel acoustic emission analyser is shown [4].

The instrumental and data processing developments of the Institute – from the preamplifiers to the main computer – gave us the potential to construct a mobile acoustic emission laboratory comprising a 32-channel acoustic emission measurement and data processing system. The other basis of this work was given by the experience gained from various acoustic emission experiments [3].

2. DESIGN CONSIDERATIONS

Design work required that the following questions should be answered:

For what sort of investigations is the laboratory to be used? Besides sensing and analysing AE signals originating from material degradation, the laboratory must be capable of leakage control and loose part monitoring.

What is to be measured by the laboratory? The measuring object can be nearly any kind of large structure (pressure vessels, piping, bridges, etc.) as well as any small specimen. The investigated materials can be metals, composites, polymers, rocks, etc.

In what way is the laboratory utilised? The laboratory is to be utilised for pressure tests, load tests (under shop, pre- and in-service conditions), corrosion measurements, fatigue tests, as well as in quality control and for supervising technological processes.

Where is the laboratory utilised? The fields of utilisation may be numerous and very different. It can be used in the chemical, oil, nuclear and power industries, and outdoor and indoor sites can equally be taken into account. In the case of research applications mostly laboratory investigations are to be performed.

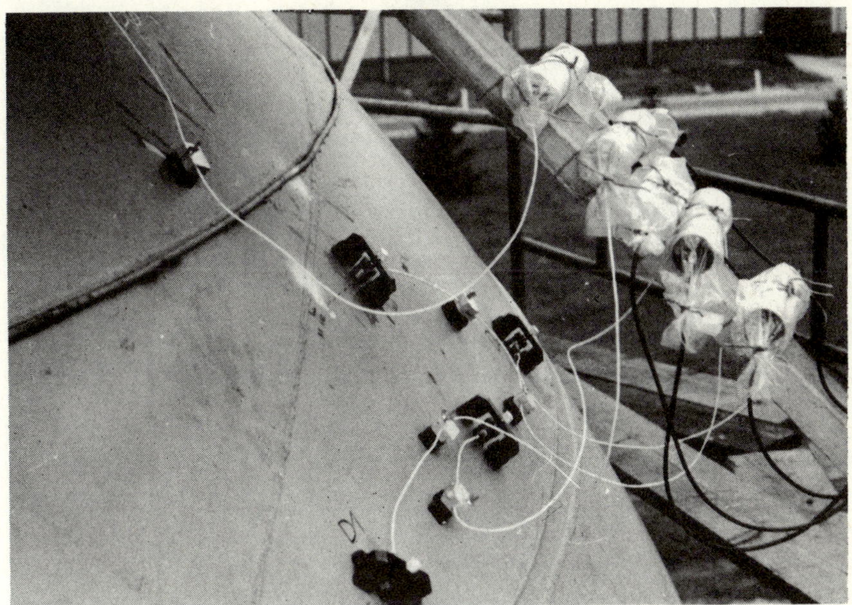

Fig. 1 – Transducers, cables, preamplifiers for the pressure test of a 400 m³ liquid gas reservoir.

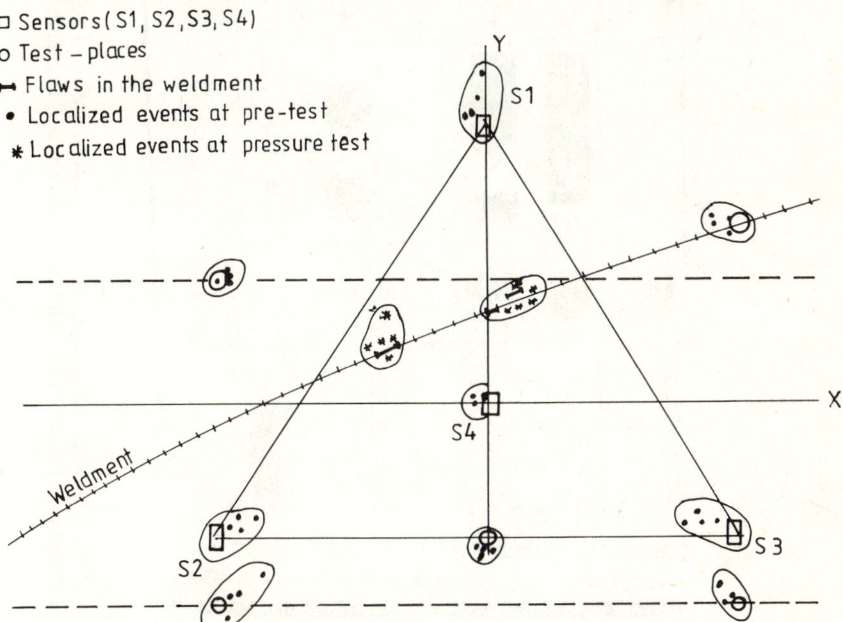

Fig. 2 – Localised events during the pre-test and pressure testing of a 400 m³ liquid gas reservoir.

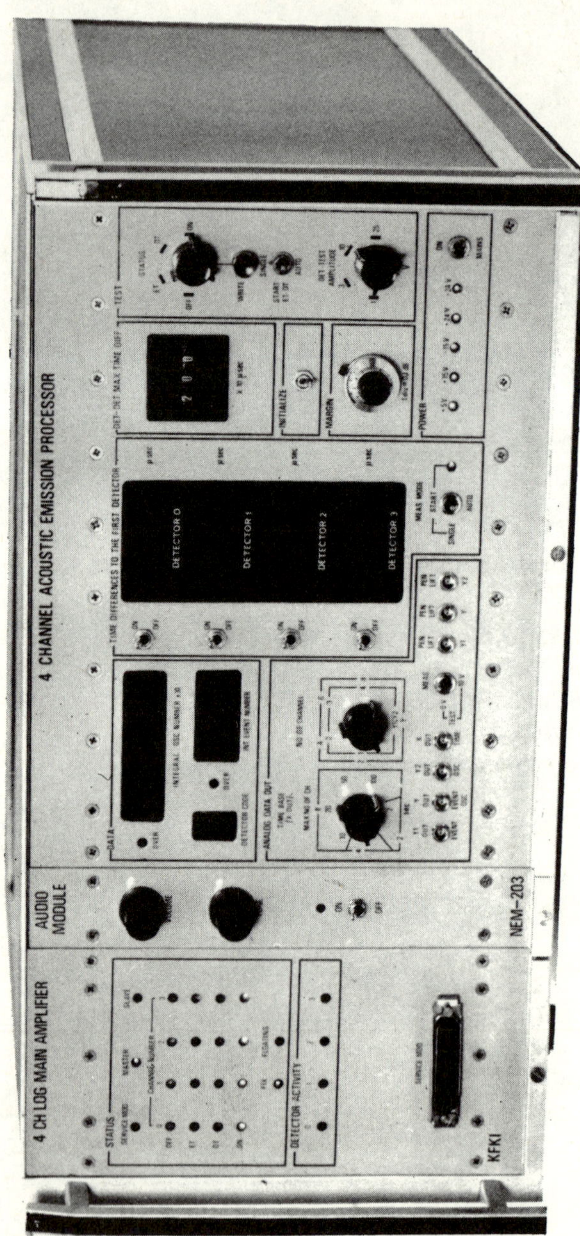

Fig. 3 – Four-channel acoustic emission analyser [Model NEZ-210] developed and used in the Central Research Institute for Physics, Budapest.

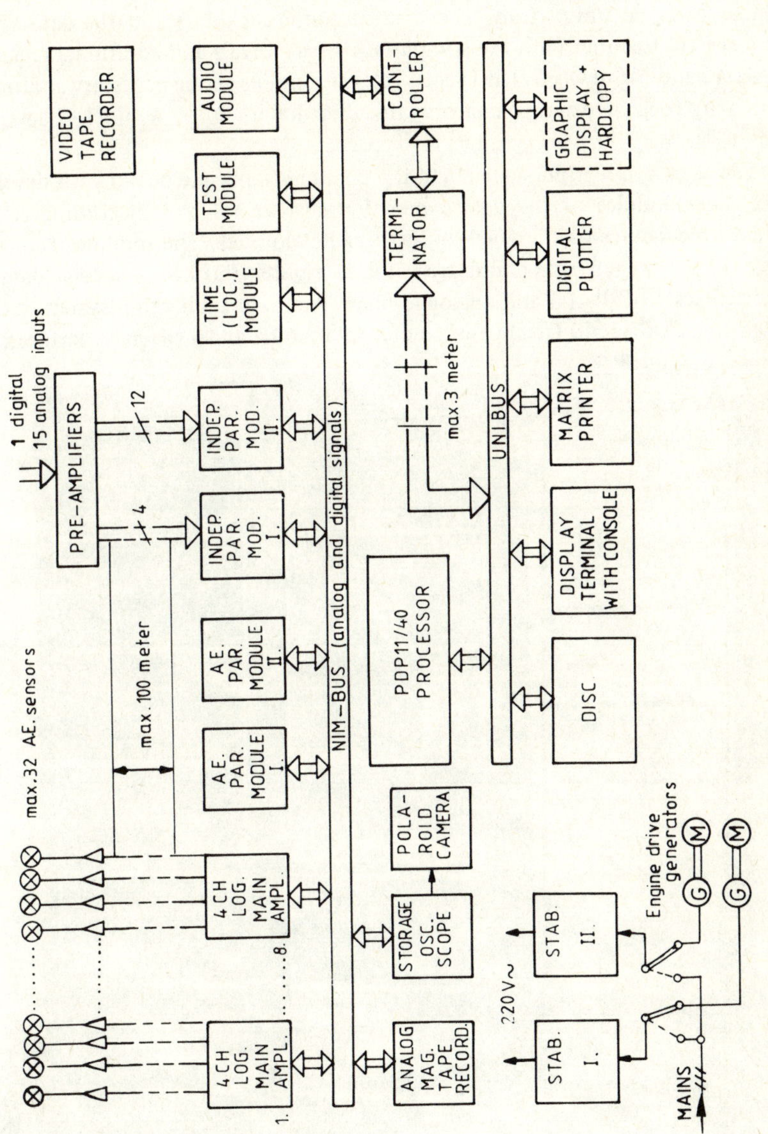

Fig. 4 – Block diagram of the 32-channel, computerised acoustic emission system.

3. THE MOBILE LABORATORY

The mobile laboratory can be considered to consist of two parts: the measuring-data processing system and the special vehicle housing the system. The former includes external units (transducers, preamplifiers, cables) and the data acquisition system together with the computer, its peripherals and additional measure- and data handling devices. The special vehicle contains all the necessary additional equipment (engine drive generators, air conditioner, etc.). A block diagram is given in Fig. 4.

The data acquisition system and the preamplifiers were of our own development. The modules of the data acquisition system are built according to the standard mechanics of the Nuclear Instrument Modules. The modules can communicate with each other and with the computer through a special databus which consists of digital and analogue lines. The data acquisition system is connected to a PDP 11/40 UNIBUS via a control module and a special interface.

Fig. 5 – The 32-channel acoustic emission fault localiser and analyser system. The data acquisition part is housed in one cabinet, the PDP 11/40 computer together with the disks, and analogue tape recorder take up two cabinets.

The a.c. power is fed to the system through two special insulator-regulators (ISOREG computer power modules); these ensure freedom from spikes for the stabilised a.c. power, even if the portable petrol engine drive generators are used. Parts of the system are shown in Figs. 5 and 6.

Fig. 6 – Acoustic emission data acquisition system. It is arranged in four 5U high 19″ width crates, and one 4U high crate is for the power supply (U = 44.45 mm).

The whole system is mounted in a special bus (Ikarus, Hungary). The internal space of the vehicle is divided into three: at the back is the entrance; the measuring compartment is in the middle; the driver's cab and passengers' compartment are in front. The inner arrangement and a photograph of the bus are shown in Figs. 7 and 8.

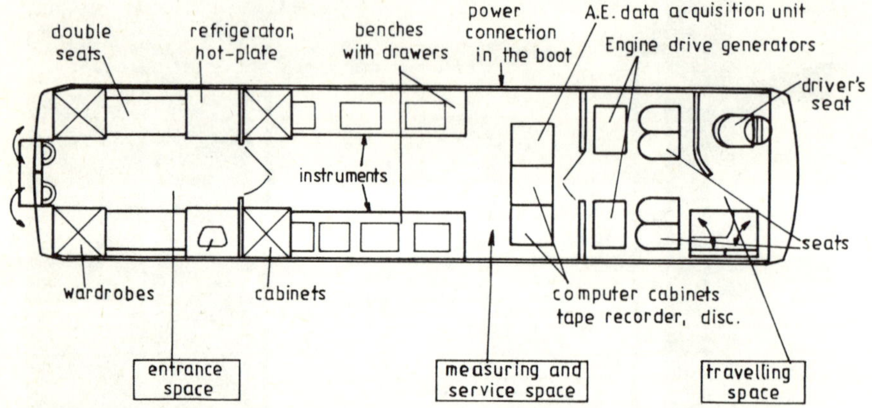

Fig. 7 – Inner arrangement of the vehicle.

Fig. 8 – Photograph of the mobile acoustic emission laboratory.

4. MAIN CHARACTERISTICS OF THE ACOUSTIC EMISSION SYSTEM

The system has 32 channels for acoustic emission signal inputs (i.e. it works with 32 transducers simultaneously) and 16 channels for input of technological parameters.

Transient (burst) and continuous acoustic signals can be sensed and analysed in the 50 kHz-2 MHz frequency band.

Many parameters of the sensed acoustic signals can be measured. (In the case of burst signals: peak amplitude, rise-time, event duration, oscillations and area of the signals of the first hit transducer and of a programmable transducer and time differences between the transducer signals for localisation. One event is localised at one time; in case of continuous signals: r.m.s. value, average value and a coefficient characteristic to the ripple of the signal).

During the measurement and processing of the acoustic signals, special effort was devoted to the high dynamic range. Accuracy was compromised, if necessary, for the sake of fast data processing.

The incoming signals can be discriminated according to their analogue properties (minimum–maximum limits can be specified for the peak amplitude, oscillations, rise-time, event duration, r.m.s. value). If the measured values exceed the specified limits, the signal is registered as invalid. Space discrimination is achieved by master–slave and coincidence–anticoincidence methods which rely on time difference measurements. Further discriminations can be made by software.

Sixteen technological parameters (force, strain, pressure, temperature, etc.) can be measured by the system. 15 analogue signals are sampled simultaneously at the occurrence of the burst signals and equidistantly in time. A single digital input is reserved for counting (e.g. for fatigue tests).

A special test method ensures the possibility of rapid, complete testing of the whole system. The external units (cables, preamplifiers, transducers, couplings between transducers and the surface of the investigated structure) can be tested via the same transducers which take part in the measurement, utilising the reversible characteristic of the piezoelectric materials.

The applications software is designed as an interactive system. The operator can conduct a dialogue with the computer through the CRT and the console (Fig. 9).

The software provided with the system consists of measurement initialisation, data acquisition, processing-displaying and test programs as well as programming aids. The programs are written in FORTRAN IV. and MACRO-Assembler (DEC language). Localisation can be performed on planar, cylindrical and spherical schemes. Direction-dependent propagation velocities can be taken into account.

The system is easily expandable (e.g. by hardware localiser, correlator, etc.).

Date 12-Jan-1983 Preliminary measurement No. 1

Event number	Arr cod	Det cod	Elapse time [ms]	Osc numb	Time rise [us]	Time width [us]	Delay times [us] 0	1	2	3
1	1	3	1000	1E2	3E1	7E3	74	120	250	0
2	1	0	1033	4E2	2E1	3E4	0	30	49	110
3	3	1	1041	3E1	1E1	2E3	14	0	49	97
4	2	2	1100	4E1	1E1	3E3	15	121	0	17
5	1	3	1150	1E2	3E1	7E3	74	125	247	0
6	1	3	1207	1E2	4E1	7E3	70	122	250	0
7	2	2	1231	4E1	1E1	3E3	15	129	0	20
8	1	3	1296	2E2	3E1	1E4	73	124	251	0

Fig. 9 – Selected and measured parameters are shown on the display terminal in tabulated form.

5. SPECIAL DATA PROCESSING FEATURES OF THE SYSTEM

Sensing the burst-type AE signals

Generally, burst-type signals are identified when they cross a reference level which is set as a fixed value or dependent on the background noise. By a single reference level a fast, direct amplitude discrimination can be achieved, but the time the signal spends above this level is heavily dependent on the reference value and so are the measured event-duration and energy. To avoid this drawback, a double reference level is used by which the event-duration and energy of the burst signal can better be determined (Fig. 10).

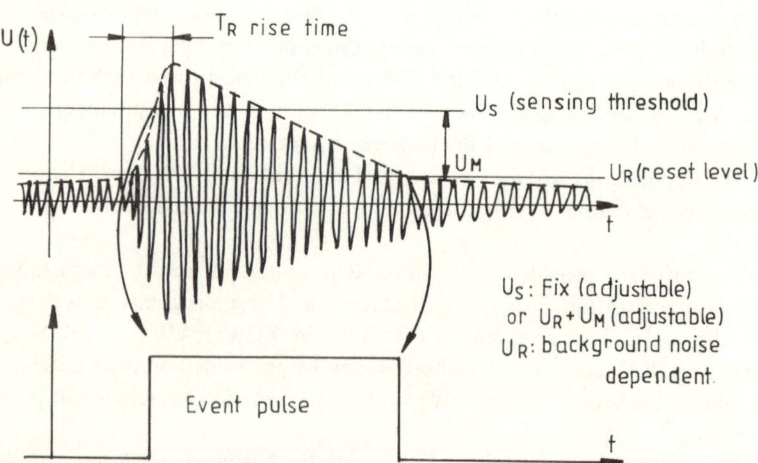

Fig. 10 – Using double reference level, the amplitude discrimination has minor effect on the width of the event pulse.

5.2 Measuring and processing the AE signals

Generally data are sent further in rounded floating point form: $Y = A \times 10^B$, where Y is the measured value, $A = 0, 1, \ldots 9$ and $B = 0, 1, \ldots 7$. In this way the accuracy of the measured values is limited but the whole dynamic range can be preserved in a single byte word-length. In this way the memory space requirement can be very much reduced. (For complete accuracy, 6 decade dynamic range needs three bytes word-length).

For the compromise between accuracy and high dynamic range, the amplifiers of every channel are of logarithmic character (after a linear amplification of 40 dB) and dispose a total dynamic range of 90 dB. In this way the cumbersome setting of the gains is avoided and the over- and underloading of the amplifiers disappear.

5.3 Measured parameters

Because of the logarithmic amplification, the energy is not measured in the conventional way but the area under the envelope of the burst signal is measured. Multiplying the area by two and dividing this value by the pulsewidth (event-duration), the result is in close relationship with the logarithm of the average power of the burst signal. Besides conventional AE parameters, it is also possible to measure the area belonging to the rise-part of the envelope and the whole area as well. As the rise-part of the burst signals characterises the signal source more than the decay part, the measured area belonging to the rise-part might give useful information about the source. With signals of continuous type and outputs of linear envelope detectors are sampled and measured. It is possible to measure the r.m.s. of the positive envelope of the signals (r.m.s.) and the r.m.s. value of the positive envelope minus d.c. part of the signal (r.m.s._R). The $\text{r.m.s.}_\text{R}/\text{r.m.s.}$ quotient is characteristic of the fluctuation of the continuous signal.

6. SUMMARY

A 32-channel, mobile acoustic emission laboratory has been designed in the Central Research Institute for Physics, Hungary. It makes possible the simultaneous, multichannel measurement of acoustic signals, with source localisation, multi-parameter recording, computer-based data acquisition and data processing. The computer-based acoustic emission system is mounted in a special bus provided with the necessary additional equipment ensuring convenience, mobility, and speed. The project is due to be completed in the spring of 1983 after which it is planned that the mobile laboratory will regularly be used for field service – primarily in Hungary, but work in foreign countries is also foreseen.

ACKNOWLEDGEMENTS

The authors are highly indebted to their director, Mr Z. Gyimesi, and to the head of department Dr P. Pallagi for continuous assistance in this project. Thanks are due to the State Committee for Technical Development for its financial support. The authors express their gratitude to Mr. J. Geréb and Mr. G. Lafranco who were responsible for a good deal of the hardware design and the software development. Thanks go to Mr A. Kömives, Mr F. Paár, Mr J. Kapros and Mr L. Késmárki for technical assistance.

REFERENCES

[1] J. C. Spanner, Acoustic emission: who needs it – and why? Conf. on Acoustic Emission, Anaheim, California, September 1979.

[2] A. Péter and P. Pellionisz, Research and development on acoustic emission technique in Hungary, Conf. on On-line Surveillance and Monitoring of Plant Reliability, London, September 1980.

[3] A. Péter, J. Geréb and G. Lafranco, Acoustic emission measurement activity of the Central Research Institute for Physics, 8th Congr. on Material Testing, Budapest, 1982, Sc. Soc. of Mech. Eng., pp. 1174–1179.

[4] A. Péter, P. Pellionisz, J. Geréb, Akusztikus emissziós kutatások a Központi Fizikai Kutató Intézetben, (Acoustic emission research in the Central Research Institute for Physics) *Mérés és Automatika,* **XXIX** (1981), 10, pp. 392–395 (in Hungarian).

Index